ROUTLEDGE LIBRARY EDITIONS:
THE RUSSIAN REVOLUTION

Volume 7

AGRICULTURAL RUSSIA

AGRICULTURAL RUSSIA

On the Eve of the Revolution

GEORGE PAVLOVSKY

Routledge
Taylor & Francis Group
LONDON AND NEW YORK

First published in 1930 by George Routledge & Sons Ltd

This edition first published in 2017
by Routledge

2 Park Square, Milton Park, Abingdon, Oxon OX14 4RN

605 Third Avenue, New York, NY 10017

Routledge is an imprint of the Taylor & Francis Group, an informa business

British Library Cataloguing in Publication Data
A catalogue record for this book is available from the British Library

ISBN: 978-1-138-21999-1 (Set)
ISBN: 978-1-315-31269-9 (Set) (ebk)
ISBN: 978-1-138-22727-9 (Volume 7) (hbk)
ISBN: 978-1-315-39626-2 (Volume 7) (ebk)

Publisher's Note
The publisher has gone to great lengths to ensure the quality of this reprint but points out that some imperfections in the original copies may be apparent.

Disclaimer

The publisher has made every effort to trace copyright holders and would welcome correspondence from those they have been unable to trace.

AGRICULTURAL RUSSIA
ON THE EVE OF THE REVOLUTION

BY

GEORGE PAVLOVSKY

Ph.D. (Lond.), F.S.S.

WITH 3 MAPS

LONDON

GEORGE ROUTLEDGE & SONS, LTD.

BROADWAY HOUSE: 68–74, CARTER LANE, E.C.

1930

Thesis approved for the Degree of Doctor of Philosophy in the University of London.

Published with the aid of a grant from the Publication Fund of the University of London.

Made and Printed in Great Britain by Butler & Tanner Ltd., Frome and London

CONTENTS

PREFACE

SOME of the ground covered by the present study has long been a favourite battlefield of conflicting schools of Russian economists and politicians. The strife over the Agrarian Problem, which had reached its climax about the time of the Revolution of 1905, far from having subsided when, in 1906, Stolypin came forward with his scheme of agrarian reforms, had actually increased in violence, and the new policy was fiercely assailed. Yet, while this struggle had been proceeding, the Russian countryside had been gradually settling down and, without paying heed to the clash of arms among the warring sections of the *intelligentzia*, was grasping at the opportunities offered it by the reform with eager hands. Forces have been, indeed, at work in the social and economic system of twentieth-century Russia, which were pulling the Russian peasant out of his rut and driving him forward, in the direction in which the way had been cleared for him by the agrarian legislation of Stolypin. It is these economic forces, which influenced the agricultural evolution of Russia in the early part of the current century that I shall endeavour to study in the present work, with a view to thus filling a gap in the extensive literature of the subject. I must be excused for dealing throughout rather with facts than with opinions. The conflicting views of such schools as the *Narodniki*, or the Marxians, are, indeed, better known to the public from numerous writings than the actual facts of the agricultural evolution of Russia and of its causes; but opinions change and are often disputed and disproved, while the forces of evolution do their work relentlessly and without fail, even when, in times of revolutionary upheavals, human will makes a determined attempt to divert them from their course.

Accordingly, I limit the scope of my work to the positive presentation of the evolution of Russian farming, as it appears to me from the study of the statistical and other records at my disposal, as well as from personal observation of the Russian countryside throughout the period dealt with below.

Various faults of omission and commission are sure to be found in this work and pointed out by its critics; and no one is better aware of its many shortcomings than its author. The necessity of writing a monograph on this subject abroad, even with the abundant resources of the British Museum Library at hand, was responsible for many gaps. Though everything has been done to avoid, as far as possible, the use of second-hand data, and the whole groundwork of this study is the result of original research, it will be seen that, on many occasions, I have been compelled to quote figures and passages on the authority of earlier writers. Errors may also have crept in, for which the writer must bear the blame.

In conclusion, it is my pleasant duty to express my sincere gratitude to all those who have so willingly helped me in various stages of my researches. To the London School of Economics and Political Science and to its Director, Sir William Beveridge, I owe the possibility of submitting this book as my thesis for the Ph.D. Degree of the London University. To Baron A. F. Meyendorff and Mr. R. B. Forrester I am deeply indebted for their assistance and valuable advice throughout the time of its writing. To Sir Bernard Pares, Director of the School of Slavonic and East-European Studies, King's College, my thanks are due for the interest and encouragement I have always had from him in my work. To Mr. A. A. Rittich, formerly Russian Minister of Agriculture, who, as Deputy Minister during the period of agrarian reforms, had been mainly responsible for the planning and the actual carrying out of the peasant enclosure movement, I wish to express my thanks for his help in reading through the chapter on enclosures. To Miss Winefride Hunt, of the London School of Economics, and Miss Dorothy Wilford, of the Stockwell Training College, I am much indebted for the excellent maps made from my rough sketches. Last, but not least, I should like to thank my wife for her assistance in the large amount of statistical work involved in the compilation of the various tables which appear in this book.

The publication of the book has been assisted by a grant made by the Senate of the University of London, for which I take this opportunity to express my gratitude.

G. P.

RUSSIAN WEIGHTS AND MEASURES

One *dessiatin*	= 2·7 acres
One *verst*	= 3,500 feet = 0·66 mile
One *pood*	= 40 Rus. lb. (*funt*) = 36 Engl. lb
One *funt* (Rus. lb.)	= 0·9 Engl. lb.

RUSSIAN PRE-WAR CURRENCY

One *rouble* = 100 *copecks* = *2s. 3d.* = *27d.*

One Arshin = 28 inches.

One Rbl. (1913) = 2s. 1½d.

RUSSIAN DATES [illegible]

[illegible]

Agricultural Russia on the Eve of the Revolution

INTRODUCTION

WITHIN the short space of some two decades immediately preceding the Great War was compressed one of the most important epochs in the history of Russian farming. Not even the period of the Emancipation of Peasants, far-reaching as have been its social, economic and political consequences, could be compared with these momentous few years in its effects on the agricultural industry, its organization and its progress.

It is, indeed, more or less generally assumed that the Act of 1861, which abolished serfdom, had inaugurated the capitalistic system in Russia. The beginning of the "industrial revolution" in Russia is accordingly, as a rule, referred to the 'sixties, and Russia, from that time onwards, is considered as living under conditions of capitalism, less advanced than those of her Western neighbours, but essentially the same. As a matter of fact, this view, by antedating capitalism in Russia, did much to confuse certain important aspects of her recent social and economic evolution and had the most disastrous effects upon the interpretation of those processes, which have been developing in the Russian countryside since the Emancipation. By the Manifesto of February 19, 1861, no such instantaneous transformation either was or, indeed, could be effected, and for several decades after the Emancipation Russia had lived under a hybrid sort of social and economic organization, in which survivals of medieval forms were rather incongruously blended with numerous anticipated elements of a future system, towards which the country was painfully feeling her way. And nowhere had the hybrid nature of the existing organization been more strongly felt than in the agricultural industry, on which the great bulk of the population depended for their living.

The basic fact of Russia's modern economic history was that when, in 1861, she crossed its threshold, she did so with only one of the elements necessary for the building-up of the capitalistic system, namely free labour. The other essential element of that system, the one, indeed, from which its very name is derived, was practically non-existent. Russia possessed no capital worth its name, and this had to be created, before the new social organization, with its inherent advantages of economic efficiency, could come into being. The bulk of the very limited capital resources of Russia, in the middle of the last century, was engaged in internal trade and, under the conditions of trade and transport characteristic of the pre-railway age, was generally tied up in the business. In the first half of the nineteenth century, a beginning had, indeed, been made in the development of large agrarian capitalism, based on serf labour, in the black-earth belt of Russia, which could eventually have resulted in the accumulation in the hands of landowners of more or less considerable balances seeking investment; but this process of capital accumulation came to an end with the Emancipation which, by abolishing serfdom, struck at its very roots. Foreign trade, perhaps the most important single source of capital accumulation in medieval Western Europe, played too small a part in the economic life of Russia before the second half of the last century to be of any real service in this respect. The credit system was only just beginning to develop and could not be relied upon for decades to come for the concentration and making available for investment of the scattered savings and reserves of the general public and of business. Thus, in the 'sixties, Russia had plentiful hands and large natural resources, to whose development these hands could be turned; but she lacked capital to set them in motion. Throughout the rest of the nineteenth century, she had been struggling, often against great difficulties, with the task of accumulating or attracting capital and of building up a national industry, based either on the exploitation of her natural wealth or on the extensive home market ensured by her numerous population. The important problem was to start the process of capital accumulation, following the maxim that, given the first hundred, the second is easy to make. Such starting, however, necessitated the immediate investment of large sums into railways, factories, mines, steamers, etc. In a country rich in capital, the latter would of itself have found profitable investment in anything for which there was sure to be a pressing need; in Russia, with no balances eagerly on the look-out for investment, and with any prospects of attracting capital from abroad only in the case of the most obviously and immediately profitable concerns, the position was different. The only means by which most developments, especially in transport and in anything pertaining to the countryside and to farming, could be started, was the direct intervention of the State, which planned them, provided the necessary funds, generally by borrowing abroad, and put the plans into execution or, alternatively, in the case of railways and of some important heavy industries, secured the co-operation of private interests, to whom it guaranteed either profits or orders. Accordingly, in the building-up of the economic system of modern Russia, the State had played a far greater part than had their own governments in the economic evolution of her Western neighbours: a phenomenon due, in the first instance, to the extreme shortage of capital in Russia, as a country up to the close of the nineteenth century almost entirely agricultural. With the constant assistance of the State, all through the second half of the last century Russia had been building up gradually, if slowly, the elements of capitalism and evolving an economic system powerful enough to support the social and political structure of a modern State. The immense size of the Russian Empire and its position in the foremost rank of the Great Powers involved financial burdens which, indeed, no purely agricultural country, devoid of the more profitable branches of national industry and trade, could possibly bear. While, however, this made industrialization imperative, it also imposed on Russia's immature economic system a financial strain she could not easily bear without detriment to the rapidity of her progress towards ultimate economic transformation. The growth of modern capitalism, therefore, has been necessarily slow, and Russia's actual transformation into a capitalistically-organized country had been delayed until the twentieth century. It was not until the last fifteen to twenty years before the Great War that the real "industrial revolution" in Russia started for good and began rapidly to transform the country and to do away, one by one, with those survivals of medievalism in the social, economic and political system which, hitherto, had appeared to some extent necessary or even indispensable. In the economic history of Russia, therefore, the early years of the twentieth century form an important chapter; a chapter, however, whose full importance in Russia's economic development would be far better known and far more appreciated, had it not been cut short by the war and the revolution, which the war brought in its wake.

The Russian agricultural industry was one of the branches of the country's economic life most profoundly affected by both the conditions of the period of transition and those of the "industrial revolution," inaugurating the final triumph of modern capitalism. Farming, in any country whose population consists mostly of peasants, is more than a mere industry. It is, indeed, rather a mode of life and, as such, is so closely and intimately interwoven with other aspects of existence, with traditions and prejudices, that it does not easily lend itself to drastic modifications, except under pressure of urgent necessity, when its existing forms fail to answer the requirements of the situation. This explains the proverbial conservatism of the peasant mass, which is so difficult to move out of the rut in which it had stuck. Any established system of cultivation, accordingly, is liable to persist for centuries, as did the three-course cropping in most parts of Western Europe, which ruled the countryside for roughly a thousand years, from about the time of Charlemagne to the beginning, and in many cases down to the middle, of the nineteenth century. That system, along with open-field tenure, involving compulsory cropping and thus hindering individual progress, throughout Western Europe, had ultimately been defeated and replaced by modern scientific rotation in one or other of its numerous forms, by the triumph of modern capitalism. Generally speaking, the influence of capital upon farming was brought to bear in one of two ways, namely either by the direct investment of capital in the agricultural industry, involving, to a greater or lesser extent, the substitution of large estates for peasant holdings and of large capitalistic farming for peasant husbandry, or by the less direct means of the influence of the growing wealth of the towns and industrial centres on agriculture in all its forms, which left the actual distribution of land and the size of farms either wholly unaffected or only partially modified. The former of these two courses was followed by the agricultural evolution of England since the later part of the seventeenth century, when the capital accumulated in trade began to seek investment in landed property and agriculture, and farming entered the most brilliant period of its technical progress. In some parts of Germany, more particularly to the East of the Elbe, in the region of predominant patrimonial organization (*Gutsherrschaft*) and of serfdom, the evolution, since the close of the Thirty Years' War, had followed somewhat similar lines and involved wholesale evictions of peasants (*Bauernlegen*) and the substitution of large farming for peasant cultivation. In various constituent States of Germany, the results of this consolidation of holdings were different, according to the extent to which their several rulers were successful in protecting the peasants against evictions, but in some of them it had gone very far and had involved the disappearance of a large part of the former peasant class. Thus, in Mecklenburg, like in England after the enclosures, the independent peasantry had been almost completely done away with, and their remnants converted into agricultural labourers on large estates. In those countries, in which the accumulated capital had been seeking investment, on a large scale, in the agricultural industry, it had succeeded, indeed, in greatly raising the standards of farming, but only at the expense of the more or less far-reaching extinction of that class of independent peasant farmers which is a source of physical and moral health for the nation. More beneficial, in the present writer's opinion, was the indirect influence of capitalistic development on the agricultural industry, exercised through an increased and more varied demand for agricultural produce, higher prices and the possibility for the farmer of investing some of his growing receipts in the improvement of cultivation. These being the natural effects of the growth of capital and of industrial and trading centres, it is by no means surprising that it was Holland, the foremost trading and capitalist country of the sixteenth and seventeenth centuries, that had been the pioneer of modern agriculture in Europe, and that it was from there that the English " spirited landlords " of the seventeenth and eighteenth centuries had imported many of the most important farming innovations. It was from Holland that Germany, in the second half of the eighteenth century, had learned the cultivation of clover and that many other improvements in arable and stock farming had spread. The early growth of trading capitalism in the industrial and commercial cities of Western Germany, on the Rhine and the Main, had certainly contributed to a great extent to the prosperity of the local peasant farmers, who, in that part of the country, formed the backbone of the agricultural industry. Wherever one turns, one can see that the stimulus that had roused the agricultural industry of Europe to new life, after a protracted period of stagnation which had lasted through the Middle Ages, originated from the capitalistic development of the countries concerned, whose growing wealth brought about a large extension of the agricultural markets and raised the prices of the products of farming. One can easily imagine, therefore, the position of the Russian agricultural producer, whether large or small, who, during the long period of transition which followed the abolition of serfdom, had been faced with conditions which neither permitted him to revert to a system of isolated natural economy, in which his ancestors had formerly lived, nor afforded any of the advantages of the still non-existent capitalistic organization, of which the principal substance, namely capital, was still lacking. It is hardly surprising, indeed, that throughout the second half of the last century Russian farming had lived in a state of utter depression, and that an acute agrarian overpopulation had developed in many a locality of Russia. The only way out of the difficulties, with which Russia in general, and her agricultural industry in particular, had been faced, lay through a final triumph of capitalism over the hybrid system in which the country had lived since the Emancipation. That triumph, which manifested itself in a rapid expansion of industry and trade, in the accumulation at home of increasing capital resources and in the influx of large amounts of capital from abroad for investment in Russia, had been the outstanding characteristic of the early years of the twentieth century. Then, and more particularly during the decade immediately preceding the Great War, the "industrial revolution" in Russia was in full swing, and Russian farming, at last, felt on itself the effects of those stimuli to development, which had hitherto been lacking. The traditional forms of tenure and cultivation, inherited from the days of serfdom and of isolated natural economy, to which the Russian peasant had been clinging during the preceding decades, but which stood on the way of progress under modern conditions, had to be discarded, and the agricultural industry, in all its branches, had entered a stage of transformation, with a view to adapting itself to the requirements of the new economic era.

This period of transformation of Russian farming, which involved far-reaching changes in the systems of tenure, in the methods of cultivation and in all the economic organization of the Russian countryside, formed, indeed, one of the most interesting chapters in the economic history of Russia. Not only, moreover, do the developments of the years immediately preceding the war present considerable interest in themselves, but they throw much needed light on the evolution of Russian farming and agrarian relations in the course of the past few decades, whose interpretation had often appeared difficult and highly controversial. Only by looking back from the twentieth into the nineteenth century and projecting the pre-war developments against the background of Russia's agrarian organization and evolution since 1861, it is possible to understand the characteristics of the Russian agrarian system, which to many observers and students have appeared as inherent peculiarities of the people, a manifestation of some essentially national spirit, but which, to the present writer, were no more than features of a passing stage of adaptation, doomed to disappearance on the coming of the modern social and economic system into its own.

It is my purpose in the present study to deal with the economics of Russian farming during the early part of the twentieth century, and to analyse, in the first instance, those economic influences which were at work and were bringing about its transformation. Starting from a sketch of the agricultural geography of European Russia, as it had been shaped by natural conditions, historical and economic factors and other influences, I shall proceed to the study of the organization and conditions of Russian farming during the period dealt with, and finally deal with Russia's agricultural production, her characteristics as an agricultural producer and the origins and disposal of her available surpluses of agricultural products.

PART I

THE AGRICULTURAL MAP OF RUSSIA

CHAPTER I

NATURAL CONDITIONS

THE agricultural geography of a country is the product of the combined influence of natural conditions, historical development and economic factors. Natural conditions impose certain territorial limitations on farming in all its branches, and though the progress of agricultural science tends, indeed, to expand those boundaries, agriculture is so essentially dependent on the forces of nature that, in the main, the limits thus fixed may be considered as practically rigid. The elaborate design of the agricultural map, on which the hand of history never ceases to work, sometimes nursing a single detail for generations or even centuries, and sometimes bringing about drastic changes in the course of a few short years, has nature for its background. In any attempt to analyse and describe the agricultural conditions of Russia, a brief outline of the natural background on which they developed, is therefore absolutely essential.

European Russia is a vast plain, intersected only by slight undulations of the ground. Changes in the relief of the territory, which exercise so marked an influence on the character of farming in most other countries of Europe, play, therefore, practically no part in shaping the agricultural map of Russia. The natural environment to which farming has to adapt itself depends mainly on two elementary geographical factors, namely latitude and longitude, which determine climatic conditions and, through the latter, to a considerable extent, also the properties of the soil.

Except for a relatively narrow fringe along its Western frontiers, European Russia possesses a climate which can be described as typically continental. This character is especially pronounced in the South and the East of European Russia, owing not only to the geographical position of these parts, far removed from the influence of Atlantic air currents, but also to the special atmospheric regime of the Russian plain. Indeed, the distribution of atmospheric pressure is such that, as a rule, the Northern and Western parts of the country are open mainly to the influence of South-Western winds, especially common during the winter, while the Southern and Eastern districts, protected from the West by the double barrier of the Alps and the Carpathians, are mostly swept by the arid winds of the Asiatic steppes. Accordingly, the severe winters of the North-West of European Russia are somewhat moderated by Atlantic air currents, as well as by the relative proximity of the Gulfstream. Both contribute to the abundance of moisture in the atmosphere, resulting in plentiful rain—and snowfall. While, on the whole, the abundant summer rains in the North-West may, to some extent, be harmful to agriculture, the presence of a thick coat of snow throughout the winter is favourable to farming in these regions, since it enables the cultivation to be extended very far North, into latitudes which, otherwise, would have been agriculturally barren. Indeed, according to the eminent Russian meteorologist and geographer, Professor Voieikov, "were it not for the cover provided by snow, the cultivation of winter crops would be impossible in two-thirds of European Russia." [1]

The farther East, away from the moderating influences of the Atlantic air currents and the Gulfstream, the colder and the drier becomes the climate in the same latitudes. Roughly speaking, the area of European Russia over which the Atlantic atmospheric regime exercises a predominant influence does not extend beyond an imaginary line drawn approximately from the White Sea near Archangel, along the Northern Dvina, to some point in the Urals, between Viatka and Perm. The country East of this boundary climatically belongs rather to Asiatic than to European Russia. The central belt of Russia, down to the steppes, on which it borders in the South and the East, is also accessible to Western air currents, whose influence, however, naturally diminishes from West to East, the continental character of the climate tending to assert itself increasingly both in the amount of atmospheric precipitations and in the extremes of temperature.

In the steppes, subject mostly to Eastern atmospheric influences, climatic conditions are entirely different. In the course of the summer months, especially from June to August, conditions here are generally favourable to the predominance in the steppes of Western and South-Western winds, but their beneficial effects are

[1] *Russland*, von Prof. Dr. A. v. Krassnow und Prof. A. Woieikow, Leipzig & Wien, 1907; p. 149.

mostly confined to the Western part of the zone. Before these air currents reach the South-East, they have practically lost their moisture in the arid atmosphere of the sun-baked steppes. Yet, they come at the right time. Indeed, in the South, approximately up to the latitude of Kiev (51–52° N.), the greatest amount of rainfall is registered in June, while about 35 % of the total annual precipitations fall in the course of the months immediately preceding the harvest. The absolute amount of precipitations, however, is far from sufficient, even in the most favoured parts of the zone. In the late spring the Southern steppes are often swept by strong North-East winds, which, originating in the cold wastes of Siberia, overrun the greater part of European Russia, but are especially violent in the vast open spaces of the South. Another enemy of the farmer in the steppes, more particularly in South-Eastern Russia, is the wind from the deserts of Central Asia and the Caspian shores which, though fortunately not frequent in summer, is disastrous in its effects on ripening crops.

The extreme South-East of European Russia, adjoining the Caspian Sea, removed as it is from Western atmospheric influences, climatically belongs rather to Asia than to Europe.

To sum up this preliminary outline of the atmospheric conditions of European Russia, whose closer consideration must be postponed till the description of the natural zones, it may be said that in every latitudinal belt, as it runs eastwards, the differences between the extremes of summer and winter temperatures tend gradually to increase, while the amounts of precipitations decrease continually. In other words, the farther East, the more continental becomes the climate.

Passing now to the properties of the soil in various parts of European Russia, it is essential to bear in mind that a division based on this principle, registering the transitions from one type of soil to another, corresponds very closely with that into belts according to climate and natural vegetation, thus giving an adequate outline of the natural background of the agricultural map of Russia. This connection between climate, vegetation and soil is due to obvious causes, but in Russia it is especially marked owing to the general flatness of the country, which contributes to the uniformity of conditions over very vast territories. In other countries, whose surface is broken by mountains, no such uniform belts would be met, horizontal and vertical zones being mingled and giving the general effect of patchwork to the map. In Russia, the natural zones stretch over thousands of square miles without interruption, going over into each other along their borders by imperceptible stages.

European Russia may be divided, according to climate, soil and natural vegetation, into the following principal zones, shown on map No. 1:

(1) Arctic zone;
(2) Zone of Northern Forests and Marshes;
(3) Zone of Mixed Forests;
(4) Middle-Russian (intermediate) zone;
(5) Black-earth zone;
(6) Zone of Southern Steppes;
(7) South-Eastern Arid Zone.

The two mountainous districts on the Southern fringe of European Russia, namely the Crimea and the Caucasus, though they certainly possess very marked characteristics of their own, are too small to be considered as separate zones. Moreover, nationally, as well as with regard to their farming, they are distinctly non-Russian, and thus fall outside the scope of the present study, dealing with the agricultural industry of Russia proper only.[1]

The *Arctic zone*, which occupies the extreme North of Russia, lies mostly beyond the limit of cereal cultivation. This limit passes approximately through the 65° N. in European Russia, the 59° in Siberia and the 51° in the Far East (Kamchatka). Only in some parts of the Murman Coast, exposed to the moderating influence of the Gulfstream, the cultivation of coarse barley is practised occasionally as far North as the 69°, actually beyond the Arctic Circle.[2] The farther East, the more severe the conditions, and the wider the zone inaccessible to the farmer.

The extreme North of the zone consists of Arctic barrens (*tundra*), with practically no vegetation, except mosses and lichens growing on a ground permanently frozen to depths varying from some 20 to about 42 inches. Owing to the frozen condition of the ground and the extreme scarcity of vegetation, the Arctic barrens have no soil in the strict sense of the word, the superficial layers of the subsoil being only slightly tinted with coarse *humus* to a depth not exceeding two inches or so.[3]

[1] The usual practice, however, is to consider these two mountainous districts as separate zones (Anisimov, Oganovsky, Viner and others). This, indeed, is the right course to take in attempting a detailed survey and dividing the country into relatively small natural zones, but does not fit into the scheme of the present work, which deals only with natural conditions in a general way, and with reference only to Great, Little, White and New Russia. Farming in the Baltic or Lithuanian provinces, as well as the native agriculture of the Caucasus, the Crimea or Central Asia, entirely different from that of Russia, is not considered here.

[2] Krasnov & Voieikov, op. cit., p. 151.

[3] P. Ototzky, section on "Soil" in the volume *Russia* of the *Russian Encyclopædia* of Brockhaus & Efron.

To the South of the Arctic Circle, the barrens change into mossy marshes with sparsely scattered undersized Northern conifers, forming the Northern fringe of the forest belt, into which they go over gradually. The cultivation of the most hardy and quick-ripening cereals, such as barley and, occasionally, rye and oats, becomes possible, and is actually practised to some extent, mainly on forest clearings, approximately on the latitude of Archangel, about the 65–66° N. Here, South of the White Sea, begins the next natural zone—that of Northern forests and marshes.

The *zone of Northern forests and marshes* is a very wide belt, stretching approximately from the latitude of Archangel down to a line drawn roughly from Petrograd through Vologda and Nizhny-Novgorod to the South of the province of Perm in the Urals. Thus, the Southern boundary of this zone runs through about the 60° N. in the West and descends to the 56–57° in the East of European Russia. The natural conditions of the zone, though they vary considerably from North to South, with regard to climate and soil, decidedly represent variations of the same type. The average temperature of the year in Archangel and the North of the zone generally is about zero C., while on the latitude of Petrograd it is about + 4° C. The period of vegetation in Archangel does not exceed 125 days; in the South of the zone it increases to about 155 days. From the agricultural point of view, this difference is one of enormous importance. The natural vegetation in the North of the zone is very poor. The forests consist mainly, if not exclusively, of conifers, growing on swampy ground covered with mosses, rough species of grass and an abundance of Northern berries. The conditions are not favourable to the formation of *humus*, and the natural fertility of the soil is accordingly low. Both climate and soil, therefore, make the North of the zone little suited to agriculture, and the latter is practically confined to the part of subsidiary occupation. In the South of the zone, namely the provinces of Petrograd, Vologda (Southern part), Novgorod (North-Eastern part), Kostroma, Nizhny-Novgorod (Northern part), Viatka and Perm, the natural conditions are more favourable to farming. Here, the area of mossy marshes is considerably smaller, while the richer vegetation and warmer climate allow the formation of more fertile soil. The predominant type of soil is, indeed, the same throughout the zone, but it differs in the proportion of *humus* and the thickness of the superficial layer enriched by it. The typical soil of the zone is known in Russia under the name of *podzol*: a word having no exact equivalent in other languages. The *podzol* is a thoroughly washed-out soil of rather coarse granular structure, varying in its mineral contents according to the nature of the underlying rock. Its superficial layer, generally whitish-grey in colour, contains distinctive streaks of darker substance representing the humified strata. Owing to rapid washing-out, the mixing of *humus* with the basic soil, which is mostly sandy, is never perfect, and most of the *humus* gets quickly sucked in and dissolved in the lower strata. The result is that the soil in the strict sense, that is the layer containing *humus*, is very thin and lacks uniformity; it is accordingly poor. In the South of the zone, where the natural vegetation is richer, the climate less severe and the ground not constantly soaked in water, the quality of the *podzol* is better, and in many places it changes into a transitional kind of soil, known as *podzolisty*, intermediate between that of the Northern forests and the sand and clay loams of Central Russia. Though, agriculturally, even the South of the Northern zone cannot be considered as favourable, its great wealth in natural meadows and pastures make it well suited for grazing and dairy farming.

It is only after crossing the Southern boundary of the Northern forest zone that one enters the really agricultural part of Russia, in which farming is more than an auxiliary occupation of the people.

South-West of the line Petrograd—Vologda—Nizhny-Novgorod lies the *mixed forest zone*, which includes the greater part of the provinces of Novgorod, Yaroslav, Kostroma and Riazan, as well as the provinces of Pskov, Tver, Moscow, Kaluga, Vladimir, Smolensk, Vitebsk, Minsk and Mogilev and the Northern districts of the provinces of Chernigov, Kiev and Volynia. The natural conditions of this zone are far more favourable to agriculture than those of the North. The climate is moderated by Western atmospheric influences. The average temperature of the year varies from + 4° C. in the North to + 7° C. in the South of the zone, and the differences between the extremes of summer and winter temperatures are much less pronounced than in the Northern and Eastern parts of European Russia. The rainfall reaches 20 to 24 inches. The period of vegetation varies from 160 to 180 days, according to locality. The country is wooded, though to a much lesser extent than the Northern zone, and the ground, especially in the Western part of the zone, is often marshy. The nature of the marshes, however, except in the North of the zone, where mossy swamps are common, is different, since here the typical marshes are really peat-bogs, which can be turned by drainage into a soil of considerable fertility. The properties of the soil within the zone vary considerably, though the most typical is still the *podzolisty* forest soil of varying *humus* contents and, accordingly, various fertility. Sands, clays and sand and clay loams of varying qualities are also well represented in this zone, especially along its Southern and Eastern fringes. The agricultural industry of this zone, favoured by the proximity of the principal markets for its produce, including both capitals—Moscow and Petrograd—is well developed and varied. Dairy farming here is developed to a greater extent than in any other part of Russia, and the standards of cultivation are superior to those of the other zones.

The next belt, bordering on the mixed forest zone in its Western half, up to about Nizhny-Novgorod, and from thence eastwards on the Northern forest zone, may be described as the *middle-Russian zone*, and possesses all the essential characteristics of a transitional or intermediate type. Here, patches of woodland lay scattered among vast open spaces, and the soil, by scarcely perceptible stages, changes from the poor *podzol* of the North, through a variety of clay and sand loams, into types approaching black-earth in constitution and fertility. The zone is not wide, especially in its Western part, where, in a narrow strip, it crosses the South of Volynia, the North of Podolia and the centre of the province of Kiev. On the left bank of the Dnieper it includes a large part of the provinces of Chernigov and Orel, the whole of the province of Tula, most of that of Riazan, the North of the provinces of Tambov and Penza, the South of the province of Nizhny-Novgorod and parts of the provinces of Simbirsk, Kazan and Ufa. At no point does its width exceed some 200 versts (about 130 miles). The natural conditions of the zone are very favourable to farming. The average temperature of the year varies from about + 8 or 9° C. in the West to + 4 or 5° in the East. The period of vegetation lasts from 180 to 200 days, the latter figure being reached in the Western part of the zone. The amount of rainfall is sufficient in the West, where it reaches about 20 inches and in the Centre, where it is about 16 inches, but in the East it does not exceed about 14 inches on the average. The typical soils are sand and clay loams, though the forest soils of Northern Russia (*podzol*) and the lighter varieties of black-earth are to be met occasionally in large patches penetrating deep into the interior of the belt. This zone, together with that of mixed forests, form the bulk of agricultural Russia outside the black-earth.

Immediately South of this zone begins the wide belt of *black-earth*, which, in its Northern half, still presents the familiar landscape of Middle Russia, while in the South it consists entirely of open steppes, without a grove of trees to relieve the monotony of the limitless plains. The black-earth runs from the Bessarabian border in a North-Eastern direction in a belt of varying width. Thus, on the meridian of Kishinev it is about 350 versts wide (approximately 230 miles); at that of Kharkov it increases to 600 versts (400 miles); at the longitude of Tambov it reaches 700 to 800 versts (about 500 miles). On the left bank of the Volga it narrows down again to some 350 to 500 versts (230–375 miles), while in Siberia it never exceeds some 150 to 200 versts (100–130 miles).

The Northern boundary of the belt passes through the extreme South of Volynia and the North of Podolia, the South of the province of Kiev, Chernigov, Orel, Tula, Riazan, Nizhny-Novgorod, Kazan and Ufa. The Southern limits run through the South of the provinces of Kherson and Taurida (above the Perekop, joining the peninsula to the mainland), the middle of the Don Cossack territory, the South of the province of Saratov and the middle of the province of Samara. The dividing line between the Northern, wooded, half of the zone and the steppes runs, approximately, from the Dniester through the North of the province of Kherson to Poltava, Kharkov, the middle of the province of Voronezh and the North of the province of Saratov to the left bank of the Volga.

Climatically, the black-earth zone, especially in its Eastern part, is typically continental, the atmosphere being very dry and the extremes of temperature very pronounced. The lack of precipitations is felt, to a greater or lesser extent, throughout the whole belt, but it is especially acute in its Eastern half, the region of the worst droughts and famines in the history of Russia. The average rainfall for the whole of the zone is about 18 inches, the maximum being in the North and West (20–21 inches), and the minimum (14–15 inches) in the South and East. Snow, in these parts, is of little use to the land, since the thaw in the open steppes, with no forests to delay it and make the process gradual, is too rapid. The rush of water from the melting masses of snow is so violent that it is not allowed time to penetrate deep into the soil and to build there a reserve of moisture. Instead of being beneficial to the land, the thaw thus rather serves to ruin the soil by washing away its most fertile strata and digging enormous ravines down the slopes. The average temperature of the year in the West of the zone varies between + 7 and + 10° C., while in the East, in spite of hot summers, it is as low as + 3° C., actually lower than that of the Northern forest zone on the latitude of Petrograd, where it is + 4°. The period of vegetation is long, varying from 180 days at Voronezh, in the middle of the zone, to 200 days in the West, and as much as 220 days in the province of Kherson.

The so-called black-earth is a variety of soil which, though naturally not uniform in constitution and mechanical properties, is distinguished by the presence in it of a very high percentage of *humus* and the thickness of the humified strata, reaching to a depth sometimes exceeding 40 inches. The richest black-earth is found along the central axis of the zone, running from the South-West to the North-East, roughly along the boundary between the wooded half of the belt and the steppes. North and South of this central strip, the soil changes gradually, until ultimately it goes over into the clay and sand loams of Middle Russia in the North and the "chestnut-coloured" soils of the Black Sea shores in the South. According to the thickness of the humified strata and the percentage of *humus* they contain, the varieties of black-earth may be classified as follows: [1]

(1) *Northern black-earth*, containing 3 to 6% of *humus* to varying depths, up to about 25 inches;

(2) *Rich black-earth*, containing 6 to 10% of *humus* in the West and 10 to 13% in the East of the belt, to depths of 40 inches or over;

(3) *Common black-earth*, with 6 to 10% of *humus* to depths reaching about 30 inches;

(4) *Southern black-earth*, with 4 to 6% of *humus* to a depth of some 25 inches.

The nature of the soil makes for an exceptional fertility of this zone, but its great drawback is the dry climate, which is responsible for the paradoxical fact that no other part of Russia is more liable to failures of crops and to famines than this belt, commonly referred to as the "granary of Europe." This is especially the case in the Eastern half of the zone, where frequent droughts combine with primitive methods of farming to keep the yield far below the level justified by the quality of the soil.

In the South, the black-earth belt borders on the zone of *Southern steppes*, with their "chestnut-coloured" soil, representing an inferior variety of black-earth. This zone stretches along the Northern shores of the Black Sea, from the South of Bessarabia, through the Southern half of the province of Kherson, the province of Taurida, the South of the province of Ekaterinoslav, the South-East of the Don territory, practically the whole of the Kuban,

[1] Prof. A. N. Chelintzev, *Agricultural Geography of Russia*, p. 26; also P. Ototzky, op. cit.

except an isolated island of heavy black-earth in the East, the Western half of the province of Stavropol and the South of the province of Samara. In this zone, the continental characteristics of the climate, present in the black-earth belt in a less marked degree, are greatly accentuated. The summer heat is intense, while the winter, though short, is extremely cold and rough, with violent winds and very little snow to protect the crops. The average temperatures of the year vary from + 6 to + 7° C. The period of vegetation is long, reaching 225 days at Odessa, in the Western part of the zone. The amount of precipitations is very small, the average for the zone being only 12 to 13 inches. The belt is of great importance agriculturally, mainly on account of its production of high-grade wheat and barley, exported in large quantities abroad, but it resembles the Eastern parts of the black-earth zone in the extensive methods of cultivation generally used.

Finally, in the extreme South-Eastern corner of European Russia, on the Northern shores of the Caspian, lies the *arid zone*, forming an intermediate stage between the steppes of European Russia and the deserts of Transcaspia and Central Asia. This belt occupies the North-East of the Terek region, the East of the province of Stavropol and the province of Astrakhan. The climate of this zone is extremely dry, the average yearly rainfall varying from 8 to 12 inches, of which only 2 to 4 inches fall in the course of the summer. The difference between summer and winter temperatures is extremely marked: the average temperature of July being + 24° C., while that of January varies from — 6 to — 18° C. The typical soil belongs to the variety known in Russia as "brown-earth," of which the principal mineral constituent is sand. The proportion of *humus* varies, but is generally rather low, owing to the scarcity of natural vegetation. The soil, however, is rich in nitrogen, which, wherever there is enough moisture, makes for great fertility and rapid ripening of plants.[1] Large tracts of land in this belt, next to the sea-shores, are covered with salt marshes. In its present condition, the agricultural importance of the belt is relatively small, but its productivity can be greatly increased by irrigation.

This short outline of the immutable natural characteristics of European Russia, as far as they affect the agricultural industry, shows how different are the conditions with which the farmer is faced in various parts of the country. It is on this natural background that the historical evolution of Russia and the play of

[1] P. Ototzky, op. cit.

economic forces which accompanied the development of the country from the ninth century of our era onwards, had drawn the contours of Russia's agricultural map, such as it was on the eve of the war and the revolution.

CHAPTER II

HISTORICAL AND ECONOMIC INFLUENCES

AMONG the historical influences to which the agricultural geography of European Russia is due, the oldest and the most fundamental one is probably that of colonization. Indeed, for centuries the history of Russia has been primarily that of the gradual colonization by the Russian people of the vast plains of Eastern Europe and, later, of Asia.

A brief reference to the outstanding points in the history of Russian colonization, which have left their mark on the agricultural map of the country, would therefore appear necessary for the proper treatment of the problem under discussion.

At the dawn of Russian history, two original nuclei served to form the beginnings of the Russian State. One, in the North-West corner of the Russian plain, close to the Baltic, was the Great Novgorod; the other, in the South-West, on the Dnieper, was Kiev. Both were situated on the " Great Route from the Normans to the Greeks "; both owed their origin and prosperity to the enterprise of the Norman warriors and traders. In the North, the Republic of Novgorod gradually extended its domains, until, by the fourteenth century, it brought under its sway the whole country up to the White Sea and the Northern parts of the Ural range. In the South, the influence of Kiev spread over the steppes, inhabited mainly by roaming tribes, and penetrated the woods of Central Russia. Both Kiev and Novgorod supplied colonists who went to settle in the centre of the Russian plain, in the triangle formed by the Volga and its tributary, the Oka, thus laying the foundation of a third nucleus, known in history as the " Rostov-Suzdal Russia." When, in the first half of the thirteenth century, Kiev, already weakened by continual feuds between rival princes, was laid waste by the Tartars, while Novgorod, spared the Mongolian invasion, was holding aloof and pursuing her own policy, it was Rostov-Suzdal Russia that became the central nerve of the nation. With the rise, in the fourteenth century, of the Moscow

Princes to a dominant position among the local rulers, it was there that the consolidation of Russian forces against the Tartars began. The power of the Moscow Princes and the comparative order and security which reigned in their dominions, led to a great increase of population in Central Russia and a considerable development of farming in that part of the country. By the end of the fourteenth century, the Moscow Princes felt strong enough to begin a determined campaign against the Tartars, gradually driving the invaders South and East into the steppes. As, in the course of the fifteenth century, the Tartars were driven back farther and the boundaries of the Russian State were extended, Russian colonization followed on the tracks of the retreating enemy, spreading in a South-Easterly direction. In the middle of the sixteenth century, when the first agrarian crisis in Russian history, due to the relative overpopulation of the Moscow region, began to assume serious proportions, the conquest by Ivan the Terrible of Kazan and Astrakhan, while striking the death-blow to Tartar power in Eastern Russia, extended Russian possessions to the middle and lower currents of the Volga and the Urals. Beyond the latter, the formal annexation of Western Siberia followed during the closing years of the same reign. Wherever the Russian frontier was moved, its extension was followed or, not unfrequently, even preceded by Russian settlers. The principal current of colonization proceeded South-East, towards the fertile black-earth of the Volga and the Don, in the wake of the Don Cossacks who had established an advanced outpost to the North of the Sea of Azov. There has been also a certain amount of migration towards the Ukraine, the Eastern half of which, on the left bank of the Dnieper, was united to Moscow in 1649. From that time onwards, the further colonization of European and Asiatic Russia has been the joint work of all the three branches of the Russian people—the Great, Little and White Russians. Steadily, Russia spread her frontiers and her population South and East, until, in the second half of the eighteenth century, she reached in the South her natural boundary—the Northern shores of the Black Sea. The colonization of New Russia, as these Southern provinces were called, proceeded practically throughout the nineteenth century, settlers being drawn from every part of the old country and even from abroad. The movement East, to the left bank of the Volga, though it began earlier, developed more slowly, since the advantages of settlement in the South, as well as the policy of the Government, anxious to populate the new territories, diverted the stream of emigration towards the Black Sea. The result was that, East of the Volga, the growth of population and the progress of farming were somewhat retarded, as compared with New Russia.

At the beginning of the twentieth century, European Russia, though still allowing, on its outskirts, sufficient scope for land-settlement on a limited scale, aiming at a partial re-adjustment in the distribution of rural population, had no more room for colonization in the strict sense of the word, as a movement *en masse*. The latter had to proceed farther East, beyond the Urals, to Siberia, the Far East and Central Asia.

The colonization of Siberia is a relatively recent development, following on the occupation of the space available in the South and East of European Russia and the great increase in the population of the older regions in the second half of the last century.[1]

The development of migration beyond the Urals in the early years of the current century, when the first attempts have been made to organize it systematically on a large scale, will be dealt with in a separate chapter.

The colonization of a large part of European Russia was so recent that, in spite of other influences which, in the course of the last few decades, had partly blurred its traces on the agricultural map of the country, its marks can still be clearly discerned in the geographical distribution of farming systems. In fact, the regions of ancient settlement, from which migration proceeded to other parts of Russia, combine a denser population with more intensive systems of farming. In the rest of the country, as a rule, the systems of farming in general use are the higher, the nearer the district is to the old centres of population, and the more distant the date to which its occupation and development can be traced. Since, as I have pointed out, the trend of migration has been mainly in the direction of East and South, the farther one advances in this direction, the more extensive are the systems of farming one comes upon, until, in the farthest South-East of European Russia, bordered by the Volga, the Ural river and the Caspian, one is actually confronted with semi-pastoral conditions of husbandry. Thus, in spite of the influences which, in the course of historical development, have come to be superimposed on that of colonization, a glance at the agricultural map of European Russia recalls to the mind, in the first instance, the story of a people which, having once overflowed the borders of the small area where it had originally grown into a nation, has been keeping on the move ever since,

[1] Professor A. Kaufmann, *Emigration and Colonization*, 1905, p. 5, where he dwells on the subject. The author is one of the best Russian authorities on the agrarian problem and, especially, on Russian colonization and its history.

spreading over the plains of Eastern Europe, as water would spread over a flat surface. Whenever the districts already settled for some time began to feel the pressure of growing population on the available agricultural land, the problem was solved partly by raising the standards of cultivation and the yield of crops, partly by the migration of a portion of population to other localities, in which land was still plentiful.

It may be said that, at least up to the nineteenth century, natural conditions and the trend of migration were the two principal forces which helped to shape the agricultural map of Russia. The country was still living, in the main, under conditions of self-sufficient natural economy. The nation had not yet grown into an organic economic entity. It was rather a mechanical combination of more or less isolated economic units linked to the central authority of the State by the bonds of taxation or service. The economic relations between such individual units were still too weak and too casual to exercise any marked influence upon their conduct. Economic factors, accordingly, could not be expected to exercise the same degree of influence that they have in modern society. Migration, unless deliberately directed to certain localities by the State, with a view to certain political objects, proceeded to those districts in which the land was most fertile and the natural conditions generally favourable. Not unfrequently, as in the case of religious dissenters after the revision of the books in use in the Russian Church in the seventeenth century, the economic advantages of situation were deliberately avoided, the settlers seeking the most remote and out-of-the-way places for their new homesteads. Under these primitive conditions, the geographical distribution of farming was determined by natural environment, on the one hand, and by the territorial expansion of the Russian State, on the other. Only gradually, as Russia was drawn out of her economic and political isolation and began to develop industry and trade, as well as to re-organize her administration and her army on Western models, economic factors began to assert themselves increasingly in her agricultural organization. Yet, until the mechanical inventions of the nineteenth century and the great economic and social transformation inaugurated by the emancipation of peasants in 1861, the progress of commercialization had been very slow and the typical form of economic life was that described as "natural economy." The bulk of agricultural produce was still consumed on the farm or in the farmer's household. The difficulties of transport, indeed, necessitated the production on every farm of most, if not all, the bulkier necessaries, irrespective of the special advantages or drawbacks of the locality. This naturally made for uniformity of production throughout the country, as far at least as the principal crops were concerned. For specialized farming there was practically no scope, except in the immediate vicinity of the towns. The latter, however, still produced a large part of their necessaries themselves and kept numerous herds of dairy cattle. The markets for agricultural produce were strictly local, protected from outside competition by the prohibitive costs or even the absolute impossibility of transport over long distances. The local and uncompetitive character of the markets was evident from the extremely wide range of variations in the prices of cereals in different parts of the country. Thus, according to contemporary authority, in 1797–1803, the prices of rye, per pood, were : 46 copecks in Astrakhan, 23 copecks at Kiev, 62 copecks at Moscow and 90 copecks at St. Petersburg.[1] Moreover, until the middle of the eighteenth century, the internal trade in cereals was subject to transit duties levied on passing from one province to another, which were a serious hindrance to commerce. They were repealed by the Empress Elizabeth in 1754, the loss to the Treasury being made good by an addition of 13 % to the custom duties on foreign trade.[2] The foreign trade in grain was still in its infancy. Besides the difficulties of transport, the condition of foreign markets, as well as the attitude of the Russian Government itself with regard to grain exports, prevented its development. In Russia, up to the second half of the eighteenth century, the attitude of the State towards the export of cereals represented a blend of medieval conceptions, similar to those which governed English commercial policy from the Plantagenets to the Tudors, and of the ideas of mercantilism which came into fashion under Peter the Great. Thus, on the one hand, Russia had not yet shaken off the medieval conception of the State as *bonus paterfamilias*, whose first concern it is to ensure a sufficiency of necessaries to its dependents. On the other hand, the Colbertian view began to gain ground, which conceived the people primarily as the principal means by which the wealth and power of the State are increased. The State now assumed an independent existence as a Leviathan. Different as the two conceptions were fundamentally, as typified in European history by the antithesis of *le bon roi Henri Quatre*, on the one

[1] Figures based on the data of Pallas and Falk, as quoted by Storch in his *Course of Political Economy*, *q.v.* P. Liashchenko, *Essays on the Agrarian Evolution of Russia*, 2nd ed., 1923, Vol. I, p. 100 (Russian).
[2] Liashchenko, *Essays*, p. 105.

hand, and *le roi Soleil*, on the other, in practice, when it came to the question of grain exports, they were both on the side of restriction, though for different reasons. Accordingly, the Russian commercial policy in the eighteenth century differed little from that of the Moscow Tzars of the seventeenth. Then, the export of cereals was directly controlled by the State, under a system virtually amounting to Government monopoly. In the eighteenth century, though permitted to private individuals, it was kept under strict Government control by means of export duties which varied in accordance with the movement of prices on the home market. It was not before the middle of the eighteenth century that the restrictive policy began to show signs of some relaxation. The change was probably due, on the one hand, to the extension of cereal cultivation into the steppes, where, in the second half of that century, large farming with serf labour began to develop with considerable rapidity. The landowners of the black-earth belt were, accordingly, in need of new markets, and the unprecedented influence exercised in Russia by the landowning gentry under the female successors of Peter the Great throughout the eighteenth century enabled them to promote a series of legislative measures aimed at protecting and encouraging agriculture. Thus, by an Ukaze of 1754, the spirit distilling industry, which was one of the most profitable productions connected with farming, was made a monopoly of the gentry. This was a profitable outlet for the surplus of cereals, since spirit in those days was distilled almost exclusively from rye. Very characteristic were the Commissariat Regulations of 1758, which insisted on the elimination of middlemen from army contracts for cereals and other foodstuffs and the direct purchase of supplies from landowners, with a view to enabling "those subjects who are directly interested in farming" . . . "to profit by it." The Government went even so far as to order that all cereals for Russian armies campaigning abroad shall be brought from Russia, irrespective of costs, "in order to provide some real encouragement to agriculture." Once having inaugurated this policy of encouragement of the agricultural interests, the Government was naturally led to a thorough revision of its policy with regard to grain exports, and in 1762, under Peter III, an Ukaze was issued providing for the repeal of export duties on cereals. In effect, however, the duties have not been abolished until the beginning of the nineteenth century, when they finally disappeared. The Government itself, moreover, since the second half of the eighteenth century, with the increase of Russia's intercourse with the outside world and the beginning of her credit operations abroad, which dated from 1769, when she raised her first foreign loan in Holland, was becoming increasingly interested in the development of exports. This, probably, had also contributed to the modification of its policy with regard to grain exports, which were still practically negligible. Indeed, in 1768–70, the export of grain from Russia did not exceed, approximately, 560–580,000 poods to the value of 114,000 roubles a year, and of the total value of foodstuffs exported abroad grain represented only 1%.[1] Apart from corn duties on foreign markets, the conditions of transport by land and by sea, before the advent of the steam traction engine, tended also to restrict the development of Russia's foreign trade in cereals. The result was that, though since the middle of the eighteenth century, grain exports have been increasing, until in the first quarter of the nineteenth they reached about 20% of the total value of Russian exports, they did not assume such proportions as could really have influenced the agricultural geography of a country like Russia. In 1778–80, Russia exported, on the average, about 2·3 millions of bushels a year; in 1806–10, just over 3 millions of bushels; while in times of war or of crop failures her exports fell to much lower figures: thus, in 1801–5, they averaged about 665,000 bushels only.[2]

It was not until well into the nineteenth century, that signs of the dissolution of the old régime of isolated natural economy began to appear, and economic factors began increasingly to assert their influence over the agricultural industry of Russia. The introduction into Russia of the mechanical inventions which revolutionized the economic life of Western Europe, and more particularly, that of steam transport by land and by water, on the one hand, and the Emancipation of peasants, in 1861, on the other, were the two outstanding events of the last century in Russia, whose importance for the evolution of farming and for the agricultural geography of the country could hardly be overestimated. By breaking up the old organization of society, based on bondage which, in the course of the eighteenth century, had degenerated into nearly complete slavery, the Emancipation opened the way to the re-organization of the country's economic system on modern competitive lines. The basis of farming, large and small alike, was entirely changed by the reform, and the importance of money, both as working capital and as measure of economic values in the

[1] Liashchenko, op. cit., p. 108.

[2] B. Pokrovsky, "Commerce exterieur" in the publication of the Russian Ministry of Finance, *La Russie à la fin du XIX siècle*, edited by V. Kovalevsky, Paris, 1900, pp. 699–700.

agricultural industry, was increased enormously, as compared with the conditions which had existed under serfdom. Both the large landowner and the peasant had to readjust their whole outlook, as far as economic relations were concerned. The immediate effect of the reform was that it placed both the large landowner and the peasant in a position in which, first and foremost, they needed ready money, the one to pay wages to those peasants who had hitherto worked on his estate as serfs and received their remuneration in the form of land, which they cultivated on their own account, and the other—to pay, besides taxes, the annual instalments of the redemption price of his holding. Thus, the Emancipation, besides making the peasant a free man and the owner of his holding in the village, also pulled him by the root from the old system of self-sufficient natural economy, and forced him to look round for some sources of ready money. To get hold of the latter, the peasant had to sell on the market either his produce, or his labour, and in those localities in which he had no sources of cash except the increase in his own production of grain and other agricultural products, he was forcibly driven into commercialization. In the districts possessing considerable local industries or other sources of earnings, sometimes paying the peasant far better than the cultivation of his own holding, farming not unfrequently became relegated to the position of an auxiliary occupation, responsible only for some part of the family's income. This was the case in most provinces of the Northern half of Russia, in which forestry, domestic industries and other outside occupations were more profitable than the cultivation of the relatively poor soil of these localities. The peasants remained on the land and continued to cultivate it, but its relative importance as a source of income was reduced by the effect of the Emancipation and its accompanying redemption of holdings, which emphasized the need for cash. In the fertile and purely agricultural Southern half of Russia, where, apart from occasional employment on the adjoining large estates, of which, moreover, many have ceased cultivation on their own account since the Emancipation, the peasant had practically no outside sources of money, the position was entirely different. The peasant here had to obtain cash from his land or, if this was insufficient to provide a marketable surplus, to lease or buy more land in addition to his own holding. Thus, a powerful impetus to the extension of cultivation had been given throughout the black-earth zone and the steppes of Southern and Eastern Russia. Besides turning into arable every available bit of their own land, the peasants leased all the land in the neighbouring large estates they could lay their hands upon, with the result that their competition for leases raised the rents far above their economic value and encouraged the landowners to let their estates to small peasant tenants, instead of cultivating them on their own account. Thus it came about that, in the Southern and Eastern steppes, which, at the time of the Emancipation, still possessed large tracts of virgin black-earth, very soon virgin soil became practically non-existent, most of the land suitable for agriculture having been ploughed up and more or less exhausted by constant cultivation, without regular fallowing or the use of manure. With the development of steam transport by rail and by sea and the rapid growth of Russian cereal exports to the industrial countries of Western Europe in the second half of the nineteenth century, this extension of arable farming throughout the black-earth and the steppes formed the salient feature of Russia's agricultural evolution since the Emancipation.

The influence of the rapid development of farming in the black-earth zone on the agricultural geography of Russia was that, in the second half of the last century, the centre of gravity of Russian agricultural production had been shifted towards the South-East, into the steppes.

The extent to which, only a quarter of a century after the Emancipation, in 1887, the available agricultural area of the black-earth belt and of the steppes had already been turned into arable, may be seen from Map No. 2. The map shows the percentages in each province of the land actually under plough relatively to the area suitable for agriculture, that is to the total area minus forests, waste and unreclaimed land and waters, according to the Survey of 1887. The map shows that, practically throughout the black-earth and steppes zone, the percentage of arable was actually higher than nearly in any province of the oldest and most densely populated parts of the country. The black-earth belt and the Southern steppes appear as the zone of extensive cereal farming, into which they have developed under the influence of the forces that have been at work since the Emancipation. It may be seen that of the virgin soil of the Russian steppes there was little left even in 1887, forty years since.

Among the influences affecting the agricultural geography of Russia after the Emancipation, the most widespread and far-reaching has probably been that of the extension of the railways. The construction of railways, indeed, had hardly begun before the Emancipation, and the total length of the lines in operation in Russia, in 1861, did not exceed some 1,400 miles. The pressing need for more railways was recognized during the Crimean War, and the problem was tackled in earnest immediately on its termination, when the "General Company of Russian Railways" was constituted and entrusted with the management of the existing lines and the construction of new ones. Hitherto, the existing railways only joined the capitals with the Western frontiers, while the agricultural areas and their connections with the home and foreign markets were practically ignored. The Imperial Ukaze concerning the establishment of the new Company charged the latter, in the first instance, with the task of repairing this important omission. The Ukaze provided, indeed, for the construction of railways "from St. Petersburg to Warsaw and the Prussian frontier, from Moscow through Kursk and the lower Dnieper to Feodosia, and from Kursk and Orel through Dvinsk to Libau. Thus, continuous railway lines through 26 provinces must join together three capitals, the principal navigable rivers, the districts producing a surplus of cereals and two harbours in the Black Sea and the Baltic, which are open almost all the year round. This will facilitate exports, secure the necessary imports from abroad and provide for the needs of the home market." [1] By the establishment of the Company of Russian Railways, a real start had been made in the construction of the first skeleton of a net which would meet the needs of the agricultural industry in Russia. In the course of the following decades, through many vicissitudes, the brunt of which fell mostly on the Treasury, the railway net in European Russia extended continually, the lines converging on Moscow as the web of a spider, thickest in the centre. The progress of construction may be seen from the following table, showing the fluctuations in the building activity at different times: [2]

	Total constructed (One verst = 0·66 miles)	Yearly average
1865–1878	17,615 versts	1,258 versts
1879–1890	7,793 ,,	641 ,,
1891–1902	24,934 ,,	2,077 ,,
1903–1912	9,855 ,,	985 ,,

In January, 1913, Russia had about 64,000 versts (43,000 miles) of railways actually open for traffic, of which about 48,000 versts were in the European part of the Empire.

The distribution of the railway net over the country played a most important part in the development of Russian farming in

[1] Liashchenko, op. cit., Vol. I, p. 145.
[2] K. Zahorsky & E. Geidanov, "Internal Transport" in *Russia, its Trade and Commerce*, edited by A. Raffalovich, London, 1918, p. 232.

the course of the last few decades. Great changes have been effected by the construction of railways in the relative economic position of various districts. Some localities which, before, had been handicapped by their distance from the consuming centres, now found themselves linked up with the latter by rail; other districts, more advantageously situated before, saw themselves deprived of their former advantages. The grain trade, which used to gravitate towards certain convenient points of shipment on the Volga or other navigable rivers, saw its needs better served by the railways and, instead of clinging to its few former centres, became scattered over an enormous number of railway stations. Yet, up to the very last, the beneficial effects of a relatively dense railway net were almost entirely confined to the part of the country South and West of the Volga. North and East of these limits the railways were still so few and far between that their influence on the agricultural conditions was very limited. The country East of the Volga, accordingly, profited relatively little by the railways; rather, it found itself put at a disadvantage relatively to some other localities, equally or even more distant from the markets, but now provided with means of cheap transport. As a result, while in other localities railways provided strong stimuli for the intensification of farming, the Eastern provinces were not encouraged to raise the standards of cultivation, local prices for cereals being the lowest in European Russia. Thus, in 1906–1910, the average spot autumn prices of the principal cereals, according to regions, were as follows:

Regions	Rye	Wheat	Oats	Barley	Average
		Copecks per pood (36 lbs.)			
Northern	118	(145)[1]	80	107	(112)
North-Eastern	76	101	54	73	76
Petrograd	113	(130)	75	94	(103)
Moscow	95	(114)	67	84	(90)
Western	95	102	69	79	86
South-Western	85	103	68	75	82
Ukraine	85	105	62	68	80
Central Agricultural	82	101	59	(70)	(78)
Middle Volga	81	108	64	77	82
Eastern	69	98	59	67	73
New Russia	82	99	72	66	80
South-Eastern	78	100	65	61	76
Caspian	83	(115)	71	67	(84)

[1] Figures in brackets represent incomplete averages.

In the Northern provinces of Russia, dependent mostly on cereals imported from the cereal-growing districts, the lack of adequate railway facilities tended, as may be seen from the table, to keep grain prices very high. Indeed, in the province of Archangel, in

1906–10, the price of rye averaged 129 copecks per pood; in the provinces of Petrograd and Olonetz it reached 121 copecks. In the Petrograd region, which could not boast a highly developed railway net, except in the immediate vicinity of the capital, cereal-growing was actually encouraged to some extent, in spite of natural disadvantages and the resulting high costs of cultivation, by the level of prices which ruled in the more remote districts.

The immediate effect of the construction of railways was that they raised prices for agricultural produce on the spot of production, thus benefiting the farmers and encouraging agricultural progress. Apart from this general effect, the railways exercised a more subtle influence on the distribution of farming systems in the localities traversed by them. This was due to their having reduced the costs of transport of produce over long distances to the ultimate markets and, at the same time, relatively increased the importance, in the farmer's bill of costs, of the expense involved in hauling it by road to the station. With the railways, it was not the distance from the station to the ultimate market, but that from the farm to the station, that became decisive for the system of farming. This new factor affecting the distribution of farming systems throughout the country had played an enormous rôle in Russia. The haulage of grain by road was estimated, before the war, to have cost, on the average, 0·15 to 0·25 copecks per pood per verst, or approximately 10 to 20 times the cost of transport by rail,[1] and the rapid rise of this item of costs to prohibitive figures with the increase in the distance had very important consequences for the evolution of farming in the districts traversed by railway lines. The effects of the railways in this respect could best be observed in the case of dairy farming and the cultivation of potatoes, which both exercise a great influence on the system of farming generally. Of the products of dairy farming, it is fresh milk that plays a decisive part in this connection, while potatoes, as a field crop included in the rotation, is characteristic because of its great bulk and its inability to bear the cost of long-distance transport. Both milk and potatoes are brought to the consuming centres by railway from considerable distances. Thus, in Russia, before the war, the greater part of the milk supplies of Petrograd and Moscow were brought from distances ranging from 100 to 200 versts (about 66 to 130 miles). Of the total quantity of milk handled by the Russian railways in 1901–6, 18·5% travelled 100 to 200 versts; 12·5%—200 to 300 versts, and 12·5% over 300 versts.[2] The great consuming centres, accordingly,

[1] B. Brutzkus, *The Economics of Agriculture* (Rus.), p. 128.
[2] Brutzkus, op. cit., p. 157.

extended their demand over very large territories, the railways converging on them acting as feeders. Yet, along the railway lines the influence of this demand actually affected a very narrow strip of only a few versts on either side of the lines, within which area dairy farming, with all its consequences for the whole system of husbandry, could be kept going by the production and marketing of fresh milk. The condition of country roads and the instruments of transport suitable for negotiating these primitive thoroughfares did not allow fresh milk to be brought to the railways from distances exceeding, at the very best, some ten miles, and even this could be considered as exceptional. Beyond this restricted area, the dairy farmer was compelled to concentrate on the production of butter, as a far more transportable article, but one requiring a greater outlay of capital and, as a general rule, considerable modifications in the whole system of farming with a view, on the one hand, to the full exploitation of the equipment and labour and, on the other, to the profitable utilization of buttermilk. The bulky nature of potatoes, too cheap at their weight to stand long-distance transport, also confines their production, as a field crop grown for the market, to within a short distance from the railways or waterways, or from local distilleries and starch factories, using potatoes as raw material. The importance of potatoes, as a regular course in the rotation of crops, both in their influence on the soil and in the great quantity of offals they yield to the farmer, enabling him to improve the feeding of his stock and, especially, to develop pig-keeping, is very great. By raising the standards of farming in those districts in which they are widely adopted, potatoes play a great part in agricultural progress, particularly so in the Northern half of Russia, where the cultivation of sugar-beet is impossible. As a general rule, since both potato-growing and dairy farming depend on convenient means of transport, they are mostly met with in the same localities or even farms, and their combination tends, as a rule, to raise the districts concerned considerably above those parts of the same neighbourhood which do not enjoy similar advantages of situation with regard to the railways.

The era of railways also brought into play another set of factors of great importance in the evolution of farming and the shaping of the agricultural map of Russia, by enabling the State to exercise a powerful influence over the economic development of the country by means of its railway-building policy, on the one hand, and by its control over rates, on the other. By the extension of the railway net, the State did, whenever it saw fit, deliberately encourage the development of certain districts. By the control of freight rates and their adjustment, it could, and actually did, intervene in the course of this development, meeting the situations as they arose. Until the eighties of the last century, the Government, faithful to the then fashionable maxim of *laissez faire*, left the fixation of rates, within the limits set by their respective statutes, to the railways themselves, and it was only in 1889 that the necessity had been recognized of concentrating the control of railway tariffs in the hands of the State, through the Railway Department of the Ministry of Finance. The leading principle to which the policy of the Government had to conform in this matter, as expressed in the resolution of the Tariff Conference of 1889, was that "each producing district should be given convenient access to the greatest possible number of points on the frontier or the sea-shore, to which it gravitates."[1] The interests of the export trade would, accordingly, appear to have played a conspicuous part in the direction of the tariff policy from the outset, and had certainly exercised a considerable influence on the agricultural development of the South and South-East of Russia. The preparation of the monetary reform, which was under way at the time, was a highly important factor in this policy, and its influence cropped up again at the Tariff Conference of 1893, which, once more, emphasized the need of encouraging the export of the surplus of cereals, though it also laid stress on the necessity of conciliating, as far as possible, the interests of foreign trade, on the one hand, and those of the home market, on the other.

One of the most important features of the tariff policy of the Government, since its inception, was the introduction of differential through rates, which encouraged the development and the commercialization of farming in the more remote parts of the country, by fixing relatively low rates for long runs. Though the principle of differential rates was not new, hitherto they had been left to the discretion of the individual railways, and, since the latter allowed rebates only on their own sections of the route, it was practically inoperative. The graduation of rebates which, at the beginning, had been very steep and, therefore, was considered by numerous experts as unduly favouring the remoter districts of the country at the expense of its more central parts, as well as encouraging exports abroad at the expense of the home market, underwent certain modifications during the subsequent general revisions of rates.

One of the special measures which had exercised a considerable influence on the development of the grain trade in Russia and on the agricultural industry, especially in the Northern parts of the black-earth belt, was the system of rebates in favour of Baltic ports,

[1] Liashchenko, *Essays*, Vol. I, p. 163.

which gave them considerable advantages over those of the Black Sea. For the agriculturist of central Russia, situated half-way between the Baltic and the Black Sea, this differentiation was a matter of considerable importance, in so far as it enabled him to produce grain for export more profitably than he could have done otherwise. The rebates, accordingly, besides increasing the trade of Baltic ports, provided, to some extent, an artificial stimulus to the exports of grain from such districts as, in the natural course, would have had to depend primarily on the home market. The system of rebates in favour of the Baltic was abolished in 1893, with the result that, on the one hand, the exports through the Black Sea increased considerably, and, on the other, that the agricultural production of the Northern black-earth belt became almost entirely dependent on the demand of the home market for the disposal of its output.

The existence of tariff rebates on long runs, which was an essential feature of the system of rates on the Russian railways, had the effect of making the influence of the foreign demand for cereals felt very far inland. The cost of transport of grain, at the rates current before the war, worked out at the following figures:[1]

1,000 versts	(660 miles)	18·88 cop. per pood (36 lbs.)
2,000 ,,	(1,320 ,,)	27·95 ,, ,, ,,
3,000 ,,	(1,980 ,,)	37·50 ,, ,, ,,
4,000 ,,	(2,640 ,,)	50·50 ,, ,, ,,

Accordingly, even provinces of European Russia as far removed from the frontiers or the sea-shores as that of Ufa, if they were in a position to produce high-grade cereals, worked to a large extent for export. In fact it was estimated that, of its total marketable surplus of cereals, mainly hard wheat and rye, the province of Ufa exported 57·7% abroad in 1909–11.[2]

Since the dependence on either the home or the foreign market for the disposal of their agricultural produce could not fail to exercise a certain influence on the development of various localities, the grouping of provinces according to the disposal of the bulk of their surplus of cereals had a direct bearing on the agricultural geography of Russia. In considering this question it is necessary to bear in mind that the gravitation of the cereals towards either the consuming centres at home, or the frontiers and ports of shipment, is dependent on a whole set of causes, partly natural, partly economic. Of the former, it is the situation and the suitability of

[1] Brutzkus, *Economics of Agriculture,* p. 128.
[2] Kondratiev, H., *The Grain Market and its Control during the War and the Revolution,* Moscow, 1922 (Rus.), pp. 22–3.

the soil and climate to the production of certain kinds and grades of cereals that plays the principal part in deciding on the ultimate destination of the produce. Among the economic factors, the decisive part is played, in the first instance, by the means of transportation and its costs. Thus, in the case of the province of Ufa, while natural conditions enable it to grow hard wheat of high quality, for which there is a ready demand on the world markets, the geographical remoteness from the frontiers is compensated for by the application of the differential grain rates on the railways. In other provinces, different combinations of conditions in the case of various cereals create a very complicated system of gravitations, certain kinds or grades of grain being produced for export, while other cereals are grown for the home market. As a result, production for the home and the foreign markets, in varying proportions, was carried practically in every province which had any surplus of cereals at all, and there were very few localities in which, before the war, either the home or the foreign demand could be said to have exercised a paramount influence, to the more or less complete exclusion of the other. The nearest approach to this could be found in the provinces of New Russia and the South-East, situated on the very shores of the Black Sea and thus well adapted for export. In these provinces, indeed, farming depended, primarily, on the demand of the foreign markets for Russian wheat and, lately, for barley. The percentage of the total surplus of the four cereals shipped abroad from these provinces in 1909–11 was estimated at the following figures: [1]

Province of Kherson	86·6%
„ „ Ekaterinoslav	64·4%
„ „ Taurida	85·6%
„ „ the Don	70·4%
„ „ „ Kuban	83·0%
„ „ Stavropol	81·1%

Here, the demand was for cereals, and more particularly for two of them, namely wheat and barley, and the whole region had grown up, in the agricultural sense, under the influence of this demand. Accordingly, it concentrated on cereal-growing to the utmost possible extent, at the expense both of live stock and of the natural qualities of the soil, which was being gradually exhausted by continuous cropping, without regular fallow or any systematic attempts at restoring the fertility of the soil by manuring.

The demands of the home market for agricultural produce in general and for cereals in particular, though also spread over the

[1] Kondratiev, op. cit., p. 22.

whole country, naturally played an especially prominent part in the agricultural provinces of Central Russia. Among these, the following disposed, in 1909–11, of one-half or more of their available surplus of cereals on the home market: [1]

Province of Riazan	75·0%
„ „ Tula	67·9%
„ „ Orel	74·2%
„ „ Kiev	61·5%
„ „ Poltava	71·2%
„ „ Kharkov	80·7%
„ „ Voronezh	70·2%
„ „ Tambov	57·6%
„ „ Saratov	55·5%
„ „ Samara	62·7%
„ „ Kursk	77·2%

In these provinces, the demand of the home market could be expected to exercise a more or less decisive influence on the agricultural industry, and this was actually the case. The influence was twofold, and affected, on the one hand, the direction of cereal production and, on the other, the whole organization of the arable and stock farming in the localities concerned. The Russian internal market for cereals, besides the cities and industrial centres, included also the greater part of the rural population of the under-producing provinces of the Northern half of Russia, which depended to a large extent on grain imported from the localities which produced a surplus. While the cities formed a market for all kinds and classes of cereals, the countryside of the under-producing provinces wanted, in the first instance, large supplies of rye, as the staple bread-stuff of the Russian peasant. The cultivation of rye, which dwindled down in other parts of the country, owing to the fall in the demand for it in Central and Western Europe, as well as in Russian cities, was, therefore, well maintained in the provinces dependent on the home market, and especially in the Middle Volga region. The principal effect of the dependence on the home market, however, and one of great advantage to the agricultural industry in the provinces concerned, was the far greater variety of demand, which prevented the excessive specialization of farming and encouraged the introduction of new courses in the rotation, as well as the general raising of the standards of cultivation. While in the South, where agriculture was mainly controlled by the foreign demand for a few staple kinds of cereals in great bulk, economic considerations induced the farmer rather to extend his cultivation than to improve it by the adoption of a scientific rotation or the development of the stock branch, in the provinces working primarily for the home

[1] Kondratiev, op. cit., p. 23.

market the agriculturist had open to him many profitable courses of development.

Thus, the gravitation towards either the home or the foreign market, according as to which of the two played the more important part, was a great factor in the evolution of farming in various localities and must, therefore, be considered as one of the forces which have been, and still are, instrumental in shaping the agricultural map of Russia. With the progress of industrial development, which was so conspicuous a feature of the years immediately preceding the war, and the consequent increase in the capacity of the home market for agricultural products, the area under the influence of internal demand had been gradually extending, to the great advantage of the agricultural industry as a whole. With the progress of industrialization, the agricultural geography of Russia had been undergoing an important process of development, the distinctive features of the several agricultural regions being gradually brought out in greater relief. The development was cut short, for the time being, by the war and the revolution ; yet, so fundamental was the nature of those causes which had called it forth in the past, that, on the economic recovery of the country, the general outline of the map will probably be found to have changed but little.

CHAPTER III

THE AGRICULTURAL REGIONS OF RUSSIA

In the two preceding chapters the influences have been outlined which contributed in making the agricultural map of Russia what it was on the eve of the war and the revolution. Natural conditions, on the one hand, and numerous other agencies, social, economic and political, on the other, were responsible for considerable local variations in the agricultural industry. According to the standards of the countries of Western and Central Europe, the range of these variations was, indeed, rather limited, since in Russia a whole set of causes made for greater uniformity of farming throughout the whole extensive territory. Among these causes, one of the most important consisted in the flatness of the country, which enabled the same systems of farming to be practised over large tracts of territory. In Western and Central Europe the broken relief of the country necessitates a much more minute adaptation of farming to local conditions, thus making for greater variety in the agricultural industry. Southern Germany, Austria and Switzerland, not to speak of numerous other examples, may be pointed out as outstanding instances of variety in this respect, due to topographical conditions. In the plains of Russia proper this cause of variation is entirely absent.

The less advanced economic development of Russia, as compared with her Western neighbours, was also responsible for the relative uniformity of her agricultural industry. The influence of the demands of the market, home and foreign, for agricultural produce, which has been gradually extending, had not yet succeeded in breaking up the last traces of isolated natural economy. In many an out-of-the-way locality and, throughout the country, in certain elements of every peasant-farmer's household, the old notions and practices of the régime of natural economy still survived, and were only being overcome by the progress of commercialization very gradually. The transformation of the countryside proceeded slowly, and in so far as the old conditions of self-sufficiency still prevailed over the modern commercial spirit, farming naturally tended to considerable uniformity. Thus throughout Russia, even in districts in which land might conceivably have been put to more profitable uses, the greater part of it was still sown with such staples as rye and oats, these two being the principal foodstuffs of man and beast. Moreover, in considering the Russian agricultural map, it is essential to bear in mind that Russia has always been a country of peasant farming, the latter accounting, in 1916, for 90·0% of the area under crops. It was small peasant farming, therefore, with its still predominant communal tenure and open fields—two most important factors of uniformity and stagnation—that, by the sheer weight of numbers, exercised a decisive influence on the contours and the colouring of the agricultural map of Russia.

Though not so varied, as the agricultural maps of her Western neighbours, that of Russia, however, presented to the eye a number of regions with distinct agricultural characteristics. The consideration of these several regions, each possessing certain peculiarities and problems of its own, is of fundamental importance in the study of Russian agricultural economics.

The actual division of the country into agricultural regions can be carried out on several principles, according to the particular requirements of the case. To take the simplest case, one may assume, as the basis of division, the ability of the various districts to supply their own needs for the principal agricultural products, namely cereals, from their own resources. This gives the familiar division of European Russia in two, by a line running approximately along the Northern boundary of the black-earth zone. The provinces North of this line, some of which are not fitted for agriculture by nature, and others too densely populated to be agriculturally self-supporting, depend, to a greater or lesser extent, on cereals imported from the districts of the South, which produce exportable surpluses of grain. The importance of this division is manifold, both with regard to a number of problems of economic policy and transport, and to farming. The agricultural industry of the under-producing provinces is in a position in many respects different from that of farming in the provinces producing a surplus of cereals. Generally speaking, the natural conditions of the under-producing region are unfavourable to arable farming, and the costs of cultivation are accordingly heavier than in the Southern half of the country. In many parts of the under-producing region, especially in the North-Western provinces, with their abundant rainfall, the costs are greatly increased by the necessity of drainage, which, besides, being effected by means of open ditches, involves a considerable waste of acreage. Then, throughout the region, the soil requires heavy manuring, which not only raises the costs of cultivation, but exercises a far-reaching influence on the whole organization of farming by increasing the importance of live stock. According to an inquiry carried out by the Department of Agricultural Economics and Statistics of the Ministry of Agriculture, in 1914, the average cost of cultivation of one dessiatin (2·7 acres) of winter rye amounted to 59 roubles 57 copecks, of which 18 roubles 36 cop., or 30·8%, was accounted for by manuring.[1] It is not surprising, therefore, that the same cereal cost to produce 60·4 copecks per pood (36 lbs.) outside the black-earth zone, and only 40 to 43·8 copecks in the Southern provinces, where no manure was used. Yet, the agriculturist of the under-producing belt had certain compensations. In the first instance, the relatively high prices of grain in the Northern half of the country enabled the agriculture of the under-producing region, to a certain extent, to hold its own even in spite of the heavy costs of arable farming. Moreover, both the natural and the economic conditions of the central and Northern parts of Russia were favourable to the growth of other branches of farming, which could make good for the relative unprofitability of cereal cultivation. By nature, many parts of the under-producing region were well adapted to the development of dairy farming, as well as to the cultivation of flax for fibre. The fact that both capitals—Petrograd and Moscow, as well as the principal centres of the manufacturing industries, were situated in the under-producing provinces of Russia, played a great part in encouraging the agricultural industry in these regions by creating an extensive and varied market for the products of farming. The economic conditions of a large part of the under-producing region, therefore, favoured greater variety in agricultural production and the raising of the standards of cultivation, in which the variety of crops and the development of the live-stock branch were very important factors. Considered as a whole, the under-producing region, with the exception of the Northern provinces, sparsely populated, far removed from the market and by nature ill-adapted to any form of farming, may accordingly be said to have been more advanced agriculturally than the black-earth belt. The Southern parts of the latter, as mentioned before, came mainly under the influence of foreign markets, which made for greater uniformity of farming and for extensive systems of cultivation. Over the Northern provinces of the black-earth belt the influence of the internal market extended gradually, bringing with it increased variety of production and raising the standards of farming gener-

[1] Brutzkus, op. cit., p. 129.

ally, especially in the Ukrainian and the South-Western provinces which were in the focus of the economic influence of several different markets, home and foreign.

The effects of the difference in the standards of cultivation, due to economic influences, on the productivity of farming, may be gauged from a comparison of the yield of the principal cereal crops in the most naturally fertile parts of the black-earth belt, on the one hand, and on the much poorer, but more efficiently cultivated soil of the provinces outside that favoured zone. The figures below refer to the provinces of the Southern and South-Eastern steppes, on the one hand, and to those of the Petrograd and Moscow regions, on the other, as the most representative of the influences which affected agriculture in the under-producing zone. In between, there was a wide belt of provinces, stretching from the South-Western region, through the Ukraine, then through Kursk, Orel, Tula, Tanbov, Penza, Voronezh, Saratov, Simbirsk, Samara, Kazan, partly within, partly outside the black-earth zone, which were subject to a mixed set of influences and occupied, agriculturally, an intermediate position. The yield of the principal cereal crops on peasant land, in the provinces belonging to the two different types, in 1911–15, averaged, in poods per dessiatin (one pood = 36 lbs.; one dessiatin = 2·7 acres):

	Winter Rye	Spring Wheat	Oats	Barley
Southern steppes :				
Kherson	52	31	56	55
Taurida	48	42	55	54
Ekaterinoslav	58	43	62	57
Don	42	38	47	50
Kuban	65	64	72	127
Stavropol	52	53	61	62
Central provinces :				
Vladimir	54	50	50	50
Moscow	57	—	60	—
Kaluga	51	—	58	61
Tver	54	—	57	56
Yaroslav	58	56	69	58
Kostroma	50	41	52	48
Smolensk	63	—	63	54
Petrograd	70	—	71	49
Novgorod	54	—	67	60
Pskov	56	—	59	41

It may be seen that, owing to extensive systems of farming and to low standards of cultivation, the yield of crops of the Southern steppes, far from exceeding that of the poorest soil of the under-producing region, was, in some cases, actually inferior to it. This comparison of the results of arable farming in the two zones conveys a fairly good idea of the effects of a relatively well-developed internal market on the character of the agricultural industry in the localities from which it draws its supplies of agricultural products.

Important as this division of agricultural Russia into two distinct zones may be for various purposes, it is far too rough to be of much service in the study of the agricultural geography of the country. Each of the two divisions includes regions entirely different from each other in every respect, except their ability to produce enough cereals. Even the causes of their inability to provide for their own needs in this respect are quite different in the case of various localities. Thus, in the Northern provinces, the deficiency is due to natural conditions. In the North-West, it is mainly the result of farming there being an auxiliary occupation, since the greater part of the income of the rural population in these districts is derived from forestry and other outside sources of earnings. In the industrial provinces of the Centre, the under-production of cereals was only relative, and was accounted for wholly by the density of population. Indeed, no two provinces are absolutely alike with regard to agricultural conditions, but, under the combined influence of their present economic environment, their history and their natural conditions, they group themselves into a number of distinct regions, clustering round certain centres of gravitation or subject, to a varying extent, to certain economic influences. No hard and fast boundaries can be drawn to define the limits of these regions. As some factors extend their influence, while other agencies gradually recede to the background, the characteristics of a region tend to overflow into the neighbouring districts, and the boundaries are shifted farther afield. The peculiar features of the regions will appear most conspicuously about their centres of gravitation and gradually fade away towards the borders, until, somewhere near the imaginary boundary, a transitional zone is reached, in which the characteristics of the two adjoining regions, sometimes but indistinctly defined, are struggling for domination.

The distinctive character of the individual regions is not necessarily bound to find expression in the uniformity of systems of farming within their limits. As a general rule, in every region, these systems will vary according to locality, following some modified form of von-Thünen's concentric zones. Looked upon from this angle, the agricultural regions are not merely geographical areas distinguished by the prevalence of certain systems of farming, but living social and economic entities, held together and controlled in their activities by definite combinations of forces akin to that of gravitation in the physical world.

Of the existing schemes of division of Russia into regions none is based on these essentially economic characteristics. Those used in the compilation of official agricultural statistics by the Central Statistical Committee and the Ministry of Agriculture have long outlived their usefulness, but had to be adhered to for the sake of continuity and comparability. Most other schemes, suggested by numerous authorities, are based on purely technical standards.[1] Here, therefore, a new scheme will be adopted, based on the principles outlined above.

The study naturally starts from Moscow, on which the eye is fixed by its central position and by the network of railways converging on it from the four quarters.[2] The railways connect the old capital and its surrounding industrial districts with the remotest corners of Russia and with her frontiers, spreading the influence of the demand of this great centre of industry and population far and wide over the country. At some distance from Moscow, the influence of its demand for agricultural produce, which gradually diminishes in intensity, first meets with other conflicting influences and then dwindles down to vanishing point, while other influences, spreading inland from the confines of the country or coming from other centres of gravitation within its frontiers, become decisive for the agricultural industry. Within a certain radius from Moscow the demands of the capital and its industrial district for agricultural products are the principal influence to which local farming is subject. This area may be referred to as the *Moscow agricultural region.* It consists of nine provinces, namely those of Moscow, Vladimir, Riazan, Tula, Kaluga, Smolensk, Tver, Yaroslav and Kostroma. The area of the region is 374,384 square versts; the density of population, according to the Census of 1897, was 37·5 per square verst; that of rural population only—31·4. The highest density of population was in the province of Moscow (83·1); the lowest—in that of Kostroma (18·8). Of the total area of the region, according to the Survey of 1887, 34·8% was arable; 17·0% under meadows

[1] Of these, the most important and recent are those of Prof. A. N. Chelintzev, reproduced in its final form in his *Agricultural Geography of Russia* (Rus.), Prague, 1924, and of Prof. Dr. G. Studensky, in *Die Grundlinien und Methoden der landwirtschaftlichen Geographie,* Weltwirtsch, Archiv, Jan. 1927. Both authors belong to the purely statistical school of agricultural geography. A critical account of the various schemes, official and other, will be found in two Russian works on the subject: B. Knipovich, *Methods of the Determination of Regions,* Moscow, 1921, and N. Nikitin, "The Economic Regions of European Russia," *Transactions of the Higher Seminar of Agric. Economics,* Moscow, 1921.

[2] For details of the statistical data quoted in the description of the agricultural regions, contained in the following pages, the reader is referred to Appendices Nos. I, II, III, IV and V at the end of the book. The boundaries of the regions are shown on Map No. 1.

and pastures; 36·9% was covered with woods; 3·0%—unused agricultural land, and 7·7%—waste and unreclaimed lands. The highest percentage of arable was in the South of the region, the provinces of Tula and Riazan accounting for 74·0 and 55·9% respectively under plough; the lowest in the wooded North, the percentage of arable in the province of Kostroma being as low as 18·5. Only two provinces (Riazan and Tula) produced a surplus of cereals, and, as a whole, the Moscow region depended on outside supplies of grain, and formed part of the "under-producing" zone of European Russia. Cereal prices in the Moscow region were, accordingly, among the highest in Russia, but the heavy costs of cereal-growing, combined with the relative advantages of the cultivation of other products, for which Moscow provided a ready market, resulted in the proportion of cereal crops being among the lowest in the country. Owing to the large and varied demands of Moscow and its industrial districts, considerable variety was introduced into the agricultural industry of the region, and its standards of cultivation raised. Among the crops, in 1916, the four principal cereals accounted for 74·1%; minor cereals for 2·0%; potatoes occupied 8·2% of the area under crops; flax—6·8%, and field grass—5·9%. Cattle-breeding, according to Russian standards, was well developed, the average number of cattle per 100 dessiatins of crops, in 1916, reaching 69·1. Dairy farming, especially along the railway lines, was generally widespread, and formed an important branch of local agricultural production. Statistically, this tendency of stock farming in the Moscow region could be seen from the low percentage of young cattle (12 to 18 months) in the herds, which, on the average for the region, amounted only to 15·2.[1] Sheep-breeding, confined to peasant farming almost exclusively, was only moderately developed, the number of sheep per 100 dessiatins of crops being only 76·7. The systems of arable farming among the peasants varied from the cruder forms of the three-course system in the North, especially in the province of Kostroma, through improved forms of the same system, with abundant manuring and the cultivation of grass or potatoes on the spring field or the fallow, to systems of four and more courses, with regular grass or root crops in the central and Southern parts, more densely populated and situated nearer to the market.

The relatively high returns of the agricultural industry, coupled with the pressure of population on the land, which was especially acute in the Southern, more purely agricultural, provinces of the Moscow region, were reflected in the prices of land, which were on the average higher than in any other region outside the black-

[1] According to the method first adopted by Prof. A. N. Chelintzev.

earth. According to land prices, the region could be divided in two different parts, the level of prices being distinctly higher in the provinces of Riazan, Tula, Kaluga, Smolensk and Tver than in the more industrial provinces of Moscow, Yaroslav, Vladimir and Kostroma, with their poorer soil and sparser population. In 1906–10, the average price of land in the region was 99 roubles per dessiatin (2·7 acres), the highest and lowest prices being, respectively, 161 roubles in the province of Riazan and 33 roubles in the province of Kostroma, on the Northern outskirts of the region.

In the North-West, the Moscow region borders on that of the other capital of Imperial Russia—Petrograd or St. Petersburg, to give it its historical name. This other great centre of population and industry, situated in the extreme North-West corner of the Empire, amid districts by nature ill-adapted to the development of farming, depended on supplies transported to it over very long distances, mostly by rail. The country over which its economic influence was paramount, and whose farming depended primarily on the demands of Petrograd and its industrial district, comprised the provinces of Petrograd, Novgorod and Pskov. The *region of Petrograd* was situated on the border line of two natural zones, namely those of Northern forests and marshes and of mixed forests. The area of the region was 181,322 square versts; the density of population, in 1897, was 25·3 per square verst; of rural population alone, 16·7. The distribution of the area, according to the Survey of 1887, was as follows: arable, 14·3%; meadows and pastures, 14·0%; other land suitable for agricultural purposes, 4·1%; forests, 48·8%; marshes and other waste land, 18·1%. The rural population, though keeping to farming as its basic occupation, derived the greater part of its income from outside sources, among which forestry was by far the most important. The costs of cultivation were very heavy, owing to the necessity of abundant manuring and, in most cases, of drainage; labour was comparatively scarce and dear, and the cultivation of cereals on a large scale did not pay, even in spite of extremely high prices, excepting the more remote districts, to which supplies could only be brought with difficulty and at high cost. The level of grain prices in the Petrograd region was, indeed, second only to that which ruled in the extreme North, the average price of rye, in 1906–10, being 1·13 rouble per pood (36 lbs.). For the greater part of the year even the rural population of the region lived on cereals imported from other parts of Russia, The country, however, provided, owing to both its natural and its economic conditions, considerable facilities for stock farming, as well as the cultivation of potatoes, flax, vegetables and field-grass. Petrograd was a good market for animal produce, as well as for vegetables and for hay, especially of the higher grades. These factors were decisive in determining the direction of farming in the Petrograd region. The percentage of cereal crops, in 1916, was only 71·0%, the lowest in Russia; potatoes accounted for 6·7%; flax for 10·4%, the province of Pskov being famous for its production of flax fibre and having as much as 15% of its sown area under flax; the percentage of field-grass, which amounted to 9·6%, was higher than in any other region. Owing to the abundance of natural meadows, however, field-grass, mostly clover, was mainly cultivated for the market and sold in Petrograd as clover hay, instead of being fed to cattle on the farm. Dairy farming was well developed, and the proportion of young cattle accordingly low (8·6 to every 100 cows). The number of cattle per 100 dessiatins under crops in 1916, reached 103·5, the region in this respect being second only to the Caspian, with its extensive grazing. Sheep-breeding was considerable, especially in the province of Pskov. The average number of sheep per 100 dessiatins of crops, in 1916, for the whole region, was 102·9. The sheep belonged exclusively to the coarse-woolled peasant types. The systems of arable farming varied considerably according to locality, the three-course system being generally typical for the peasants; its simplest forms could be met in the outlying wooded districts of the provinces of Petrograd and of Novgorod, where even more primitive forms of cultivation could be met with occasionally. In the more densely populated parts of the region, however, improved three-course cropping and, not unfrequently, varieties of some more advanced forms, mostly involving multiple grass crops, and belonging to the type known under the German name of *Feldgrasswirtschaft*, were common.

The quality of the soil in the region is very poor, which fact, combined with the sparseness of population, was responsible for land prices being generally very low. This applied more particularly to arable land, since natural meadows and forests, with their much higher net yield, fetched generally better prices. In this connection, indeed, it must be pointed out that the Petrograd region was extremely heterogeneous in this respect, land varying in value enormously according to situation, mode of exploitation and the quality and condition of the soil, with the result that average prices there were generally far less representative than in other localities, more homogeneous as to the soil. Not unfrequently, in an estate of some few hundred dessiatins one would find, alongside with each other, areas worth 25 roubles per dessiatin along with land worth anything up to 250 roubles, especially if the estate contained rich natural meadows flooded in spring or good building timber. In 1906–10, the average land prices in the Petrograd region worked out at 71 roubles per dessiatin, with a maximum of 103 roubles in the province of Pskov and a minimum average in that of Novgorod (44 roubles).

To the West of the Moscow region, and adjoining that of Petrograd, lies the *Western agricultural region*, consisting of the White-Russian provinces of Mogilev, Minsk and Vitebsk. The area of the region is 133,927 square versts; the total population, in 1897, was 32·4 per square mile; the rural population alone—28·9. The region, situated entirely within the zone of mixed forests, had, according to the Survey of 1887, a high proportion of woodlands (37·3%); 26·7% was arable; 13·0%, meadows and pastures; other land suitable for agricultural purposes accounted for 3·5%, and 1·8% was waste and unreclaimed land. The region was to a large extent dependent on cereals imported from other parts of Russia. Cereal cultivation was not expedient economically beyond a limited extent, and, in the course of economic development, as conditions of transport improved and supplies of grain could be obtained cheaper relatively, cereal crops tended to lose in importance, as compared with other plants and, more particularly, potatoes. The cultivation of the latter was encouraged, especially in the provinces of Mogilev, and Minsk, by the numerous spirit distilleries, which provided an extensive local market for potatoes. In 1916, the distribution of the area under crops in the Western region was as follows: 75·5% were sown with the principal cereals; 6% with minor cereals; 8·5% with potatoes; 4·0% with flax; 4·1% with grass, and 1·2% were accounted for by other crops. The Western region, situated relatively near both capitals and their industrial districts, as well as Poland and the Western frontier of Russia, and not naturally suited to extensive cereal cultivation, concentrated to a considerable extent on stock farming. Dairy farming and meat cattle-raising were well developed throughout the three provinces. The number of cattle per 100 dessiatins under crops was 65·5, which, considering the size of the cultivated area, was actually a denser cattle population than that of the region of Petrograd. The relatively high percentage of young cattle, 16·4 to every 100 cows, pointed to the meat-producing branch being more prominent than in the Petrograd Region, as this, indeed, could well be expected, considering the longer distance from the markets. In pig-breeding, the Western region, which exported considerable numbers of live pigs, as well as pork, to Germany, was easily first in Russia, with 71·2 pigs per 100 dessiatins of crops. Sheep-breeding, on the other hand, was only moderately developed, the number of sheep per 100 dessiatins of crops being only 77·5.

The systems of arable farming, though they varied considerably according to locality, were mostly based on the improved three-field cropping, with grass and potatoes on the spring field or fallow, while four-field rotation was also common. The systems of cropping differed from those of the Moscow region mostly by the preference generally given in the rotation to potatoes, as against grass crops.

The average price of land in the Western region, in 1906–10, was about 92 roubles per dessiatin, the land being cheapest (77 roubles) in the province of Minsk and fetching approximately the same prices (100–101 roubles) in the other two provinces.

Farther South, one enters the agricultural region of the *Ukraine*, on the left bank of the Dnieper, and the *South-Western*, on its right bank. Inhabited by the same branch of the Russian people, namely the Little Russians or Ukrainians, and having Kiev for their common cultural and economic centre, the two parts must be considered as distinct entities owing partly to economic and partly to historical causes. While the Ukraine was united to Moscow in 1649, the South-Western provinces remained under Polish rule until the close of the eighteenth century. Their agrarian relations, accordingly, developed, until the nineteenth century, and to a large extent even through it, along different lines. Economically, the two regions, except for their common local interest in Kiev, as by far the most important urban centre of the South-West of Russia, as well as their position as the centre of sugar-beet cultivation in Russia, were different in that, among the numerous cross-currents of influences to which their farming was subject, the Ukraine was mostly affected by the demands of the Russian internal market, while the South-Western region, situated, as it was, along the Western frontier of Russia, with its South within easy reach of the exporting harbour of Odessa, and its North adjoining the Polish provinces, was situated in the focus of various competing influences, mostly coming from outside. A very important difference between the two regions is also that of the density of population, since the South-West had, in 1897, 72·9 inhabitants per square verst as against the Ukraine's 54·9. To this greater density of population in the South-West, which actually exceeded that of any other region in Russia, corresponded also a higher intensity of farming.

The principal agricultural characteristics of the two regions were as follows. The area of the Ukraine, which includes the provinces of Kharkov, Chernigov and Poltava, is 137,771 square versts; that of the South-Western region, comprising the province of Kiev, Podolia and Volynia, 109,373 square versts. The rural population of the Ukraine, in 1897, was 48·7, and of the South-West 65·5 per square verst. The distribution of the total area, according to the Survey of 1887, is given below:

	The Ukraine	The South-West
	Percentages	
Arable	66·9	53·9
Meadows and pastures	10·0	9·9
Forests	12·2	18·8
Unused agricultural land	4·7	9·3
Waste and unreclaimed area	8·1	7·2

The distribution of the area under crops, in 1916, in the two regions is shown in the following table:

	The Ukraine	The South-West
	Percentage of total Crops	
Principal cereals	75·7	71·2
Minor cereals	10·5	12·5
Potatoes	3·7	4·4
Sugar beet	2·5	5·1
Field grass	4·0	3·4

These two regions were the principal sugar-beet-growing localities in Russia, and this particular crop, especially in the South-West, played a most important part in the organization of farming. Indeed, the combination of sugar-beet and potatoes, which supplied the raw materials respectively for the sugar and the spirit industries, highly developed in that region, were the basis of a system of cropping in which cereals, roots and grass, following each other in regular succession, greatly raised the standard and the yield of farming. Agriculturally, therefore, the South-West was certainly the most technically advanced part of Russia. The Ukraine, though in the distribution of crops it resembled its Western neighbour to some extent, lagged considerably behind the South-West in its standards of farming, especially in the South, where it bordered on the extensively-farmed steppes of New Russia.

With regard to live stock, with the exception of sheep, of which there were more in the Ukraine than in the South-West, the position was much the same in the two regions, as this may be seen from the figures below, referring to the year 1916:

	The Ukraine	The South-West
Cattle per 100 des. of crops	49·7	50·6
Young cattle per 100 cows	23·2	19·9
Sheep per 100 des. of crops	54·6	30·8
Pigs " " "	31·5	29·9

It may be said generally that, in their agricultural development, the Ukrainian provinces followed on the footsteps of the South-Western region, though lagging slightly behind.

The prices of land in the two regions, in 1906–10, were approximately the same, the exact averages being 217 roubles per dessiatin in the Ukraine and 216 roubles in the South-West.

Now, starting again from the central region of Moscow and proceeding in the direction in which Russian colonization had been spreading originally over the black-earth towards the South and the East, one enters two of the oldest territories colonized by the Russian people in the course of its expansion. In the twentieth century these territories formed two distinct agricultural regions, namely the Central Agricultural region and the region of the Middle Volga.

The *Central Agricultural region*, situated due South of Moscow, consists of the provinces of Penza, Tambov, Voronezh, Orel and Kursk. It covers an area of 232,421 square versts and had a population of 47·7 per square verst in 1897; its rural population was 43·8 per square verst. The small difference between the density of the total population, on the one hand, and that of the rural population alone, on the other, is in itself a characteristic of this region, which was almost exclusively agricultural. These provinces, which belonged to the belt of oldest settlement and were situated in the very heart of the country, close to its old capital, supported a very dense population, depending almost entirely on the excellent soil of the region for its living. The proportion of arable in the Central Agricultural region was, accordingly, very high. In 1887, arable accounted for 67·0% of the total area, while according to provinces it varied from a minimum of 60·6% in the province of Orel to a maximum of 72·8% in that of Kursk. The area of meadows and pastures, already in 1887, was reduced to 9·0% of the total by the continuous encroachments of the plough. Since in these localities there were no adequate facilities of forest grazing, the forests accounting only to 13·6% of the area, this involved, under the prevailing three-course system, a severe restriction of the numbers of cattle, with its unavoidable effects on the agricultural industry generally. The unused reserve of land suitable for agriculture, in 1887, represented 3·3% of the total area; waste and unreclaimed land accounted for 4·9%.

The distribution of crops in the region, in 1916, pointed to a great concentration on the production of the principal and minor cereals, the former accounting for 75·6%, and the latter for about 12% of the total area of crops. Potatoes accounted for 3·9%;

sugar-beet for 0·9%; hemp and grass for 1·9% each. The number of cattle per 100 dessiatins of crops was only 39·5; that of sheep, on the contrary, was very large and amounted to 117·7 per 100 dessiatins of crops. Moreover, the cattle here, as in the Ukraine and the South-West, as well as in other parts of the black-earth belt, included draught oxen, which circumstance tended to raise the relative numbers of cattle, as compared with other districts of Great Russia.

For the disposal of its surplus of agricultural produce the Central Agricultural region depended mainly on the internal market. The greater part of its surplus of cereals, mostly consisting of rye and oats, was exported to the under-producing provinces of Russia, while some of it, as well as a considerable part of the available surplus of potatoes, were used within the region by the numerous spirit distilleries. Among all the regions of Great Russia, the Central Agricultural, in its Southern parts, was the only one in which sugar-beet-growing was developed to any extent, its cultivation being encouraged by climatic conditions and the cheapness of labour.

The average prices of land in the region, in 1906–10, were relatively high, owing mainly to the acute "land-hunger" among the peasantry of these provinces, with their relative overpopulation, and reached 169 roubles per dessiatin, with a minimum of 123 roubles in the province of Penza and a maximum of 205 roubles in the province of Kursk.

The *region of the Middle Volga* comprises the provinces of Nizhny-Novgorod, Kazan, Simbirsk and Saratov, of which the first belongs to the under-producing group, while the other three have a surplus of cereals. The area of the region is 218,727 square versts, and its population is not so dense as that of the Central Agricultural region. In 1897, it amounted to 35·1 per square verst, while the rural population alone was 31·8. The proportion of the total area under the plough, according to the Survey of 1887, was 53·1%; and, though meadows and pastures did not exceed 10·9% of the whole, the considerable percentage of forests, which amounted to 25·7%, afforded sufficient facilities of forest grazing. 3·1% of the area, though suitable for farming, was not actually used, and 7·2% was waste and unreclaimed land. In the arable farming of the region, the cereal-growing side was strongly emphasized, the principal cereals alone accounting, in 1916, for 82·1% of the crops; minor cereals represented approximately another 9%; potatoes accounted only for 2·1%; flax and hemp for 0·8% each; grass only for 0·7%. Cattle-breeding was somewhat better developed than in the Central Agricultural region, the number per 100 dessiatins of crops amounting to 41·4. The proportion of young cattle did not exceed 22·9 per 100 cows, and was thus only slightly higher than in the Central Agricultural region. Generally speaking, in its live stock, as well as in its arable farming, though the latter is as typically representative of the three-course system as that of the Central Agricultural region, the region of the Middle Volga forms an intermediate stage between the intensively cultivated Central and Western parts of Russia, possessing relatively large numbers of cattle, and the extensive cereal-growing districts of the East and South, either having very little cattle, or keeping large herds under primitive semi-pastoral conditions of husbandry. Sheep were numerous, their number reaching 106·4 per 100 dessiatins of crops. A characteristic feature of the Middle Volga region, due partly to its natural conditions and partly to its situation on the borders of the under-producing zone, was the great part played in the cultivation by rye, which, in 1913, accounted for about 43% of the total area sown.

The average prices of land in the region, in 1906–10, worked out at about 105 roubles per dessiatin, the minimum being 91 roubles in the province of Nizhny-Novgorod and the maximum—118 roubles—in that of Saratov.

On the left bank of the Volga lies the *Eastern agricultural region*, which, with its extensive cultivation and large surpluses of cheaply produced cereals, formed the agricultural background of European Russia. The region comprises the provinces of Samara and Ufa and covers an area of 239,934 square versts. Its population, in 1897, numbered 20·6 per square verst; the rural population alone—19·5. The region was traversed only by two main railways connecting it with the Volga and the denser railway net on its right bank. With their few branches and the navigable tributaries of the Volga and the Kama, these railways were the principal means of transport in the region. The local prices of cereals and other agricultural products were, accordingly, low, and production was carried on extensive lines. The proportion of arable was relatively large and amounted, in 1887, to 41·4% of the total area; meadows and pastures represented 31·5%; forests 23·1%, mostly in the province of Ufa. Arable farming had a strong cereal-growing bias, the principal cereals, in 1916, accounting for 86·3% of the crops; minor cereals for 9%; potatoes for 1·6%; grass for 1·4%. The systems of arable farming varied from the extensive types of cultivation, without regular rotation or fallow in the Eastern and South-Eastern parts of the region, to the three-course cropping, which was common in the Western districts, near the Volga. Stock farming was also extensive, and though the number of cattle was not large, as compared with the enormous arable area, and worked out at no more than 43·1 heads per 100 dessiatins of crops, it was large relatively to the population, while the high proportion of young cattle, which reached 30·6 per 100 cows, pointed to a distinct meat-producing tendency in the local cattle-raising. Sheep-breeding was also developed to a considerable extent, the number of sheep per 100 dessiatins of crops reaching 98·0.

While, by reason of its geographical position, the Eastern region could be expected to depend almost exclusively on the home market, as a matter of fact, a considerable part of the surpluses of cereals produced, namely 57·7% in the province of Ufa and 37·3% in that of Samara, according to an estimate referring to the period 1909–11, actually were exported abroad, where the high-grade hard wheat of the Eastern steppes was in good demand.

Land prices in the Eastern region, in 1906–10, averaged 77 roubles per dessiatin, or 87 roubles in the province of Samara and 67 roubles in the province of Ufa.

In the Southern steppes one is confronted with another zone of extensive cereal-farming, conducted mainly with a view to export abroad. Here, on the Northern shore of the Black Sea, lie the two principal cereal-exporting regions of Russia—New Russia and the South-Eastern region. The distinction between the two is one rather of degree than of kind, and, in its agricultural evolution, the South-East has been generally following on the footsteps of its more advanced Western neighbour—New Russia.

The agricultural region of *New Russia* consists of the provinces of Kherson, Taurida and Ekaterinoslav, and occupies an area of 170,973 square versts. Its total population, in 1897, was 36·8 per square verst; its agricultural population—28·5. Though the region is one of relatively very recent colonization, its natural and economic advantages, as well as the policy of the successive Governments since the reign of Catherine II, brought about its rapid development. The region is practically all steppes, only 3·0% of its area, mostly on the outskirts, being covered with forests, according to the Survey of 1887. Already then, 70·3% of the area were arable; meadows and pastures accounted for 15·4%; other agricultural land—5·0%, and waste land—6·3%. Originally, on its first annexation to Russia, the region afforded exceptional facilities for extensive cattle-raising and sheep-breeding, and these branches of farming were extremely important in the rural economy of New Russia down to the second half of the last century, when the plough began definitely to gain ground on the herds. The cereal cultivation of New Russia grew under the influence of the rapid industrialization of Western Europe, until most of the land was turned into arable, and the region, from the first rank of grazing districts in Russia, changed into one of the localities in which cattle was definitely scarce. The land was cultivated, as a rule, in a most primitive manner, reliance being placed entirely on its natural fertility. Even at the beginning of the twentieth century, vast tracts of land, especially in the Eastern part of New Russia, were still sown in an haphazard way, without regular fallowing or the use of any kind of manure, and when the soil, in the course of time, began to give out, the custom was to abandon it and turn it to grazing for a number of years. Before the war, however, some form of regular fallowing or other, according to the crops sown, was becoming common in the older and more highly developed Western parts of the region. The distribution of crops, in 1916, in New Russia was characteristic of the type of farming. Indeed, 90·2% of the area under crops was sown with the four principal cereals; 5·3% were accounted for by the minor cereals; 1·2% by potatoes; 0·2% by sugar-beet (province of Kherson); 1·0% by grass crops. The number of cattle was very small, there being only 27·7 heads per 100 dessiatins of crops, and the herds included a high percentage of young animals, amounting to 36·1 to every 100 cows. The cattle, indeed, was being raised rather for meat than for the purpose of dairy production, which was confined to the neighbourhood of the large towns. Sheep-breeding, which in old days was probably the most important branch of farming in New Russia, by the close of the nineteenth century had been reduced to relative insignificance. In 1916, sheep in New Russia numbered 29·4 per 100 dessiatins of crops, which was the lowest proportion in the whole country, with the only exception of the extreme North.

The prices of land in New Russia, in 1906–10, averaged about 190 roubles per dessiatin, the maximum being paid in the province of Kherson (216 roubles) and the minimum in the Taurida (165 roubles).

The *South-Eastern region*, adjoining New Russia in the East, consists of the Don and Kuban Cossack territories and of the province of Stavropol. It occupies an area of 274,983 square versts and its population, in 1897, averaged 19·5 per square verst; the rural population alone being 17·0. Agriculturally, as pointed out above, it has much in common with New Russia, as the latter was a few decades ago. In the Western parts of the region, the systems of farming are similar to those of the Eastern districts of New Russia. Before the war, in both, the so-called *pestropolie*, that is the continuous raising of cereal crops, without any regular rotation, which allowed the land only a few months' rest between successive crops, was practically general. The same system was common in the greater part of the Kuban as well, while in the province of Stavropol the land was cultivated more primitively and extensively still, crops being sown continually until exhaustion of the soil and the land then abandoned for a long term of years, in which to recover its natural fertility. Of the total area under crops, in 1916, 88·6% were sown with the four principal cereals; 4·6% with the minor cereals; 0·8% with potatoes; 0·9% with grass; the rest included various other crops, each with a fraction of 1%. Cattle in the South-East was much more numerous than in New Russia, there being, in 1916, 56 heads per 100 dessiatins under crops. These, however, included an especially large proportion of draught oxen, which, in these provinces, were more widely used for work than in any other part of the country. The presence in the herds of as many as 47·1 young animals per 100 cows was the result of this circumstance, as well as of the strong meat-producing tendency of stock farming in this region. Sheep-breeding also was considerable, the number of sheep per 100 dessiatins of crops amounting to 81·8. In the South-East, indeed, was concentrated by far the greater part of what still remained, in the twentieth century, of the formerly considerable Russian merino-breeding. Yet, of the 7·5 millions of sheep in the South-Eastern region, only under 1·5 millions belonged to the fine-wool breeds, generally referred to under the name of "merinos."

The average price of land in the South-Eastern region, in 1906–10, worked out at 154 roubles per dessiatin, the highest being 182 roubles in the Kuban and the lowest, 130 roubles, in the province of Stavropol.

The extreme South-East of European Russia, namely the *Caspian region*, which consists of the provinces of Astrakhan and Terek, occupies, in respect of both natural and economic conditions, an intermediate position between Europe and Asia. While the Terek Cossack Territory forms a transition between the plains of Russia proper and the mountains of the Caucasus, with their peculiarities of farming and of life generally, the province of Astrakhan may be looked upon as the entrance hall into the Transcaspian and Central Asia. The population is very sparse: 5·3 per square verst in the province of Astrakhan and 14·7 on the Terek. Farming in each of these two provinces possesses peculiarities of its own, that of Astrakhan being of a very extensive, semi-pastoral type, with little land under the plough and large herds of cattle and sheep; that of the Terek, on the other hand, differing greatly in various parts of the province, in accordance with the configuration of the locality. Thus, in the Northern and Western parts of the province, it resembles the extensive cereal farming of the Kuban and Stavropol, which it adjoins, while in the mountainous districts of the South and the East, it is akin to the native husbandry of the Caucasian tribes. The two provinces, therefore, though they are joined together in one agricultural region, have little in common, except their geographical position and their pronouncedly transitional semi-Asiatic character. Since, however, neither of them can properly be attached to any of the other regions, it would appear best to consider them in this combination.

The province of Astrakhan, in 1916, had only 680,300 dessiatins under crops, out of a total area of 207,193 square versts. Of the total area under crops, 89·9% represented the principal cereals; about 5·5%—the minor cereals; 1·6%—potatoes; 1·1%—flax; and the rest—various minor crops. The number of cattle was very large, there being 200·3 heads per 100 dessiatins under crops. The same applied to sheep, which numbered 363 per 100 dessiatins of crops. In the twentieth century, it was still largely a country of extensive wild grazing.

The Terek was more agricultural. In 1916, the area under crops was 628,400 dessiatins, of which 47·5% were sown with the principal cereals—rye, wheat, oats and barley, 26·3% with maize, 19·0% with millet, and the rest with a number of minor crops. The number of cattle per 100 dessiatins reached 151·5; that of sheep—313. Thus, the Terek also was a locality of extensive cattle—and sheep-raising. Indeed, no other region in European Russia could compete with the Caspian in the relative number of cattle, with reference to the cultivated area.

The average price of land, in 1906–10, in the Terek territory, reached 126 roubles per dessiatin. Of the province of Astrakhan there are no corresponding records of the Peasants' Bank. Yet, during the same period, the Bank has been selling land there out of its own reserve at extremely low prices, ranging from 15 to 25 roubles per dessiatin.

To the North of the Moscow region lies the *Northern region*, which comprises the provinces of Archangel, Olonetz and Vologda. This region occupies an enormous territory, its total area exceeding 1,200,000 square versts, and agricultural conditions in it, accordingly, vary greatly. As a whole, however, with the exception of the province of Vologda, and more particularly, of its Southern part, where the standards of farming are raised by the development in the course of the last few decades, of dairy production, the Northern region, mainly by reason of its extremely severe natural conditions, is agriculturally undeveloped. The population is extremely sparse, the average density for the whole region being only 1·7 per square verst. In the province of Archangel it is as low as 0·5; in that of Olonetz—3·2, and in that of Vologda—3·8. Most of the ground is occupied by forest and marshes, the distribution of land according to modes of exploitation, in 1887, being as follows:

	Archangel	Olonetz	Vologda	Region
		Percentages		
Arable	0·1	5·0	2·3	1·3
Meadows	0·2	5·2	5·9	2·4
Forests	45·4	60·5	86·1	59·2
Suitable land	—	17·1	0·4	1·8
Waste	54·3	12·2	5·3	35·2

In 1916, in the whole vast territory of the Northern region, there were, accordingly, only 798,700 dessiatins under crops. Of this area, 64,400 dessiatins were in the province of Archangel; 117,200 in that of Olonetz and 617,100 in that of Vologda. Of the principal cereals, only rye, barley and oats succeed in the rigorous conditions of the North, and of the total area of crops in the region, in 1916, they occupied 89·4%. Potatoes accounted for 3·1%; flax for 5·1%; other crops, including grass, were confined to fractions of 1%. Relatively to both the population and the area under crops, the number of cattle in the region was very large, namely 153·7 per 100 dessiatins under crops. Dairy farming was well developed in the Southern parts of the region, especially in the province of Vologda, where it was the mainstay of the local peasant farmer. Sheep-breeding was insignificant, the number of sheep per 100 dessiatins of crops being only 11·3.

The prices of land were highest in the province of Vologda, where, in 1906–10, the average was 54 roubles per dessiatin. For the provinces of Archangel and Olonetz no real average price can be arrived at, owing partly to the great variety of conditions, and partly to the extremely small number and casual character of transactions. In 1906–10, the average yearly prices recorded by the State Peasant Bank in the province of Olonetz ranged from 5 to 60 roubles per dessiatin. For the province of Archangel, there is only one record, in 1907, the price per dessiatin being 12 roubles. Speaking generally, only in the province of Vologda land could be treated as possessing any definite value in money, since in the two other provinces farming is too casual and too undeveloped to allow of such estimation, except in a few particular districts.

Far more developed agriculturally, in spite of its remoteness and its severe climate, is the *North-Eastern region*, comprising the provinces of Viatka and Perm and extending over an area of 425,188 square versts. The density of population in this region, in 1897, was 14·1 to a square verst; that of the rural population alone—13·5. Of the two provinces, that of Viatka is more densely populated and more advanced agriculturally, than the province of Perm, the Eastern part of which lies in the Northern Urals, but, as a whole, they form a distinct entity with economic characteristics of its own. Of the total area of the region, according to the Survey of 1887, 19·5% were arable; 9·3% represented meadows and pastures; 54·5%—forests; 10·8% formed a reserve of land suitable for agricultural purposes and 5·9% were waste and marshy ground. Of the area under crops, in 1916, 90·0% were sown with the principal cereals; about 2·0% with minor cereals; 0·9% with potatoes; 3·6% with flax; 1·8% with grass. The system of cropping in common use was the three-course cultivation, though more primitive forms of husbandry were still widely practised in the outlying parts of the region. Cattle was moderately numerous, the number per 100 dessiatins of crops reaching 61·1 heads. In stock farming, the dairy branch predominated in the province of Viatka, which, along with that of Vologda, belonged to the Northern dairy-farming area, supplying the two capitals with a considerable part of their milk and butter, while meat-cattle-raising was more peculiar to the province of Perm. Sheep-breeding was more developed in the province of Viatka. On the average for the whole region, sheep numbered 91·8 per 100 dessiatins of crops, which was just above the average for European Russia (88·1).

Among the under-producing provinces of Russia, those of the North-Eastern region were distinguished by having the lowest grain prices. Indeed, in 1906–10, the average prices of the four principal cereals in the region were 76 copecks per pood for rye, 101 copecks for wheat; 54 copecks for oats and 73 copecks for barley; a level of prices which did not exceed that of the black-earth belt, from which surpluses of grain were exported to the home and foreign markets. The region, indeed, was a large producer of cereals, whose output was very nearly sufficient to meet local requirements, so that only a small balance had to be imported, for the needs of the towns and of the industrial population of the metallurgical areas in the Urals.

The average price of land in the region, in 1906–10, worked out at about 38 roubles per dessiatin.

To conclude this general survey of the agricultural map of European Russia on the eve of the war and the revolution, the reader is now asked to throw a last glance over the whole. On the map, as spots of heightened colour, he will then perceive the old centres of population in the regions of Moscow, Petrograd, White Russia and the Ukraine, East and West of the Dnieper. To the South-East of Moscow the colouring begins to fade slightly, as these old centres merge with the regions of earliest colonization—the Central Agricultural and the Middle Volga. Here, in the course of centuries, the population had grown to a relatively great density, and its complete dependence on agricultural production for a living had been responsible for some of the most difficult social and economic problems which Russia had been called upon to face in the course of the last few decades. To the North and the East of the Volga, as well as in the steppes of the South, lie vast regions, less densely populated, in which the colours fade gradually away. As one advances farther afield, away from the centre, the variety of farming diminishes. The predominance of cereal crops, noticeable throughout Russia, becomes ever more pronounced; instead of regular rotation, wanton cropping, without fallowing or manure, becomes more and more general; stock farming changes completely, dairy farming being superseded by the raising of cattle for meat and extensive grazing in the Eastern steppes. Proceeding roughly in a South-Easterly direction, these changes can be observed in the agricultural industry all along, until, on the Caspian shores, European farming is actually left behind, and the semi-nomadic conditions of the steppes of Asia assert themselves in the typical forms of local husbandry.

PART II

THE ORGANIZATION AND CONDITIONS OF FARMING IN RUSSIA

CHAPTER IV

THE RUSSIAN AGRARIAN PROBLEM, ITS ORIGIN AND DEVELOPMENT

AMONG the issues of Russian politics on the threshold of the twentieth century the agrarian problem stood easily foremost. In the programmes of all the advanced political parties Land was placed in a position of honour, enjoying precedence over Freedom, which could never appeal to the peasant's heart in the same way as the promise of an addition to his holding of a few acres. In the political strife the agrarian problem has been exploited for all it was worth as a revolutionary factor. As a ferment of the revolutionary movement, its value, indeed, could hardly be overestimated. Presented in its simplified form of "land-hunger," the agrarian problem thus formed the principal plank of most political platforms, and it was this simplicity of presentation, which made a very clear-cut issue of a highly complex economic problem, that enabled politicians to use it to the greatest advantage.

Since the Emancipation a copious literature on the subject had been accumulated, the literary activity having been especially marked at the close of the last and the beginning of the current centuries, when the agrarian crisis reached its climax and broke out in the widespread agrarian disturbances which preceded and accompanied the revolution of 1905. The bulk of this literature could be definitely identified with certain political parties or currents of thought, among which, since the 'nineties, the two main branches of Russian Socialism, namely the "populists" (*narodniki*), on the one hand, and the Marxian Social Democrats, on the other, were the most conspicuous. Conservative opinion and several nondescript varieties of liberal eclecticism were also represented. The attention was focused on the sizes of peasant holdings, the relations between the peasants and the large landed proprietors, the economic position of the peasantry generally and, above all, on the tendencies of the social evolution of the peasant class in Russia, but seldom, if ever, had the agrarian situation been considered in its relation to the economic development of the country as a whole. Whenever the problem had been approached from this angle, even by highly competent writers, their work was mostly marred by the fact that, being staunch partisans of certain theories, they were not unfrequently blind to such points as did not fit into their own particular schemes.[1] Meanwhile, though the "land-hunger" of a large section of peasants was real enough, and the unsatisfactory condition of the Russian countryside could not honestly be denied, the agrarian problem implied much more than the mere insufficiency of peasant holdings, and to cope with it the situation had to be grasped as a whole.

Though, strictly speaking, the origin of the agrarian problem and its development since the Emancipation of peasants are beyond the scope of the present study, reference to this page of Russia's economic history cannot here be avoided. Without pretending to deal at all exhaustively with this subject, which could fill volumes by itself, I shall endeavour in this chapter to give a general outline of the problem in what, to my mind, appears to be the right perspective. In doing so, I shall have of necessity to traverse a good deal of ground already covered by numerous Russian and foreign writers in greater detail, in order to gain clearer insight into the conditions and problems forming the subject of the following chapters.

To appreciate the nature and origin of the Russian agrarian problem, as it presented itself in the second half of the nineteenth century, following the Emancipation of peasants in 1861, it is essential to bear in mind that, at the time of the great reform, Russia was practically a purely agricultural country, whose well-being depended, in the first instance, on the prosperity of the peasant class, which ultimately supported the whole structure of the State and society. This characteristic of Russia, which distinguished her profoundly from such countries as England, Germany or France, where even in the Middle Ages industrial activities and

[1] Among the best works of this class it is necessary to mention the following: Vl. Ilyin (Lenin), *The Development of Capitalism in Russia*, 1899; P. Maslov, *The Agrarian Problem in Russia*, 1903; P. Liashchenko, *Essays on the Agrarian Evolution in Russia*, 1908; all three representative of Russian Marxism at its best. V.V. (V. Vorontzov), *The Destiny of Capitalism in Russia*, 1882, and *The Fate of Capitalist Russia*, 1907, are equally representative of the "populist" outlook. I. Hourvich, in his *Economics of the Russian Village*, 1892, holds views akin to the Marxians'.

urban centres were highly developed, was responsible for certain important peculiarities in the attitude of the State to the peasants. Throughout the history of Imperial Russia, with the exception of the greater part of the eighteenth century, when, under the Empresses Anne, Elizabeth and Catherine II, the landowning class acquired great influence on the policy of the State, the latter consistently upheld the interests of the peasants. Under serfdom, as well as at the Emancipation, this attitude was partly dictated by humanitarian considerations, partly by the necessity, in the interests of the State, to keep the economic condition of the rural population, as taxpayers, unimpaired by the excessive greed of their masters. Accordingly, in the eighteenth century, while the great majority of the serfs had to acquit themselves by the payment of certain fees (*obrok*), since labour services (*barshchina*) on the land cultivated on the landowner's account were still rather exceptional, the Government had been constantly engaged in efforts to keep these payments down. When, later on, since the second half of the eighteenth century, the landowners began to extend their home farms, especially in the black-earth belt, by the use of the labour of their serfs, the State intervened by legislation with a view to limiting the number of days of such work (*barshchina*). Such attitude, in itself, was certainly not peculiar to Russia alone, and at one period or other of their history, when they were still mainly agricultural and the well-being of the peasantry loomed large in the eyes of their governments, other European countries have had somewhat similar experience. In England, indeed, a similar note can be traced in the attitude of the early Tudors to the development of enclosures and the accompanying eviction of peasants. In France, the same solicitude for the cultivators of the soil transpired sporadically from Louis XI onwards, the outstanding points of the later history of that country being marked by *le bon roi Henri Quatre* with Sully, and the unfortunate Louis XVI, by whom the peasants of the royal domains were set free in 1775. In various parts of Germany the same attitude was apparent, with intermissions, ever since the close of the Thirty Years' War in 1648, when the process of eviction of peasants (*Bauernlegen*), which assumed threatening proportions, met with opposition on the part of the State, more particularly in Prussia. The reign of Frederick the Great marked an epoch in the history of peasant protection in that country, just as that of his contemporary, Joseph II, did in Austria. Yet, throughout Western Europe, this attitude manifested itself more or less sporadically and had never been consistently adhered to for long periods, the influence of the powerful landowning aristocracy having generally proved too strong in the long run.

With the rising tide of individualism, which followed on the French Revolution and the spread of the doctrine of *laissez faire*, propagated by the disciples of Adam Smith on the Continent, the policy of peasant protection was definitely abandoned throughout Western Europe. It was in this atmosphere of economic liberalism, which insisted on the removal of all restrictions on the freedom of trade in land, as well as under the strong pressure of the vested interests of large landowners, that the last traces of medieval bondage had been abolished in the greater part of Europe in the first half of the nineteenth century. The policy of peasant protection, which had been characteristic of the period of enlightened paternal government of the eighteenth century, and had only been successful in so far as the power of the respective Sovereigns was sufficient to override the opposition of the landed aristocracy, was entirely out of place in the liberal State of the nineteenth century. The final abolition of bondage, as well as the working-out of the legal position of the peasants with regard to the land on which they were settled, were brought about by a process of social evolution, through a long series of conflicts and compromises between the interests concerned. The peasants, accordingly, on emerging from the long-drawn struggle, succeeded in keeping as much of their land as they have been able to save from the fray.

In Russia, events have followed a different course. There, the essential conditions of a consistent policy of peasant protection were present throughout the nineteenth century, while the whole economic organization of the country rendered such policy in many respects imperative. Historically, Russia had not built up an organized class of landed aristocracy, bound by feudal traditions and strong enough to assert its influence, as a class, on the Monarchy. Russian absolutism, accordingly, was in a position, whenever it saw fit, to effect what were practically social revolutions. By Peter the Great, such a revolution had been effected in the whole social and political life of the Russian nation. When the time came to do away with serfdom, the problem was solved by Alexander II in the same drastic manner, with an eye rather to what he considered the needs and advantages of the State and the people as a whole, than to the vested interests concerned in the reform. Indeed, the history of Europe had no precedents of emancipation carried out in so generous a spirit with regard to the peasantry, as was the reform of 1861 in Russia. The actual expropriation of part of the landowners' estates for the settlement on them of the former serfs, though it was not exclusively peculiar to Russia, had nowhere been carried out on the same scale or according to the same principles. The classical instance of peasant emancipation with the transfer to them, on the title of property, of their actual holdings, was that of the Edict of 1771 in the Principality of Savoy. In the nineteenth century, the most notable case of this kind was the land-settlement of the former bondmen in Prussia under the Hardenberg legislation. There, however, by the Edict of 1811, the peasants were only allowed to keep from one-half to two-thirds of their former holdings, according to the conditions of their former tenure, the rest being retained by the landowner as compensation for the losses inflicted on them by the reform. Moreover, by a subsequent enactment of 1816, the group entitled to benefit by the original law was restricted to the relatively small number of the better-to-do peasants (*spannfähige Bauern*), the rest being excluded from its operation. As a rule, everywhere, while the peasants have been freed from all the remaining traces of personal dependence on their landowners, they have been left to settle the question about their holdings by private agreement with the proprietors, and it was not until after the revolutionary movement in Europe in 1848, that, in Germany, facilities have been provided to enable them to purchase the land on which they were settled. The Russian Act of 1861 went very much further by recognizing the right of every peasant, in Great, Little and New Russia, as well as in part of the White-Russian provinces, to a certain area of land, representing a normal peasant holding for the given locality. If his actual holding, at the time of the reform, happened to be below the minimum fixed for the locality, the peasant was entitled to an addition, out of the rest of the landowner's estate, which would bring it up to the norm. State credit facilities for the redemption of the holding, which would make the peasant independent proprietor of his land, were provided from the beginning, and, within some twenty years, the redemption was made compulsory, because experience had proved that to leave it to private agreement between the parties concerned was inexpedient. Though, in certain cases and with regard to certain details, of which more will be said later, the influence of the landowning class succeeded in modifying the original scheme of the reform to the detriment of the peasants, on the whole, even in its final form, the land-settlement of the peasants at the Emancipation was unparalleled in history for the generosity with which they have been treated. This point has to be particularly emphasized, because in dealing with the Russian agrarian problem most writers are disposed to lay too much stress on the land-settlement at the Emancipation as the principal cause of the misery of the Russian peasants at the close of the last and the beginning of the current centuries. This view of the problem cannot be accepted without very important qualifications. To a large extent it is based on the failure to apply to the problem the test of comparison with other countries and their agrarian evolution. Such comparison shows at a glance that, whatever the shortcomings of the land-settlement at the Emancipation, it is not to them, in the first instance, that one has to turn for the causes of Russia's agrarian distress. A comparison of the Russian peasant's position with that of the peasantry in almost any country of Western Europe in the second half of the last century would show that, in spite of the far less favourable conditions of their land-settlement after the final abolition of feudal agrarian relations, they were comparatively prosperous, and no agrarian problem, such as that which arose in Russia, had been heard of in those countries. If in Russia, in spite of the generous terms of land-settlement at the Emancipation, the agrarian situation became so bad as to threaten the whole social and political structure of the State, it was surely not to the great reform, but to other causes that one should look for explanation.

This is why, in the present study, while assigning their due share of responsibility for the agrarian distress to certain features of the Emancipation, I shall endeavour to describe the agrarian situation as a whole, in the general economic setting of nineteenth-century Russia, which alone can account for the agrarian problem as it faced Russia on the threshold of the current century. It will be seen that, as one of the aspects of the general economic situation, the problem ultimately resolved itself into a question of a protracted agricultural crisis into which the country had been involved in the course of the decades following the Emancipation. This crisis affected the whole of the agricultural industry, and it was not before the beginning of the twentieth century that signs began to appear of the cloud being gradually lifted.

A. The Peasants after the Emancipation

The agrarian organization of Russia, such as it was at the beginning of the twentieth century, dated from the Emancipation, though in certain important details the reforms of 1861, while they abolished serfdom, sanctioned local customs of tenure and certain traditional features in agrarian relations. The necessity to adapt the new organization of the countryside to local requirements, resulting in the publication of four separate *Statutes of Peasants* in 1861, as well as in numerous local amendments in details, was one of the causes of certain differences in the aspects of the agrarian problem as it faced Russian political thought at the close of the century.

For the purpose of Emancipation, Russia was divided into four principal areas, according to peculiarities of tenure and existing agrarian relations. For each of these areas a special Statute of Peasants had been issued, adapted to local conditions, in addition to the General Statute containing the principles of the reform. Thus, there was, first, the *Local Statute of Peasants for Great Russia*, which, besides the 29 Great-Russian provinces proper, applied to three provinces of New Russia (Kherson, Taurida and Ekaterinoslav) and to part of White Russia (province of Mogilev and part of that of Vitebsk). With some modifications of detail, it applied also to the Don, Stavropol and Siberia. The outstanding characteristic of this Statute was that the typical form of tenure sanctioned by it for this part of the country was the *mir* or rural commune, involving redistributions of holdings among its members and joint liability for taxes. Villages with holdings in perpetuity in open fields, common in the rest of Russia, were exceptional in these provinces. The allocation of land to the former serfs in this part of the country was made on the basis of uniform local norms, according to the number of male peasants in the commune.

The *Local Statute of Peasants for Little Russia*, which applied to the provinces of Poltava, Chernigov and part of Kharkov, was characteristically different in that it sanctioned agrarian relations on which Western customs, through the medium of Poland, had exercised considerable influence in the past. In the matter of the allocation of land, this influence was evident in the Statute providing for two classes of holdings, corresponding to the existing division of serfs into *tiagly* peasants, who possessed oxen (the equivalent of the *spannfähige Bauern* in Prussia, who were placed in a privileged position by the Hardenberg legislation), on the one hand, and *pieshy* peasants, who had no oxen and performed their services on foot, on the other. Accordingly, the two classes of holdings provided for by the Statute were, first, the "basic" or *pieshy* holding, representing a minimum to be given to every peasant, and, second, the "additional" holding, which it was optional for the peasant to claim. Thus, in the allocation of land in Little Russia, a principle had been adopted entirely different from that which found application to Great Russia. While, in the latter, the sizes of the holdings to be allocated were made to depend entirely on the size of the family and on its needs, without regard to its economic position, in the case of Little Russia, following local custom, the law provided for a certain degree of adaptation of the size of holdings to the productive resources of the peasants. The form of tenure, typical of the locality, was that of hereditary holdings in open fields. Common rights of the whole village extended only over the so-called *mirskaia* or *gromadskaia zemlia*, that is the part of the land allocated to the village community at the Emancipation left over on the allocation out of it of holdings to the landless peasants of the village. This land, indeed, formed a complete parallel to the German *Gemeinde* or the English *common*, and while, for their individual holdings in the open fields of the village, the peasants were responsible severally, the liability for taxes and other dues from the common land rested with the village community as a whole.

The principal difference of the *Local Statute of Peasants for the provinces of Kiev, Podolia and Volynia* from that of Little Russia was that the former did not fix a statutory minimum for the holdings to be allocated to the peasants. In these provinces, as well as the Western, in 1846 a general survey had been made of peasant tenures, and the land held by them entered into official registers or *inventories*, with a view to protecting the serfs against any attempts on the part of the landowners to reduce their holdings. In 1861, these *inventories* provided the basis for the allocation of holdings to former serfs, and the latter were permitted, in the course of six years after the Emancipation, to claim back any land of which they may have been deprived by their masters in contravention of the *inventories*. The provisions concerning "basic" and "additional" holdings differed from those applying to Little Russia in that the latter were given only to the former *tiagly* peasants, possessing oxen. The allocation of common land, in addition to the individual holdings of the separate villagers, depended entirely on agreement between the peasants and their former master, and the element of communal tenure was, accordingly, present here in a still lesser degree than in Little Russia.

The *Local Statute of Peasants for the provinces of Vilno, Grodno, Kovno, Minsk and part of the province of Vitebsk*, which also contained no provisions concerning a minimum size of holdings, left to every peasant family the land held by it according to the *inventories*. The element of communal tenure here was entirely absent, the land being held in perpetuity and no provision having been made for any common areas.

From this short outline of the characteristics of the agrarian relations sanctioned by the Local Statutes it may be seen that, in the case of by far the greater part of Russia, to which applied the first two of the four enactments dealt with above, the peasants, when set free, had been given holdings of a size fixed according to certain norms. These norms differed according to locality and indicated, in every case, the minimum and the maximum per male peasant. If, at the Emancipation, the peasants of a given village were actually holding less than the minimum to which they were entitled, the deficiency had to be made good to them out of the rest of the landowner's estate. If, on the other hand, they held more land than the area to which they were entitled, the landowner could retain the balance over and above the minimum. The maximum served a less obvious purpose and found practical application mostly in those localities in which the land as such, without peasants, was of little value to the proprietor, and the latter could be tempted to give the former serfs more of it than they could conveniently manage, with a view to getting for it either rents, until it was redeemed, or its value on its redemption by the peasants. Thus, the Local Statute for the Western provinces, though it did not fix any minimum, fixed a maximum of 20 dessiatins per family (*dvor*). In the provinces of the West and the South-West, to which referred the two last of the four Local Statutes, it was not considered necessary to fix minima for the sizes of holdings, seeing that, only a few years before the Emancipation, peasant tenures had been regulated by the Inventory Commissions.

As I have pointed out above, it is usual to refer to the land-settlement of the peasants at the Emancipation as the principal cause of the "land-hunger" and of the generally miserable condition of the Russian peasantry after the reform. In dealing with the actual effects of the reform, it is essential to distinguish between the influence on the economic position of the former serfs of the provisions concerning the allocation to them of land, on the one hand, and that of financial liabilities imposed on them in consequence of such allocation, on the other.

In the first instance, there is the origin of insufficient holdings and of the consequent "land-hunger" among certain groups of the peasantry, which must be considered. To do so, it is necessary for a moment to look back to the agrarian conditions which prevailed in various parts of Russia under serfdom, in so far as they affected the extent of the peasants' own cultivation and, therefore, the actual sizes of their holdings, before the Emancipation. Up to the second half of the eighteenth century, while the landowning gentry was still, as a rule, in a position of tenants for service and spent most of their lives in the Army, the Navy or Government offices, large farming with serf labour was almost unknown, the land being left to the peasants settled on it as serfs to cultivate for a fixed annual rent. This payment, which varied in amount according to the estate and depended, primarily, on the requirements of the owners and the paying capacity of the serfs, was called *obrok*. It differed from the customary fees of the medieval English manor in that it was rather an impost on the earning capacity of the serf than a form of rent for the land on which he was settled. Accordingly, a serf who, being a skilled artisan, lived in town on business of his own and held no land in the village to which he belonged, also paid an *obrok*; and, in fact, such serfs, engaged in profitable occupations away from their villages, were the financial mainstay of many a landowner, especially in the provinces outside the black-earth belt. With the abolition, in 1762, of compulsory service for the landowning gentry, followed by Russia's territorial expansion in the Southern steppes, large farming began to develop. The exploitation of the virgin black-earth of the steppes made farming on the landowner's account, with serf labour, a paying proposition, and the area of the landowners' home farms began to extend. Instead of paying the *obrok*, the serfs had in such cases to provide a certain number of days' labour on the fields of their masters. At the close of the eighteenth century, the number of days of *barshchina*, as such labour was called, was limited to three per week, but this legal limit, failing adequate control, had not been universally respected. In the course of the first half of the nineteenth century farming became very fashionable among the Russian gentry. Agricultural associations were established, and foreign improvements studied and imitated. Generally speaking, in Russia it was a period somewhat similar to that of the English "spirited landlords" of the eighteenth century, with that essential difference, however, that the progress of farming in Russia rested entirely on the foundation of unpaid labour supplied by the serfs. As a result of this development in the black-earth belt, by the time of the Emancipation the conditions of the serfs in the Northern half of Russia, outside the black-earth, and those which prevailed within the latter, were very different. In the former, the old system of *obrok* was still predominant, large farming and *barshchina* being rather exceptional. In the black-earth zone, on the contrary, the *barshchina*, which left the serfs little time for their own farming, and tended, therefore, to reduce their actual holdings, was more common. Thus, it was estimated that, at the time of the Emancipation, in the thirteen provinces of Great Russia situated outside the black-earth the proportion of serfs on the *obrok* was 55%, while in the seven provinces of the black-earth belt it did not exceed 26%.[1] In the Western, South-Western and Little-Russian provinces and in New Russia, the *obrok*, on the eve of the Emancipation, hardly existed at all, practically all the serfs being on the *barshchina*, as the statistics of these provinces show: [2]

	Percentage of Serfs on *obrok*
White Russia:	
Mogilev	3·0
Minsk	2·6
South-West:	
Kiev	1·6
Podolia	3·7
Little Russia:	
Kharkov	1·5
Poltava	0·7
Chernigov	0·2
New Russia:	
Kherson	0·1
Ekaterinoslav	0·2
Taurida	—

Thus, in the newly-colonized Southern provinces the *barshchina* was practically universal. In the South-West and in part of White Russia, Western influences, which reached them through the Polish landowning gentry, were strongly felt. There, as well as in Little Russia, large farming was of long standing. The peasants were not unfrequently turned out of their holdings and made to work for their masters as agricultural labourers, and even the measures taken by the Government to prevent such evictions were not always successful.

On the position of the peasants after the Emancipation this difference in the agricultural organization of various localities exercised a very great influence. Since, in the fixing of the maxima and minima sizes of holdings at the Emancipation, the sizes of their average holdings as serfs were taken into account, the fact that in the provinces where large farming and the *barshchina* were common their holdings were generally small, had an important effect on the norms governing the allocation of land. The sizes

[1] Estimated by V. I. Semevsky.

[2] Statistics of the *Redaktzionnia Kommissii*, which drafted the legislation in connection with the Emancipation, quoted by Prof. Miliukov, art. "Peasants," in the *Encyclopædia* of Brockhaus.

of holdings allocated to the peasants of the black-earth belt were, accordingly, generally small, as compared with those of the former serfs of the rest of Russia, who, under the *obrok* system, had practically the free run of most, if not the whole, of their masters' estates. Moreover, as far at least as the black-earth was concerned, in which land was rapidly rising in value, the landowners were loath to part with it and, by exercising strong pressure in the Commissions which drafted the reforms, endeavoured by all means to reduce the area that had to be ceded to the peasants. In this they succeeded to a certain extent, and reductions have been effected in two ways. In the first instance, the local minima and maxima, according to which the actual allocation of holdings had to take place, have been reduced, as compared with the original draft; secondly, it was provided that, by agreement between the parties concerned, the landowner could give his former serfs so-called "gratuitous holdings" of one-quarter the normal size, in which case he forfeited his claim for payment for the land. In consequence of the reduction in the local norms, considerable areas of land, formerly used by the peasants, after the Emancipation have been cut off the holdings of many villages and retained by the landowners. The actual extent of such reductions varied greatly according to locality and was not easy to ascertain statistically, and its estimates vary considerably. The best modern authorities put it, however, at about one-fifth of the total area held by the peasants before the Emancipation.[1] The effect of the reduction on the economic position of the peasants concerned was, certainly, most injurious in those localities in which the holdings were smallest, and thus was generally more painfully felt in the black-earth belt of Russia. In the Northern half of the country, where meadows were the most valuable parts of the estates, the reductions mostly affected this essential element of the holdings, which made them especially obnoxious to the peasants. Whatever their economic influence, these reductions were a constant source of bitterness and quarrel between the peasants and the landowners, the former considering themselves, not without justice, done out of their own land. To the more unscrupulous among the landed proprietors these "cut-away lands," as they were called (*otriezki*), provided an opportunity for the exploitation of their neighbours, who were often vitally concerned in keeping the land for their own use, even at the cost of exorbitant rents, for fear of having otherwise to change the whole of their farming or of utter ruin. The evil consequences

[1] Prof. A. Kaufmann, *The Agrarian Problem* (Rus.), 1908, p. 26.

of such reductions were so obvious that even conservative writers, as, for instance, A. S. Yermolov, a former Minister of Agriculture and one of the best authorities on farming in Russia, considered it only fair in such cases to have recourse to compulsory expropriation of the land in question and its restitution to the peasants.[1]

The "gratuitous holdings" (*darstvennyi nadiel*) were another source of trouble, dating from the Emancipation. In those districts in which land was rapidly rising in price, it was obviously to the advantage of the proprietor to forego payment for it at its present value, rather than to part with four times the area. The peasants, especially in those localities where land was still plentiful and could easily be obtained on lease—a condition which, at the time of the Emancipation, still prevailed throughout the greater part of the black-earth belt—were often quite willing to adopt this course, with a view to escaping payment for the land. As a result, "gratuitous holdings" were fairly numerous in the black-earth belt, especially in the South and the East of Russia, still rich in virgin land. On the whole, they involved 640,000 male "souls" or about 6% of the total number of former serfs. In the provinces of Ekaterinoslav, Voronezh, Simbirsk and Kazan they accounted for 20 to 30%; in those of Taurida, Saratov, Viatka and Perm—30 to 40%; in the provinces of Ufa and Samara—40 to 45%; in that of Orenburg—as much as 70 to 75%.[2] Other provinces with a large percentage of "gratuitous holdings" were those of Kherson, Poltava, Kharkov, Kursk, Chernigov, Tambov and Penza. In the rest of the country, the proportion was negligible. The consequences of this measure were generally injurious. In this case, the peasants started life on obviously insufficient holdings and became entirely dependent on leases or outside earnings. So long as land in the black-earth belt was still plentiful and cheap, they did not fare much worse than the rest; sometimes they may even have fared better than their neighbours who had to pay for their holdings. But the development of the steppes proceeded at a rapid pace, and the rents, fostered by the peasants' keen competition for leases, increased by leaps and bounds. Employment for so large a mass of people, tied to the locality, could not be found, especially in the purely agricultural provinces, in which the "gratuitous holdings" were most common, and in a very few years the peasants who had succumbed to the temptation became, as a rule, the synonym of the village poor.

[1] A. S. Yermolov, *Our Land Problem* (Rus.), 1907, p. 113.
[2] Prof. A. Kaufmann, op. cit., p. 28.

Apart from these reductions in the sizes of holdings, there existed two classes of former serfs whose treatment at the Emancipation was different from that of the rest. In the first instance, there were the *dvorovy*, or serfs employed about their masters' households in various capacities, from common servants to artificers, musicians or even actors on the stage, etc. These, if they held no land in the village, which they usually did not, were set personally free, but were not entitled to the allocation of holdings. They formed the nucleus of the real proletariat, but as many of them were highly skilled, while others remained in the service of their masters for the rest of their lives, their position did not give rise to any special difficulties. The position of the other class, namely the serfs of small landowners, whose estates would have been completely wiped out or rendered useless by the allocation of holdings to the peasants, was worse. This class, in 1858, numbered 325,000 male "souls," of whom 137,000 were actually landless. Those of them who had any holdings at all, were permitted to keep them on the Emancipation, but were not entitled to any additions to bring them up to the local norm. Those who had neither land nor a house in the village were treated as the *dvorovy* and received no land at all. Finally, those who had only houses in the village were entitled to Government assistance for emigration and settlement on State-owned land elsewhere.

In the Western and South-Western provinces, in which the former serfs have, as a general rule, received the holdings on which they have been actually settled before, there existed a numerous class of peasant small-holders, whose holdings were not large enough, and who have always been mostly dependent on their earnings as agricultural labourers in large estates. In the three Lithuanian provinces and the Western districts of White Russia (province of Minsk and part of Vitebsk), in 1863, the peasants who have been turned out of their land before the year 1857, when the *inventories* came into force, were permitted to claim back 3 dessiatins (8·1 acres) per family (*dvor*), while those evicted since 1857 could claim the restitution of their holdings in their entirety. The 3-dessiatin holdings, however, were admittedly too small, and their owners formed, from the start, a semi-proletarian element in the villages. The peasants of the Western and South-Western provinces have, indeed, profited by the privileged treatment accorded them by the Government in consequence of the Polish Insurrection of 1863, but the difference between rich and poor among the peasantry of these provinces, more pronounced than in any other part of the country, had not been affected by the efforts of the Government to ensure to Russia the adherence of the peasants of these localities, in which a very large proportion of the landowning class was of Polish origin.

It may thus be seen that, from the start, various groups of former serfs were placed in a different position with regard to the sizes of their holdings, and that some of them began to feel the "land-hunger" from the day of their emancipation.

The peasant population of Russia, however, in the middle of the nineteenth century, was very far from being composed entirely of serfs. According to the tenth and last Census of poll-tax payers (*revisia*) taken in 1858, the serfs numbered about 10 millions of male "souls," while another 10 million "souls" was made up of peasants settled on State-owned land (about 9·2 million "souls") and of those of the appanages of the Imperial Family (about 860,000 "souls"). Both these classes have emerged from the period of bondage in a far better economic position than the serfs of private landowners.

The peasants settled on the appanages and the domains of the ruling Emperor (*kabinetskia zemli*) were the first to be set free. The Imperial Ukaze of June 6th, 1858, set personally free the peasants of the appanages; that of August 26th, 1859, did the same for those of the domains. The Ukaze of March 5, 1861, settled on these classes of peasants all the land hitherto held by them. By the Statute of June 26th, 1863, they were made proprietors of their land, the annual rents paid by them for their holdings being converted into redemption payments for a period of 49 years, at the end of which they were to hold their land free of any liabilities.

The peasants settled on State-owned land were provided for by the Statute of November 24th, 1866, according to which they kept all the land hitherto held by them. They were given the option of purchasing their holdings, but very few availed themselves of this opportunity before 1886, when the redemption was made compulsory.

Thus, the Russian peasantry, after the Emancipation, was divided into three classes, namely, the former serfs of private landowners, the peasants of the appanages and domains, and the peasants settled on State-owned land. Of these three groups, the serfs were placed in the worst position with regard to the size of their holdings. The peasants of the appanages and domains, whose holdings had not been reduced, were far better off as to land. But the real patricians among the peasantry were the State peasants, whose holdings, on the average, were much larger than those of any other group. Before the Emancipation, they have enjoyed far greater freedom and independence than the serfs, and their position had especially been improved by the reforms in their tenures and organization, introduced in the thirties of the last century by Count Kiselev, the first Minister of Agriculture and Domains. The State peasants, indeed, were the strongest and most independent element in the Russian countryside, both economically and morally. Before 1861, their legal position was that of *glebae adscripti*, and among the rest of the peasantry, reduced to a condition approaching slavery, they have always been referred to as *volny*, which means free men. A large admixture of the sober, hard-working religious dissenters—adherents of the "Old Faith" —all of whom belonged to the number of State peasants, also added to their economic and moral strength. When they were set free, they thus easily occupied the top of the social scale of the peasantry, and have kept the position to the present day.

The average sizes of holdings in the three groups, after the Emancipation, were as follows: [1]

State peasants . . .	8·7	dessiatins per male soul
Appanage peasants . .	5·75	" " " "
Serfs	3·5	" " " "

Local variations in the sizes of holdings of each of the three groups were very great, the holdings being generally smallest in the Northern half of the black-earth belt, from the South-Western provinces through the Ukraine and the Central Agricultural region, as well as the adjoining parts of the steppes in the South and of the Moscow region in the North. This was the part of the country in which, at the time of the Emancipation, farming was most developed and the prices of land highest, and where, therefore, the pressure of the landed interests for a reduction in the sizes of holdings to be given to the former serfs was strongest. Here it was also that the agrarian over-population and the consequent agrarian crisis had assumed an acute form at the earliest.

The sizes of holdings, however, were only one aspect of the land-settlement of former serfs at the Emancipation. Another very important element of the problem was the redemption by the peasants of the land allocated to them. This, though at first envisaged by the Government as optional and entirely dependent on voluntary agreement between the parties concerned, was made compulsory, in the case of the Lithuanian, South-Western and part of the White-Russian provinces, with their mainly or largely Polish

[1] Prince A. I. Vassilchikov, *Landownership and Farming in Russia and other European Countries* (Rus.), 2nd ed., 1882, Vol. I, pp. 448–9.

landowning class, in 1863, after the Polish Insurrection. In 1881, the principle of compulsory redemption was extended to all the former serfs, and in 1886 it was applied to the former State peasants as well. Indeed, from the very start, it has been the intention of the Government to make the peasants proprietors of their holdings, but the object in view had been approached gradually by easy stages. Thus, the *General Statute of Peasants* of 1861 provided originally for the allocation to the peasants of land, within the fixed limits per male "soul," in perpetual usufruct, against the payment of a certain rent (*obrok*) to the titular owner. According to this enactment, the redemption of the holdings was made optional and depended on agreement between the peasants and the landowners. The *Statute of Redemption*, in 1863, issued with a view to facilitating the purchase by the peasants of their holdings, made a distinction between the actual homestead (*usadebny uchastok*) and the rest of the holding, which included arable, meadows, etc. In the case of the former, the redemption was made compulsory, and the peasant had to pay for it himself, without credit assistance from the State. As to the rest of the holding, its redemption depended on agreement between the parties, and should such agreement be reached, the State undertook to provide financial assistance to the purchasers to the extent of 75 or 80% of the purchase price. The balance of 20 or 25% the peasant had to find himself. If no agreement could be reached and the parties came to a deadlock, the landowner could claim a compulsory settlement, in which case he had to forego the 20–25% balance of the full price of the land, otherwise payable by the peasant, and content himself with 80% of the amount if the peasants were given their holdings in full, or with 75% if their holdings have been cut down to the statutory minimum. The payment was actually made in Government *redemption bonds*. The price of land was worked out by capitalizing the amount of the annual rent (*obrok*), payable for its usufruct, at 6%. The arrangement provided for by the *Statute of Redemption* in its original form, however, proved exceedingly cumbersome, and the land-settlement of the former serfs was greatly delayed by the inability of the parties to come to agreement. Accordingly, in 1881, the redemption was made compulsory in all cases. The peasants became proprietors of their land, either collectively, as village communes (*mir*), which was the common form of tenure in Great Russia, or individually, the latter being most common in the Ukraine, the South-West and the Western provinces. Their land was virtually mortgaged to the State, as security for the advance of the purchase price, for a period of

49½ years, during which they had to pay 6% p.a. in interests and repayment.

The influence of the redemption payments, to which I have briefly referred in a preceding chapter in one of its important aspects, was great and manifold. Not only did they increase the need for ready money among the peasantry generally, and thus contribute to the commercialization of peasant farming, but, in the case of certain large groups of peasants they proved a very heavy burden indeed. Since the valuation of the land transferred to the peasants was arrived at by the capitalization of the rent payable for its usufruct, and that rent depended primarily on the *obrok*, which the serfs in the district concerned used to pay before the Emancipation, in those localities in which the *obrok* used to be high, the payments, in spite of the relatively high rate of capitalization, were heavy. Moreover, while, under serfdom, in the more agricultural provinces of Russia the majority of the peasants were kept on the *barshchina*, thus paying their dues in labour, and the *obrok* was mainly confined to those of them who, having some trade or other source of money income, could pay it more easily, now the system of money payments had been extended to the peasantry as a whole. The result was that, though, as compared with the *obrok*, the redemption payments were, as a rule, rather lower, those peasants who used, before the Emancipation, to be on the *barshchina*, and could not easily reorganize their whole husbandry now, so as to get hold of the necessary cash, found themselves faced with a very difficult problem indeed.

The position of the former State peasants, in this respect, as in others, was better than that of the former serfs. Unlike the latter, before the Emancipation, the State peasants, as a general rule, had all been on the *obrok*, and the latter, fixed under Count Kiselev on the basis of a general cadastre of State-owned land and moderate in amount, was left in force after the reform. When, in 1886, the redemption of their holdings was made compulsory, it served to fix the capital value of their holdings. Accordingly it was estimated that, on the average for European Russia, the amount of annual redemption payments per dessiatin in the case of State peasants was 0·83 roubles, while former serfs paid as much as 1·31 roubles: a difference of about 51%.[1] According to the statistics of the Agricultural Commission of 1872 and of a Fiscal Commission of the same year, the total payments of the State peasants in thirty-seven provinces of European Russia (i.e. exclusive of the nine Western and South-Western provinces) amounted to 92·75%

[1] Article "Peasants" in the Russian *Encyclopædia* of Brockhaus.

of the estimated net yield of their holdings, while those of the former serfs reached 198·25%.[1] Both these comparisons, however, though true enough in so far as they show the advantages generally enjoyed by the State peasants, can hardly be accepted unreservedly. In the case of the first of them, it must be observed that the State peasants in their great bulk were concentrated in the remoter parts of the country, mostly outside the black-earth belt, and that, accordingly, their land being of less value, it was only natural that their redemption payments per dessiatin should be lower than those of the serfs, most numerous in the agricultural provinces of the black earth. The second comparison rests on the very unreliable basis of net yield, which is extremely difficult to ascertain statistically, and too much importance should not, therefore, be attached to its actual figures.[2] That the conclusions of the official and other inquiries, which pointed out the burdensome nature of the redemption payments, however, were substantially correct, was confirmed by the accumulation of large arrears in the instalments, as well as by the generally precarious economic condition of the peasantry after the Emancipation. Whether the payments were absolutely too heavy, as numerous writers contend, and actually exceeded the utmost possible limits of the yield of the land under any conditions, it is hardly possible to decide, but, in any case, they proved too heavy for the great bulk of the peasants of nineteenth-century Russia, with her limited agricultural markets and her few facilities for the profitable disposal of either labour or produce. The fact was recognized by the Government, and by the Ukaze of December 28, 1881, the payments were reduced considerably. The reductions varied according to provinces, but for the whole country they reached 27% of the amount hitherto paid annually.[3] From the 'eighties onwards, they could hárdly be looked upon as a heavy burden, and, had the general economic development of the country in the second half of the last century been more rapid, they would not have exercised any real influence on the progress of the peasantry. That, in many cases, even after their reduction, the payments were still felt as a burden, only served to show that the commercialization of peasant farming, its transition from natural to money economy, had been proceeding very slowly indeed, and still left large areas and large groups of rural population practically unaffected by the influence of the market.

From what has been said above, one can see that, indeed, some

[1] *Ibidem.*

[2] *Q.v.* Kaufmann, op. cit., part I, pp. 51–3.

[3] Prof. L. Khodsky, article on "Redemption Payments" in the Russian *Encyclopædia* of Brockhaus & Efron.

of the causes of the "land-hunger" among certain groups of the peasantry, as well as of the impoverishment which was characteristic of the decades following the Emancipation, could be traced to the great reform. When, in 1861, the Russian peasant emerged from the long-drawn ordeal of serfdom, in some cases he suffered a reduction of his holding; and that reduction, whenever it had taken place, had certainly been responsible for the development of "land-hunger" among the groups concerned. In some cases, though there had been no reduction, the holdings were too small at the time of the Emancipation, and have not been increased in the course of the reform. According to Professor L. Khodsky, an eminent Russian economist and statistician, whose estimate is accepted by other authorities on the subject as the most reliable,[1] the position of the peasantry in Russia, after the Emancipation, with regard to the sizes of holdings, was as follows. Taking as standards the average sizes of holdings of State and appanage peasants in every province, which he considered as sufficient for the full employment and support of a peasant family, he proceeded to the study of the grouping of peasants according to the sizes of their holdings, as compared with that standard. The results arrived at were that, of the State and appanage peasants, 50·7% had received holdings in excess of this norm; 35·6% had holdings of the average size, and 13·7% had holdings below the standards of sufficiency. In the case of serfs, the proportions respectively were 13·9%, 43·5% and 42·6%. Considering the peasantry as a whole, without making a distinction between the State peasants and the former serfs, and applying to it the results of the above computation, it will be found that the proportion of peasants with insufficient holdings in Russia, on the morrow of the Emancipation, could be put at about 30%. This percentage, though certainly considerable, included all holdings below the standard of sufficiency, from the "gratuitous" plot of a dessiatin or so to anything just falling short of the fairly generous size of the average holdings of State and appanage peasants. All the rest—70% of the peasants—received holdings either sufficient or in excess of the norm. This calculation helps to reduce to measurable terms the extent of the influence exercised by the land-settlement of the peasants after the Emancipation on the origin and development of the agrarian problem, in so far at least as its most familiar aspect—the "land-hunger" of the rural population—is concerned. After the Emancipation, with 70% of the peasantry in possession of holdings either sufficient or more than sufficient, under existing conditions of

[1] *Q.v.* Kaufmann, op. cit., part I, pp. 48–9.

cultivation and yield, to make them self-supporting, Russia could hardly be considered as having laid the foundation of the agrarian distress which had developed in the course of the following decades. Even among the rest, whose holdings were in a varying degree insufficient for their support, possibilities of economic progress were not entirely excluded. Indeed, even the minute size of "gratuitous holdings" had not always proved an insuperable obstacle on the way to prosperity, and in some cases, especially in the vicinity of large markets, their owners have made good by the development on them of vegetable gardening, fruit-growing or some other form of specialized production. In the case of families with relatively large, though still not self-supporting holdings, the problem could mostly be satisfactorily solved by the practice of some auxiliary occupation or casual employment in the neighbourhood of the village; sometimes by outside employment of some member or members of the family. In fact, all these means of eking-out an insufficient income from their land were largely practised in the greater part of the under-producing belt, with the result that, on the whole, the "land-hunger" and the agrarian crisis there, in spite of the poor quality of the soil, was less acute than in the black-earth zone. Yet, as a matter of fact, the agrarian distress involved ever-increasing masses of peasants all over the country, and especially in the more purely agricultural districts.

The development could be described, in general terms, as a rapid growth of rural population unaccompanied by a simultaneous progress in the yield of the agricultural industry and in the opportunities of employment of the excess of hands in the countryside. The forms assumed by the phenomenon, as well as the intensity of its manifestation, varied according to locality, but essentially it was the same everywhere, and was produced by a combination of causes. Of the latter, certain were rooted in the organization and conditions of the countryside and of farming, while others, and the most potent by far, had to be sought in the character and the evolution of the economic structure of nineteenth-century Russia.

Among the causes of agrarian overpopulation a prominent place must be given to the conditions of peasant tenure. It will be remembered that the bulk of the peasantry in Great Russia, as well as some of the peasants of White, Little and New Russia, held their land in common, subject to periodical or occasional redistributions of holdings among the members of the commune, under the system of communal tenure (*mir, mirskoie vladenie*). The alternative system, predominant in the Western, the South-Western and the Little-Russian or Ukrainian provinces, involved tenure in perpetuity (*podvornoie vladenie*). Under both systems the individual holdings of the peasants were scattered in open fields in more or less numerous strips. The number of strips varied greatly, according to the nature of the soil, the lay-out of the village land and the sizes of holdings. The principle was that, in every field, each peasant must have a share equal to that of any other. The cottages and farm buildings of the peasants were all built in the village, mostly in two rows facing each other on either side of a central street or main road. In the Northern provinces of Russia, where the soil is of very uneven quality and suitable land is relatively scarce, as well as in the agricultural centre, with its dense population and small holdings, the strips were often extremely narrow and very numerous.[1] In the South and South-East, where the land in the steppes is more uniform, the number of strips was generally much smaller and they were wider, but here, owing to the often very large size of the villages, clustering round the sources of water supply or in other convenient places, this advantage was often counterbalanced by the important drawback of the remoteness of some of the land from the village and the consequent difficulty of its proper cultivation. Here, some of the strips, lying miles away from the village, not only necessitated a great waste of time in reaching them and practically excluded all possibility of manuring in the usual way, but had to be left unattended, except for occasional visits, from sowing-time to the harvest.[2] The mixing-up of strips, their often extremely small width and their remoteness from the village—features of open-field tenure familiar to the student of English farming before the enclosures of the eighteenth century—need no enlarging upon as obstacles to the progress of the agricultural industry. In Russia, by the most ardent opponents of the commune, the *mir* had often been saddled with the responsibility

[1] Thus, according to figures quoted by P. Pershin in his article on "The Forms of Tenure" in the Russian publication *O Zemlie*, Moscow, 1921, part I, p. 54, in the provinces of Archangel, Vologda, Yaroslav, Olonetz, Novgorod, Pskov and Vitebsk, for which data were available, 10·5% of the peasants held their land in 11 to 20 strips; 32·9% in 21 to 40 strips; 25·6% in 41 to 60 strips; 19·6% in 60 to 100 strips, and 7·7% in over 100 strips.

[2] In the provinces of Podolia, Kiev, Chernigov, Poltava, Kherson, Taurida, Ekaterinoslav, Kharkov, Don, Voronezh, Tambov, Penza, Saratov, Simbirsk, Samara, Astrakhan and Stavropol, according to Pershin (op. cit., p. 54), 5·3% of the peasants had their remotest strips at a distance of 1 verst (0·66 mile) from the village; 7·4% at 1·1 to 3 versts; 11·5% at 3·1 to 5 versts; 38·7% at 5·1 to 10 versts, and 37·1% at a distance of over 10 versts (just under 7 miles). The influence of such distance on the standards of cultivation and the yield of the land can easily be imagined.

for these evils, which, however, in fairness, should be attributed not to it but to any form of open-field tenure.

Indeed, about the part played by the system of tenure in the development of the agrarian crisis in Russia much has been written. The village commune (*mir*), involving redistributions of land among its members, had been vigorously assailed by numerous opponents, ranging from Conservatives to Marxian Socialists, and no less vigorously defended by its partisans, from the Slavophils and other sections of the Nationalist Right to the most advanced groups of the *narodniki*. Here, the *mir* must be taken for granted, and its influence on the development of the agrarian problem in Russia considered dispassionately from the economic standpoint. In the view of the present writer, the main count in the indictment against the *mir*, or redistributory rural commune, as such, was that it contributed to the development of agrarian overpopulation by relieving the individual peasant from responsibility for the excessive increase of his family. The task of finding room for any new member devolved on the commune, which did it at the expense of the whole village, at the cost of relatively small individual sacrifices. As to the rest, it may be said that, while it is perfectly true that communal tenure hinders agricultural progress by restricting individual initiative, this fault is not peculiar to the *mir*, but is shared by all forms of open fields, either in Russia, or in England before the enclosures and in Germany before the agrarian reforms of the nineteenth century. It must even be admitted that in a village commune with effective redistributions the evil of small scattered strips lends itself more easily to correction, while in open-field villages without redistributions it is unavoidably perpetuated and grows gradually worse.[1] The special objection directed against the redistributory commune or *mir* that it hinders agricultural improvements by making the position of the peasant with regard to his land insecure, really applies only to capital improvements, which, however, it is hardly possible for any individual to undertake on his strips, either in the *mir* or, for that matter, in any open-field village with holdings in perpetuity. As to other improvements, such as the use of artificial fertilizers or improved implements and seeds, they can equally well be used under both systems, subject to the general limitations involved in open-field cultivation. The introduction of improved rotation encounters the same obstacles in both cases, compulsory cropping (*Flurzwang*, as it is called in

[1] *Q.v.* Prof. A. Manuilov, "Notes on Communal Tenure," in the *Essays on the Peasant Problem* (Rus.), 1904, Vol. I, p. 256. Attention to this would appear to have been drawn first by Prof. Posnikov.

Germany) being an unavoidable attribute of open fields. Moreover, open-field tenure, either under the communal system or in perpetuity, has that great drawback that, with it, the peasant farm, as an economic and technical entity, does not exist. The division of the peasant's holding, even if carried to extreme, does not involve the breaking-up of an efficiently constituted, well-balanced farm into a number of inconvenient fragments: a fact which effectively helps in eliminating all checks on the growth of rural population and contributes to the excessive subdivision of holdings and the consequent pauperization of the peasantry.

Thus, in the systems of tenure in general use among the peasants in Russia there were elements present which undoubtedly contributed, on the one hand, to the unreasonably rapid increase in the rural population and, on the other, to the slowness of agricultural progress. Besides, the sufficiency of the holding depends not only on its actual size, on the fertility of the soil and on the form of tenure. From the purely agricultural point of view, there is the constitution of the holding which has to be taken into consideration as an important factor. In other words, whether or not a holding of a certain size is sufficient for supporting its owner in reasonably comfortable conditions, depends on the harmonious combination in it of the various elements essential for its successful exploitation. Thus, especially under a three-course system of cropping, in which no grass crops are cultivated, besides a sufficient area of arable, there must be a corresponding area of meadows and pastures for the maintenance of a certain number of cattle. This was probably the weakest spot in the land-settlement of the peasants after the Emancipation, and with the increase in population and the extension of arable at the expense of grass, it has been growing worse ever since. In a large number of cases, sometimes throughout whole districts, it was not the sizes of the holdings, but the relative shortage in them of this essential element that was the root of the problem. Here, indeed, one could see the application of the "law of the minimum," as enunciated by Liebig with regard to the constitution and fertility of the soil, to the economic aspect of farming. On the average, under the three-course system, the area of meadows and pastures must be approximately equal to that of arable: a proportion of which, in most localities, the Russian peasantry fell very far short. By the close of the last century, the ratio, in the great majority of cases, did not exceed one-third or one-half of the arable. In the most densely populated agricultural provinces of Russia, there was only a fraction of a dessiatin of grassland for every dessiatin of arable; at its worst, in the province of Kursk that fraction did not exceed one-thirtieth.[1] In the parts of Russia outside the black-earth, where land required heavy manuring, the extent of the available meadows was the principal factor in deciding on the area actually under plough. In most parts of the North and North-West of Russia, the abundance of forest pastures provided sufficient food for the live stock in the summer, but the feeding of cattle in the course of the long winter presented great difficulties. The position of the Russian peasant in this respect was very similar to that of the English open-field farmer of the period preceding the enclosures of the later seventeenth and the eighteenth centuries and the spread of grass and root crops. In the Northern half of the country, the shortage of meadows resulted, on the one hand, in the incomplete exploitation of the available arable; on the other, it did not permit to develop dairy farming or other branches of production based on live stock, without a radical change in the system of cropping, which would allow the introduction of grass in the rotation, but was extremely difficult under the existing system of tenure. In the black-earth belt, peasant farming was also crippled in the branches depending on live stock, and especially on cattle, while the land was being exhausted by the insufficiency or the complete absence of manuring. In all cases, the yield of peasant farming was reduced to below what it could have been, even under the prevalent systems of cropping and cultivation, had the constitution of holdings been well-balanced. The lack of balance, which, in many cases, was itself the result of the growth of rural population and the consequent extension of arable at the expense of grassland, in certain localities could be definitely traced to the conditions of the peasants' land-settlement on their Emancipation. This was the case particularly in the Northern half of Russia, where arable land was generally less valuable than meadows, which formed the principal source of the landowners' income and, moreover, required a minimum of labour for their exploitation. It was accordingly the grasslands that the landowners, in transferring part of their estates to their former serfs, were most anxious to keep. The reductions, therefore, in the sizes of peasant holdings at the Emancipation, which in some provinces of this region were very considerable, mostly involved grasslands. In the province of Novgorod, for instance, the total area of peasants' holdings was estimated to have been reduced at the Emancipation by roughly one-third (from 1,600,000

[1] Prof. A. I. Chuprov, *Small Farming and its Principal Needs*, first published in 1907 (Rus.); references here to the 1921 edition, published in Berlin, p. 100.

to 1,045,000 dessiatins),[1] most of the difference being accounted for by meadows and pastures. The result of this was that, though the peasants' holdings in the province of Novgorod were among the largest in Russia, and averaged, in 1905, no less than 13·5 dessiatins per family, with only 4·8% of small holdings below 5 dessiatins, the peasantry was greatly handicapped by the difficulty of providing sufficient hay for the cattle in winter, and thus often prevented from making the best use of their land.

It would, indeed, be impossible even to attempt, in this brief outline of the origins and development of the agrarian problem in Russia, to enumerate and describe all the numerous factors which, to a greater or lesser extent, were responsible for the difficulties of the Russian countryside. The problem was exceedingly complex, and the important point was that, while, strictly speaking, the insufficient size of the holdings lay at its root for a considerable section of the peasantry, practically the whole of the rural population was involved in it in some way or other, irrespective of the area of land in their possession. In fact, whatever the sizes of holdings, the real trouble was that the land was not used to the best advantage, and that the yield of peasant farming was miserably low.

With the population increasing with great rapidity throughout the second half of the nineteenth century, two things were absolutely essential to prevent the development of an acute agrarian crisis. In the first instance, an outlet was needed for the growing surplus of rural population. Russia stood urgently in need of that exodus from the villages which, throughout Western Europe, with the development of capitalism, kept on depopulating the countryside and swelling the ranks of industrial workers at the expense of the peasantry. This exodus, helped by emigration to new regions on the confines of the Empire, was the only possible means for relieving the growing pressure of the population on the land and the rapid parcellation of holdings into units below the economic size. Then, secondly, in order to provide for the increased demand of the growing population for agricultural produce, the standards of cultivation had to be raised, and the yield of the land increased gradually. The two developments, as a matter of fact, depended to a large extent on each other, since the rural exodus due to industrial growth would have of itself provided the necessary stimuli for the improvement of cultivation by extending the market for agricultural produce. Here it is that the student of the Russian agrarian problem comes to face those general economic conditions

[1] Kaufmann, op. cit., part I, p. 27.

of nineteenth-century Russia which, more than anything else, have been responsible for the agricultural and agrarian situation on the morrow of the Emancipation.

When, in 1861, Russia emerged from its old social and economic régime, essentially based on serfdom, she did so rather under the pressure of moral and political considerations, than because the change had been brought about by the course of her economic development, which had already evolved the necessary elements of a new system. Indeed, as a Great Power, Russia could not go on living on the meagre resources of a medieval economic organization and was urgently in need of substituting for that inadequate basis a modern economic system, able to bear the financial burdens of a great nineteenth-century State. Accordingly, the old system was done away with, but the new capitalistic organization, which would eventually put to profitable uses Russia's large natural wealth and make the country more prosperous and powerful than ever, had still to be evolved. This proved a laborious task, and the development of modern capitalism in Russia proceeded slowly. Starting practically without capital, which was still, in the main, to be either accumulated or imported from abroad, it was only natural that Russia, in the course of the second half of the last century, should suffer from a surfeit of hands in the country, and that instead of rural exodus she should develop agrarian overpopulation. The increase in the volume of foreign and internal trade, which accompanied the construction of railways and the growth of steam shipping, had been the first to manifest itself, as a sign of new times. But it was not to trade that one had to look for an outlet for the surplus population of the countryside. Only a rapid growth of industry, organized on capitalistic lines on a large scale and giving employment to an increasing number of hands, could solve the problem satisfactorily. Industrial capitalism, combined with an extensive development of large, capitalistically-organized farming, would have provided remunerative employment both for the absolute surplus of rural population and for those peasant-farmers, who, being settled on insufficient holdings, were dependent on local earnings for their livelihood. Yet, until the close of the nineteenth century, neither industrial capitalism, nor large farming, had developed to a sufficient extent to relieve in any marked degree the pressure of population on the land. The accumulation of capital in a country still almost entirely agricultural proceeded slowly, while the sources of foreign capital available to Russia at the time had to be used almost exclusively for the purpose of providing the country with at least a skeleton of the necessary network of railways. Moreover, during the first two or three decades after the Emancipation, which coincided with the reconstruction and industrial development of the United States after the Civil War, Russia had a powerful competitor on the foreign capital market. Neither did the conditions of the Russian currency before its stabilization and final reform in 1897 encourage investment there in preference to countries with stable exchange. Accordingly, though foreign capital was being invested in Russia to some extent, and more particularly in her railways, mining, textiles and metallurgical industries and oilfields, since the 'seventies, the scarcity has always been painfully felt. It may be said that, in the second half of the last century, industry and agriculture in Russia were locked together in a kind of vicious circle. The countryside stood in need of the rapid growth of industry as a market for its produce and an outlet for its growing surplus of hands. Without such outlets for its production of both foodstuffs and men, rural Russia, overcrowded and poor, could not aspire to prosperity. On the other hand, Russian industries, which were naturally confined to the home market for the disposal of their products, depended for their development on the progress of the peasantry and the gradual increase in their purchasing capacity. The combination of circumstances with which Russia had been faced after the Emancipation could not be described as favourable; in fact, the conditions were such, that progress in either respect was delayed. The vicious circle had to be broken through at some point, and this, as will be shown later, actually did take place at the close of the last and the beginning of the present century, when the forces of growth, hitherto stinted, were gradually released.

The agrarian crisis reached its climax by the close of the nineteenth century, and from the later 'nineties down to the middle of the following decade several important official investigations of the condition of the peasantry had been carried out, whose results are of great value to any student of the problem, since they give a fairly comprehensive idea of the situation at that time.

In the first instance, in order of importance, the Survey of Landed Property in European Russia, referring to the year 1905, should be mentioned. This inquiry, to which I shall have to return on more than one occasion in later parts of this study, showed the distribution of landed property in European Russia in 1905, and had been undertaken with the express purpose of providing the necessary statistical basis for any possible measures of agrarian policy. Here, it is mostly interesting in so far as it supplies information concerning the total area of land in the possession of the peasants and the sizes of peasant holdings in various localities at the beginning of the current century. In this connection, it is necessary to point out that the landed property of the peasant class in Russia consisted of two distinct categories of land, namely, on the one hand, of the original holdings allocated to the peasants at the Emancipation and referred to as "allotment land" (*nadiel*) and, on the other hand, of land bought by the peasants, either individually or collectively, subsequently to the Emancipation, by private treaty. With the latter category, which involved a very considerable and rapidly growing area, I shall have to deal later. In the present connection, it is the former category, the original *nadiel* of the peasants, forming the principal basis of their farming, as a class, as well as the principal support of the great mass of the Russian people, that is of interest.

According to the Survey of 1905, in the forty-five provinces of European Russia, as represented on the Agricultural Map, the number of peasant families (*dvor*), the total area of their allotment land (*nadiel*) and the average sizes of their holdings in various regions were as follows:

Regions	Number of Families (dvor), in 1,000	Area of the "nadiel," in 1,000 des.	Average sizes of Family Holdings, in dessiatins
		(One dessiatin = 2·7 acres)	
Northern	352·5	7,894·1	22·4
North-Eastern	1,003·0	16,055·5	16·0
Petrograd	479·7	5,378·5	11·2
Moscow	2,077·3	16,508·7	7·9
Western	550·8	5,158·4	9·2
South-Western	1,114·2	6,159·9	5·5
Ukraine	1,178·3	7,187·8	6·1
Central Agricultural	1,658·0	12,873·1	7·8
Middle Volga	1,235·2	10,282·9	8·3
Eastern	655·8	12,964·3	19·7
New Russia	698·0	6,779·8	9·7
South-Eastern (Don only)	130·7	9,847·4	75·3
Caspian (Astrakhan only)	67·9	2,354·9	28·4
European Russia (exclusive of the provinces of Stavropol, Kuban and Terek)	11,211·4	119,445·3	10·7

According to the sizes of their holdings, the peasants in various regions of European Russia were distributed as follows:

Regions	Percentages of peasant families (dvor) with Holdings of		
	Up to 5 dessiatins (13·5 acres)	5 to 10 dessiatins (13·5–27 acres)	Over 10 dessiatins (27 acres)
Northern	17·0	21·0	62·0
North-Eastern	10·0	11·6	78·4
Petrograd	5·5	48·5	46·0
Moscow	18·8	60·9	20·3

Regions	Percentage of peasant families (dvor) with Holdings of Up to 5 dessiatins (13·5 acres)	5 to 10 dessiatins (13.5–27 acres)	Over 10 dessiatins (27 acres)
Western	8·9	60·9	30·4
South-Western	57·6	32·9	9·5
Ukraine	45·6	43·1	11·3
Central Agricultural	22·5	53·9	23·6
Middle Volga	18·7	55·0	26·3
Eastern	7·6	18·3	74·1
New Russia	20·4	44·0	35·6
South-Eastern (Don only)	69·0	26·7	4·3
Caspian (Astrakhan only)	1·9	4·4	93·7

It may be seen that both the average sizes of peasant family holdings and their distribution according to size varied greatly from one locality to another. As a general observation, however, it may be pointed out that the largest percentage of small holdings, not exceeding 5 dessiatins, was to be found in the South-Western and the Ukrainian provinces (57·6 and 45·6% respectively), in which the increase in population since the Emancipation combined with the smallness of the original holdings to produce this effect. The Central Agricultural region, New Russia, the Moscow region and the Middle Volga came next, in the order named. Here, the large proportion of small holdings was due, primarily, to the rapid growth of rural population. The economic effects of the subdivision of holdings, however, were not felt to the same extent in all these localities. The crisis, indeed, was at its worst, by the close of the last century, in the Central Agricultural and the Middle-Volga regions, which were almost entirely agricultural, with the only exception of the province of Nizhny-Novgorod, and provided few local sources of earnings to the rural population. In the Moscow region, and more particularly in its central and Northern parts, the peasants found an outlet for the excess of labour in domestic industries, which were highly developed in that part of Russia. In New Russia and the South-East, the large local demand for seasonal labour helped the peasants with insufficient holdings to make both ends meet.

Though, as a general rule, it may be accepted that the holdings up to 5 dessiatins were, under Russian conditions, too small for the support of a peasant family, even in the most fertile parts of the country, except when they could be used for vegetable gardening or other highly intensive forms of cultivation, there is very little else that can be gained by the study of the distribution of peasant holdings according to size. The study of such statistics, though necessary, throws but little light on the nature and causes of the agrarian crisis in Russia. Without, therefore, denying or even minimizing the importance of the distribution of holdings according to size in any study bearing on the economic position of peasant farmers, I should think that, in the vast literature of the subject, the attention had generally been far too exclusively concentrated on this particular aspect of the problem. In the Russian agrarian situation at the close of the last century, the dominant fact was not so much that there existed a large section of the peasant population whose holdings were insufficient to support them, but that the yield of the land, owing to general economic conditions, was so low that an average holding was insufficient to keep its owner in conditions of reasonable well-being, while the countryside was overcrowded with an excess of people whom, under any circumstances whatever, it could not possibly employ.

In this connection, the results of some other investigations referring to the same period, are exceedingly interesting.

By the Imperial Ukaze of November 16th, 1901, an official Commission was appointed with a view to "investigating the question of the progress, during the period from 1861 to 1900, in the prosperity of the rural population of the central agricultural provinces of Russia, as compared with other parts of the country." The terms of reference of the Commission were an outstanding example of the official art of putting things nicely, since the proposed investigation had to deal with the progress of misery rather than prosperity. The Commission, generally known as the *Commission of the Centre*, did a considerable amount of useful work, which brought to light the real condition of the areas most affected with the agrarian crisis. Among the vast collection of its statistical materials, there was a compilation of data bearing on the question of the gross and net yield of peasant holdings in various parts of the country.[1] These figures revealed a situation so bad as to be hardly credible. They referred to 27 provinces of European Russia, or 273 administrative districts (*uiezd*), involving a total of 72,786,500 dessiatins of peasant allotment land (*nadiel*), of which 60·9% were arable, 13·4% under permanent grass, and 25·7% under forests, etc. The compilation was based on statistics collected by the Zemstvos of the provinces concerned during a period extending over 25 years, from 1877 to 1901: a fact which detracts considerably from the comparative value of

[1] As the original "Transactions" of the Commission were not available to the present writer, the figures are quoted from A. Yermolov, *Our Land Problem*, pp. 36 ff., where they are given in detail. *Vide* also P. Sokovnin, *The Standards of Peasant Farming on Allotment Land* (Rus.), St. Petersburg, 1907.

the figures. Allowance being made for the difference in the moments to which the data refer, some idea can be obtained, however, of the insignificance of the income derived by the peasants from their land in the last quarter of the nineteenth century. The average gross yield of arable per dessiatin in the twenty-seven provinces was 11·78 roubles, with a minimum of 8·8 roubles in the province of Samara (1894–1900) and a maximum of 21·88 roubles in the province of Petrograd (1880–82). The gross yield of meadows averaged 12·66 roubles, ranging from 5·63 roubles in the province of Viatka (1884–93) to 25·63 roubles in that of Voronezh (1885–87). The yield of all other land averaged 54 copecks per dessiatin. The cost of cultivation and harvesting worked out, on the average, at 7·65 roubles per dessiatin of arable, 4·02 roubles per dessiatin of meadows, and 15 copecks per dessiatin for the rest. The average gross yield of all peasant allotment land, irrespective of the mode of exploitation, was 8·99 roubles per dessiatin, while the cost of exploitation averaged 5·22 roubles. The net yield per dessiatin of arable, accordingly, averaged 4·13 roubles; that of meadows—8·64 roubles. The net yield of all other classes of peasant land did not exceed 39 copecks per dessiatin. The net yield of all peasant land per dessiatin was only 3·77 roubles on the average.

As mentioned above, the real comparative value of these figures, interesting as they are, is very limited. Yet, when the data for separate provinces are considered, due account being taken of the time to which they refer, they are highly illuminating. Indeed, it is a striking fact that, in the province of Moscow, in the last decade of the nineteenth century (1892–1900), the average net yield per dessiatin of peasant land did not exceed 5·29 roubles, with the result that the total net income from an average-sized peasant holding, which, in that province, was 7·5 dessiatins (about 20 acres), was just below 40 roubles a year, equivalent to £4 in English money. On the rich black-earth of the Southern steppes, in the province of Kherson, during the same decade (1892–1900), an average holding of 7·8 dessiatins yielded a net income of 42 roubles a year, or just over £4. The range of variations in the net income of average-sized holdings extended, for the whole period from 1877 to 1901, from a minimum of about 22 roubles a year in the province of Vladimir (1896–9), where the average holding was 8·9 dessiatins, to a maximum of 66 roubles in the province of Novgorod (1886–98), where the average holding reached 13·5 dessiatins. It may be seen that the yield of peasant land worked out at exceedingly low figures.

It must be borne in mind, however, that the net yield of the peasant's holding, which he gets in his joint capacity of proprietor and manager, does not account for the whole of his income from his land. Since, with the assistance of other members of his family, he performs either the whole, or at least most, of the work on his land himself, he has to be credited with that part of the produce which otherwise should be paid out in wages. Exactly how much this part of the peasant's income amounts to, it is practically impossible to ascertain with any degree of precision, the more so that the outlay of labour per unit of area varies greatly according to the conditions and standards of farming. In order to obtain an approximately correct idea of the peasant's total income from his land, it may be more expedient, therefore, to start from the gross income, in which labour costs are included, and then try and eliminate the capital outlay.

In this connection, the work of P. Sokovnin, at one time Director of the Department of Agriculture, on the *Standards of Peasant Farming on Allotment Land*, published in 1907, is very interesting. In this instructive statistical study, an estimate is made of the actual gross yield of peasant land (*nadiel*), at the close of the nineteenth century, in every province of European Russia. It has that important advantage over the statistical compilations of the Commission of the Centre, dealt with above, that it is based throughout on uniform data and can, therefore, be considered as affording suitable material for comparison.

Arranging Sokovnin's figures according to the geographical scheme adopted throughout the present study, the following amounts of gross money yield of average-sized peasant holdings in various parts of European Russia are obtained: [1]

Northern region	about	135	roubles	a year
North-Eastern region	,,	196	,,	,,
Petrograd region	,,	190	,,	,,
Moscow region	,,	146	,,	,,
Western region	,,	185	,,	,,
South-Western region	,,	127	,,	,,
Ukraine	,,	118	,,	,,
Central Agricultural, ex. Penza	,,	145	,,	,,
Middle Volga region	,,	118	,,	,,
Eastern region	,,	155	,,	,,
New Russia	,,	204	,,	,,

The total outlay involved in the exploitation of the holding, exclusive of labour, comprises seeds, manure, as well as capital expenditure in all forms, and though it varies greatly according

[1] Sokovnin, op. cit., Table XV, p. 25, in which figures are given according to provinces. No estimates have been made for the South-Eastern and Caspian regions.

to conditions and standards of farming, it may be taken to amount, on a rough average, to some 25 or 30% of the gross yield. Putting it at the lower figure, so as to avoid underestimating the peasants' incomes from their holdings, it may be seen that the average income derived by a peasant family from its land varied, at the close of the last century, somewhere between a maximum of 150 roubles and a minimum of about 90 roubles a year. This did not include anything derived from other branches of farming, besides arable cultivation, but these, as a rule, provided a very small addition to the basic income from the land. According to the few available data concerning these branches of agricultural production, which could only be obtained from peasant household budget statistics referring to a few provinces, Sokovnin puts this part of the peasants' gross income at about 47 roubles a year per family on the average under the head of live stock, and to some 1 to 6 roubles a year for vegetable gardening, etc. On the whole, these sources may have provided the peasant family, on the average, with an additional net income of some 40 roubles a year. In other words, the total actual income of an average peasant family from all branches of farming on allotment land, at the close of the last century, would work out at some figure between 130–150 and 180–200 roubles a year on the most liberal estimate—the equivalent of 13 to 20 pounds in English money. The estimate thus arrived at is avowedly a very rough one; but the fact that, as a general rule, peasant families with average-sized holdings were dependent, throughout the country, on outside earnings to a more or less considerable extent would appear to confirm the substantial correctness of these figures. To quote Sokovnin, "the peasants' income from their allotment land, on the whole, falls one-half or three-quarters short of the amount absolutely necessary for their existence, and the deficiency has to be made good, as far as possible, by outside earnings and by the lease of land from landowners." Not only the smallholder, but even the owner of an average-sized holding of twenty to thirty acres of excellent agricultural land in the black-earth belt, had to look for outside sources of earnings to eke out his resources. Practically throughout the black-earth belt, during the winter months, when agricultural activities were reduced to a minimum, a large proportion of the rural population was unemployed, and many of them left the villages and went to look for work in the towns and industrial centres. Yet, while this seasonal migration of labour helped a section of the peasants, it only involved, on the average, some million or so of people, and the demand of the industry was not sufficient to absorb the amount of labour wasted every winter. This annual waste of working power, which, if employed, could probably double the national income of Russia, was brought out with striking effect by an inquiry carried out by the Direct Taxation Department and referring to the year 1900. According to this investigation, in European Russia, at that time, 52% of the available working power was being absolutely wasted. In the Northern half of the country, with relatively well-developed domestic industries and other outside occupations, among which forestry was especially important, the proportion was considerably lower than in the purely agricultural black-earth belt, in which it reached 64 to 68% of the available labour.[1] These few figures are, perhaps, the best diagnosis of the agrarian trouble which affected Russia; but their real purport, in so far as they have been discussed at all, would appear to have been to a large extent overlooked. Indeed, while it was recognized that there is an appalling waste of labour in the countryside, the inference drawn was mostly either that the remedy lies in the increase of the area of land in the peasants' possession, which would provide them with more work on their own holdings, or that the problem can and must be solved by the raising of standards of cultivation, which would involve a far greater outlay of labour on the existing holdings, while at the same time greatly increasing their yield. Both these views, while true to a certain point, were singularly lacking in their grasp of the situation as a whole. With regard to the first of these views, while it was perfectly right in so far as a certain section of the peasantry, whose holdings were altogether too small, was concerned, as a solution of the whole problem it failed to meet the situation entirely. Indeed, it has been pointed out that, under existing conditions, even the peasant with an average-sized holding could not, in most cases, make both ends meet, and was dependent to a large extent on outside earnings. Accordingly, had it even been possible by any means whatsoever, including the wholesale expropriation in favour of the peasants of all the land belonging to other classes of proprietors and to the State and suitable for cultivation, to raise the size of all peasant holdings to the average of about 11 dessiatins per family, the problem would have remained unsolved, since, though the holdings of the poorer peasants would have been increased, they would not have become self-supporting. Besides, as a large proportion—roughly about 30%—of all the agricultural land of other proprietors was

[1] As the original work was not available, the figures had to be quoted according to A. I. Chuprov, op. cit., p. 112.

already held on lease and cultivated by the peasants, this measure would have largely involved robbing Peter to pay Paul, while some important sources of local earnings for the peasantry would have been done away with by the liquidation of the little large farming there existed in Russia. Characteristically, in this connection, most writers on the agrarian problem in Russia, who adhered to this point of view would appear to have overlooked the cardinal fact that the peasants were already, as a class, in possession of the greater part of the land in Russia, and of by far the greater part of the agricultural area in particular, and that any possible additions to their holdings at the expense of other classes of proprietors could only have a very limited effect.[1] Indeed, in 1905, in forty-seven provinces of European Russia the peasants owned 128 millions of dessiatins of allotment land (*nadiel*) and, with other land privately bought by them, had nearly 160 millions of dessiatins. The gentry had 46 millions of dessiatins and other classes of proprietors—about 18 millions of dessiatins; of this privately-owned land about 20 to 25 million dessiatins were held on lease and cultivated by the peasants already, though not owned by them. The State had 145 millions of dessiatins, and the appanages just under 8 million dessiatins; but of the lands of the State about 75% were in the Northern provinces and represented woodlands and marshes, while of the relatively small area of agricultural land the greater part was leased to peasants regularly (*obrochnia statii*). Thus, besides failing to provide a solution of the problem, the proposed increase of the holdings of the peasants generally at the expense of other classes of owners to an extent which would raise them to the existing average size, would have been hardly possible physically.

According to the other school of thought, the solution lay in the raising of the standards of cultivation, which would greatly increase the yield of the land, while at the same time providing remunerative employment for the rural population. The would-be reformers of peasant farming pointed to the example of Western Europe, where the peasants, though they possessed, in most cases, smaller holdings, enjoyed relatively considerable prosperity, entirely due to highly intensive cultivation, or even to that of China, where the exploitation of land was carried to such perfection that it could maintain several people on an area which, in the best-farmed parts of Europe, could scarcely feed a single person. As most such arguments, this contention was a mixture of both undeniable truth and most palpable error. In fact, it would be idle to

[1] This was amply proved by the results of the expropriation in 1917–18.

deny that Russian peasant farming was in urgent need of technical improvement, and that the standards of cultivation were generally quite out of date. The yield was, accordingly, very much lower than it might have been, had the methods of farming been more efficient. Yet, most comparisons, either with Western Europe, and more particularly with such countries as France, Italy or the Rhineland, in which peasants can live on very small holdings, or with China, were based either on ignorance or on the deliberate neglect of the important differences which distinguish Russian farming generally from that of the countries referred to for comparison and emulation. The natural conditions of Russia provide, in the greater part of the country, few opportunities for specialized intensive cultivation on small holdings. The latter is confined to the vicinity of the markets, in the case of vegetable gardening or poultry farming, or to certain districts, specially favoured by nature, in the case of fruit, vine and tobacco-growing. The great mass of the Russian peasantry, as a result of natural and economic conditions, have to depend on arable farming as their mainstay. This, whatever its standards, requires a relatively large area of land for the maintenance of a peasant family, and if the sizes of Russian peasant holdings are compared with those of independent peasants in those parts of Western Europe in which farming is predominantly arable, and concentrates to a large extent on the production of cereals, it will be found that, there also, the holdings are not very small. Another fallacy involved in the argument is that it omits to take due account of the economic factor, which enables the peasant of Western Europe to raise the standards of cultivation to a high level, namely the relative density of the non-agricultural population and the consequently widespread influence of the demand for the finer products of farming, which ensures a greater variety in the rotation and a higher yield per unit of area. In Russia, wherever such influences have been strong enough, owing to proximity to large consuming centres, or to good means of transport, as in the vicinity of Moscow, Petrograd and other large cities, they did not fail in raising the standards of peasant farming and increasing its returns. But such centres were few and far between, and on the huge territory of Russia their influence could only produce some local improvement, leaving the bulk of the country, with its millions of peasant farmers, either practically unaffected, or only affected but very slightly. Thus, while the general raising of the standards of farming in Russia was certainly needed and would go a long way in helping the solution of the agrarian problem, it depended, first and foremost, on the industrialization of the country and the consequent extension of the agricultural market, which would call it forth automatically.

B. The Agrarian Evolution of Nineteenth-century Russia

Another factor in the agrarian situation of Russia, which had played a considerable part in shaping the course of further evolution, was the position of the landowning gentry on the morrow of the Emancipation. At the time of the great reform, the advocates of the measure were wont to contend that, since free labour had been generally proved to be more efficient than that of serfs or slaves, not only moral considerations, but economic expediency as well, necessitated the abolition of serfdom. Yet, neither the closer study of the conditions of Russian farming on the eve of the reform, nor the precarious position in which the large farming landowner found himself after 1861, would appear to bear out this theoretical contention. Monstrous as Russian serfdom was in the middle of the nineteenth century, considered from the moral and social view-point, economically it was the only system under which progressive large farming could exist in Russia in those days. The landowning gentry, as a rule, had no capital to invest in farming. The whole economic organization of the country, from top to bottom, was still only just feeling its way towards commercialization, of which large farming, based on serf labour, and the grain trade, intimately connected with these beginnings of agrarian capitalism, were the first pioneers in Russia. Large farming with serf labour, especially in the black-earth belt, had been growing until the eve of the Emancipation. Many important agricultural improvements could be traced to the progressive large farmers of the early nineteenth century ; suffice it to point here to the development of merino-breeding in the South ; the pioneer work of Count Bobrinsky in the sugar-beet industry ; the development of improved breeds of live stock and the first introduction into Russia of scientific rotation, based on foreign models. The progress of the agricultural industry accomplished in the first half of the nineteenth century had been due entirely to the existence of serf labour, which was always at hand and reduced the capital expenditure of the landowner to a minimum. In the black-earth belt, at that time, it was generally conceded that farming on the landowner's own account, on the *barshchina* system, was far more profitable than leaving the land to the peasants to cultivate on the *obrok*. Accordingly, as I have had the opportunity to point out, the *barshchina* was prevalent, and in some provinces practically universal, in the South of Russia. The agricultural literature of the day was entirely in its favour, as by far the more profitable way of exploiting the available working power of the serfs. In the course of the year 1809 alone, three essays advocating the advantages of large farming with serf labour have been awarded prizes by the Imperial Free Economic Society. Wilkins, one of the foremost agricultural experts in Russia in the thirties and forties of the last century, proved by calculations based on his extensive practical experience that the best thing the landowners could do was to turn their serfs out of their holdings and, undertaking to keep them, to equip their estates with the necessary live and dead stock and make them work exclusively on their masters' account.[1] In actual life, something very much like this had been taking place in the South-West, the Ukraine and the Western provinces, where peasant evictions have been common before the Emancipation, in spite of all the efforts made by the Government with a view to protecting the serfs.

Thus, when the Emancipation came, for the landowning class generally, and for the farming landowners in particular, it was nothing short of a social cataclysm. For many of them, who could carry on, somehow, in the old traditional ways, but possessed neither the capital, nor the knowledge and initiative necessary to re-organize their farming on new lines, it spelt ruin. Indeed, the qualifications now required of a large farmer were very different from those of the old-fashioned Russian country squire (*pomeshchik*). In the case of the latter, one of the principal items of costs did not count, since the necessary labour was supplied by the serfs, and the latter were paid for it not in money, but in land—a commodity still plentiful and possessing in itself, apart from the peasants' labour, very little value. The cost of equipment was also reduced to a minimum, the cultivation of the fields and the cartage of produce being done by the peasants with their own stock. With costs thus reduced to an absolute minimum, which left a large margin for what the landowner could look upon as his net profit, the commercial problems of farming were easily solved, even when, in the second quarter of the last century, the excessive concentration on the production of cereals brought their prices down to a very low level. Since the Emancipation everything changed completely, and the problems of labour and of capital assumed an enormous importance, especially in those localities in which the population was relatively sparse, the peasants' holdings more or less sufficient or easily supplemented by leases, or where other

[1] P. Struve, *The Economics of Serfdom* (Rus.), 1913, p. 59.

sources of earnings diverted the peasants from employment on the land. Moreover, the peasants had still to get used to employment as hired labour, while the landowners had to learn to handle free labour over which they did not possess the disciplinary powers which made the management of serf labour relatively easy. With regard to agricultural labour, the position, indeed, as will be seen later, remained very difficult for the farming landowner until the close of the nineteenth century, in spite of the increase in rural population and of its growing need for employment. According to the Report of the Agricultural Commission of 1872, the landowners complained, in the first instance, of the shortage and high cost of labour: "The shortage of labour, indeed, puts farming in a position of uncertainty, since agricultural labour cannot be obtained, and the available hands cannot be relied upon. There is also a shortage of farm servants."[1] For farm servants, in the main, the landowner had to rely on the dregs of the village or on young men and girls who, no sooner they were getting proficient in their work, married and left their employment to settle on their own in the village. It is hardly surprising, therefore, that the first attempts of Russian landowners after the Emancipation to organize the exploitation of their estates by means of hired labour should meet with great difficulties, disappointments and losses.

The extreme shortage of capital was another difficulty, only partly mitigated by the issue to the landowners of *redemption bonds* in payment for the land allocated to their former serfs. In the first instance, since, in most cases, amicable settlement of the question of land proved impossible, and the landowners had to apply for compulsory redemption, they only obtained 75 to 80% of the assessed value of the land ceded. Then, as they were not paid in ready money, which most of them stood in immediate need of, they were pressed with the disposal of the bonds, and by flooding the market with these securities brought their price down to a point reaching, at times, as much as 30% below par. Moreover, the estates of many landowners, by the time of the Emancipation, were already burdened with mortgages to the old credit establishments which, by 1860, have all been put into liquidation.

By 1881, the aggregate of these debts, deducted from the amount of redemption payable to the debtors, reached 302,666,300 roubles, as against 748,531,300 roubles due for the land transferred to the peasants.[2] Thus, approximately two-fifths of the total amount

[1] Quoted according to P. Maslov, *The Conditions of the Development of Farming in Russia* (Rus.), 1903, p. 478.
[2] Prof. L. Khodsky, "Redemption Payments" in Brockhaus.

due for the land ceded to the peasants had never been actually paid to the landowners, having been written off for debts.

The situation was greatly aggravated by the absence, at the time of the Emancipation, of organized credit facilities for the landowners. The 'sixties, during which the whole of Russia was being feverishly reformed, saw also the first attempts at the organization of land banks on mutual credit principles. But, on the one hand, the attempts mostly proved unsuccessful; on the other, they were too few to meet the situation. The old credit institutions, of which the principal was the *State Loan Bank*, founded in 1786, were all closed down in 1859, and the *State Bank*, established in 1860, did not transact business in land mortgages. The first mutual credit land bank, opened in 1864, was the *Zemsky Bank of Kherson*; its importance, however, was purely local. In 1866, the *Mutual Land Credit Society* was founded in St. Petersburg, which, after a chequered career, was put into liquidation in 1891, its outstanding accounts being transferred to the Special Branch of the *State Land Bank for the Nobility*. Joint-stock land banks, to the number of eleven, all appeared in 1871 and 1872 and, with the only exception of the *Bank of Saratov-Simbirsk*, they all survived into the twentieth century. The activities of these institutions, which transacted business in mortgages on land and buildings in town and country, were confined in each case to a certain locality. Thus, there were the banks of *Kharkov, Poltava, Petersburg-Tula, Moscow, Bessarabia and Taurida, Nizhny-Novgorod and Samara, Kiev, Vilna, Don, Yaroslav and Kostroma*, and *Saratov and Simbirsk*. Their establishment eased the situation to some extent, but could not do much, because, owing to the general dearth of money, their terms were too heavy, considering the depressed condition of the agricultural industry. It was not until the middle of the 'eighties that, urged by the gentry and the Zemstvos of several provinces, the Government, by the establishment, in 1885, of the *State Land Bank for the Nobility*, came to the assistance of the landowning class and provided it with facilities of cheap long-term credit. The need for short-term credit, however, still remained unsatisfied, in spite of the system, adopted by the *State Bank* since the 'eighties, of discounting promissory notes bearing the signature of landowners. All these facilities, of which more will be said later, developed gradually and slowly, as the agricultural depression of the last quarter of the century made the needs of the farming landowners more urgent. During the first twenty-five years or so after the Emancipation, even the slender credit resources which had been provided later, were not available, and the landowners had to fight their own battles as well they could. The struggle was a very hard one, and the losses suffered were heavy indeed.

The conditions of the agricultural market in the second half of the nineteenth century could by no means be described as favourable to the farmer. In the course of the 'sixties and the earlier 'seventies, the prices of cereals in Russia, in paper roubles, have, indeed, on the whole, been rising, but since the close of the 'seventies, when Russia began to feel the full force of the agricultural depression which had set in in Western Europe since 1874, they have been seized in a headlong decline, which continued until the middle of the 'nineties. The only short-lived break in the fall was caused by the disastrous failure of crops of 1891, after which the decline resumed its course. At first, when the construction of railways had been proceeding with great rapidity, and their penetration into fresh districts of the agricultural provinces produced a rise in spot prices of cereals over extensive areas, this local improvement tended to some extent to counteract the effects of the general depression, but in the 'eighties, with the aggravation of the agricultural crisis and the slowing-down of the railway-building activity in Russia, these influences, which had hitherto helped to mitigate the situation, have weakened very considerably. The depression descended on Russia with all its consequences for the agricultural industry. The continuous decline in agricultural prices, however, did not affect all the different groups of the farming community in the same way: a circumstance which had a far-reaching influence on the agrarian evolution of Russia during the second half of the last century. The large landowner, indeed, with his complete dependence on the market for the disposal of his produce, and the relatively very narrow margin of profit, left to him now, when he had to include labour among his items of costs, had probably been hit harder than the other groups of producers. Among the peasants, the influence of falling grain prices was felt in a variety of ways, according to locality and to the group of peasants actually concerned. In the case of all peasant farmers, however, there was one circumstance which enabled them to bear the decline in prices easier than did the large farmer, dependent on hired labour. Indeed, the margin of the peasant's gross income, which remained to him on the deduction of his actual costs, was larger than that of the capitalistic farmer by the amount of wages, since he supplied the labour himself. Accordingly, when prices declined to an extent which left nothing above the actual costs to the large farmer, and the latter, economically, had no more inducement to go on with his cultivation, to the peasant this only meant that he will have to work for a rate of remuneration lower than before, but still equal to that he would get as a labourer. A further fall, which would involve the large farmer into actual losses, to the peasant would only mean a reduction of his pay more or less below that of the agricultural labourer: a condition which he would readily put up with in the circumstances, seeing that he has no other source of earnings. In other words, with regard to his costs, the peasant was, and generally is, in a position somewhat analogous to that of the old farming landowner under serfdom, who also was not faced with the necessity of including labour in the costs, and whose net income from farming was, accordingly, larger and more elastic. As a result the peasant farmer, generally, is better able to bear the effects of agricultural depression, and though he certainly suffers and gets poorer, he can stand for a long time a price level at which the large farmer has to abandon cultivation. Thus, the period of depression, in the last quarter of the nineteenth century, has been generally less severely felt by the peasants than by the large agricultural producer. Moreover, it must be borne in mind, that, in Russia, though every peasant is an agricultural producer to a greater or lesser extent, there are many peasants who do not produce sufficient grain for their own needs and who, on the balance, are rather buyers than sellers on the agricultural market. These, besides the small-holders throughout the country, include practically all the peasantry of the under-producing Northern half of Russia, and for all this numerous class the fall in grain prices, far from being a cause of loss, was rather profitable in so far as their cost of living was reduced and their earnings went farther.[1] Finally, there existed, in those days, considerable groups of peasants, especially in the more remote districts of Russia, whose relations with the market were still so slight and so casual, that the changes in prices hardly affected them at all in either direction.

The effect of the agricultural depression on the general position of the Russian agricultural industry, accordingly, was that it discouraged the large farmer and often compelled him to give up cultivation altogether, while it left the peasant farmer either relatively little affected or, in the case of the under-producing group, rather better off than he would have been under different conditions. It will be remembered that the Emancipation, by

[1] *Q.v.* Shcherbina, F. I., "Peasant Household Budgets and the Influence on them of Crops and Grain Prices," in the *Influence of Crops and Grain Prices on various Aspects of the Agricultural Industry in Russia* (Rus.), Vol. II, p. 74 ff.

increasing the need of the peasants for ready money, forced them, in those localities in which there were no other sources of earnings, to extend their cultivation by ploughing up their own land to the utmost possible limits and by leasing as much land as they could from the neighbouring large estates. Thus, while the large landowners, on the one hand, were being driven out of cultivation by the fall in grain prices, which was now added on top of their original difficulties, on the other hand, especially in the purely agricultural provinces of the black-earth belt, the demand of the peasants for leases provided them with a profitable alternative use for their land. As a result, while cultivation on the landowners' account diminished gradually, peasant leases have been extending ever since the Emancipation, and by the 'nineties it was estimated that, in the agricultural provinces of Russia, there was hardly any land left available for a further extension of such leases. There are no statistics of peasant leases covering the whole country, but at the beginning of the current century they were estimated by the best authorities at about 20 to 25 millions of dessiatins.[1] In the black-earth belt, the peasants leased mostly arable land; in the Northern half of Russia—mainly meadows, for which the rents were highest. The land was let on lease either direct by the landowner to the peasants, either individually or in whole villages or specially formed partnerships (*tovarishchestvo*), or through middlemen, the latter having been frequently the case in the Ukraine and some other parts of the black-earth belt. Such middlemen made large profits, since the rents for whole estates were much lower than those for the plots leased to peasants. Thus, in the province of Poltava, for instance, rents for large leases, in 1890, did not exceed 6·56 roubles per dessiatin, while for small peasant leases they varied from 8·61 to 9·73 roubles per dessiatin.[2] The arrangement, from the landowner's standpoint, had that important advantage over cultivation on his own account, that, with practically no trouble and expense, the land paid far better than it would have paid under the most skilful management and the most favourable conditions. The advantage for the peasant was that, though mostly at the cost of uneconomically high rents, he thus found employment for the surplus of labour which he could not use profitably otherwise. For the country as a whole, from the agricultural as well as the social point of view, however, the growth of peasant leases appeared in an entirely different light. From

[1] A. S. Yermolov, *Our Land Problem*, p. 44 ff., and A. Kaufmann, *Agrarian Problem*, p. 99.

[2] A. Kaufmann, op. cit., part I, p. 104.

the agricultural point of view, the prevalence of short-term leases in small plots, without any definite contracts or a possibility of control and enforcement of their stipulations over the vast areas and the millions of transactions involved, was a great evil. The one-season tenant had no interest in the land, except that of snatching from it as much as he could with the least effort and outlay on his own part, with the result that the land was soon ruined for a number of years. Instead of developing in the peasants a tendency towards improving their methods of cultivation, such leases rather encouraged them in extending the area sown in preference to intensification of farming. No less detrimental were the social consequences of this development. Apart from the serious evil of the appearance in certain localities of numerous middlemen, with no interest in the land except the profits they could make by the unscrupulous exploitation of the peasants, the spread of peasant leases involved another, more subtle, but extremely important, source of social danger. Russian history had not taught the peasant to reverence landed property as such. Deep in his soul was ingrained the notion that land is the free gift of God to those who toil on it in the sweat of their brow. Barring a very few exceptions, and that mostly in the case of monasteries, large farming was practically unknown in Russia proper until the eighteenth century, the only cultivators of the soil being peasants. The agrarian organization of Russia, based, until the second half of the eighteenth century, on service tenures, involving, in the case of the gentry, the right rather to the income from a landed estate, than to the land itself, which remained the property of the State, was responsible for certain peculiarities in the popular notions of the right to the land, which found expression in the old formula, with which the serfs addressed their masters: "We are yours, but the land is ours." And, in fact, until at least 1762, the land, to whomsoever it happened to be granted by the State in service tenure (*pomestie*), was, as such, rather the concern of the peasants settled on it, than of the legal holder of the estate, whose interest in it was confined to the collection, through a bailiff, of the *obrok*, on which he lived, and to a certain extent of general responsibility for the maintenance of law and order among his subjects, the regular payment by them of taxes, the supply of recruits for the services and other public duties. Originally, service tenures were confined to the lifetime of the actual holders, but since the sixteenth century the land once granted had often been allowed to remain in the family so long as there were male members available to take up the service. This practice became more or less general in the course of the seventeenth century, but this change had little effect on the legal position of the tenant for service (*pomeshchik*) in respect of his land and of the serfs settled on it, which still preserved the character of conditional usufruct, until, in 1762, service tenures were finally abolished and the estates handed over to their actual tenants on the title of private property. This measure, at the time, had caused much trouble among the serfs, whose conception of normal relations between the classes and the State had been rudely shaken. In their minds, they reasoned that, since, hitherto, they had to work for the gentry so as to enable the latter to serve the State, and now the gentry was released from this liability, there was no more reason why they, too, should not be set free: a recurring motive in the peasant risings under Catherine II. The historical construction of the gentry's rights to their estates was shattered, and the landowning class had to find some other justification of their legal title to the land. This they could do, and actually did in some cases, by assuming the part of leaders in the agricultural development of the country. But the lease of their estates struck at the very root of the peasants' ideas about the "right to the land," for which, in their eyes, apart from any title-deeds, a certain ulterior justification was needed. When the peasants, besides, were treated unfairly in connection with leases, bitterness was being added on top of the moral doubt, and the feeling spread in the discontented atmosphere of the Russian countryside. Thus, both economically and socially, the development of short-term peasant leases in general and of leases through middlemen, in particular, had the most undesirable effects.

In 1881, according to information collected by the Central Statistical Committee, the proportion of the land of private landowners, leased to peasants, was 35% in the black-earth belt and 29% in the rest of European Russia.[1] The regions in which the practice was most widespread, were, at the close of the last century, New Russia, the South-Eastern provinces, the Ukraine and the Western districts, along the Dnieper, in which the available area for leases was greatest. In the Central Agricultural region, in spite of the fact that the need for leases was greater than in most other districts, their extension was limited by the shortage of available land. In the black-earth provinces on both banks of the Volga peasant leases were also extensive. The extent to which the leases depended on the relative sufficiency or otherwise of the peasants' own holdings, may be seen from the following calcula-

[1] Quoted from A. Chelintzev, *Agric. Geography*, p. 120.

tions made by Professor Karyshev in his standard work on peasant leases in Russia. According to him, the percentages of peasant families which leased land about the close of the last century, in each of the principal groups of peasants distinguished by the average sizes of their holdings, were as follows:[1]

	Regions with few Leases	Regions with widespread Leases
	Percentages	
Former State peasants	21·9	36·6
" appanage peasants	—	50·3
" serfs with full holdings	39·3–46·6	55·9–61·7
" serfs with "gratuitous" holdings	66·6	65·1

These figures point to the existence of an intimate connection between the sizes of peasant holdings and the percentages of families having recourse to leases. Yet, it would be wrong to conclude, as this is not unfrequently done, that the majority of leases were accounted for by the poorer peasants. The percentage of the latter among the tenants was, indeed, relatively high, but they generally leased small plots, while the larger leases, which accounted for the greater part of the area leased, were in the hands of the better-to-do, who could afford to take the risks and possessed the stock necessary for the cultivation of the land leased.[2]

According to the system of payment of rent, peasant leases could be divided into three principal classes. The first, which was especially common immediately after the Emancipation, and has been gradually falling into disuse later, was the payment of rent in labour (*otrabotki*): a form most widely used in the case of leases by whole villages. Instead of a money rent, the village community contracted to supply, at a specified time, a certain number of days' work to the landowner: an arrangement which, in those days of semi-natural economy, suited both parties well, since the peasants were not compelled to find money for the rent, and the landowner, mostly short of labour during the critical seasons, secured in advance a certain number of hands, though mostly recruited from the worst workers in the village. Next came the system of payment in kind, fairly widespread in the Central Agricultural region and the adjoining provinces, on the principle of share-tenancy, the share of the landowner varying, according to local conditions, from one-half to three-quarters of

[1] Table quoted according to Chelintzev, op. cit., p. 121, the original work of Professor Karyshev being unavailable. A short article on leases by him appeared in *La Russie à la fin du XIX siècle*, pp. 125–7, to which I refer below.

[2] Prof. N. Karyshev, Article "Amodiations" in *La Russie*, etc., p. 126.

the produce. These two systems, however, were gradually being discarded, along with the development of money economy in the Russian countryside, and money rents were becoming more general until, in some localities, by the close of the last century, they supplanted other forms of payment almost completely. It was estimated that, in the later 'nineties, in New Russia and the Ukraine no less than 80% of the rents were paid in money, while elsewhere the practice was spreading gradually. How far this mode of payment was preferable for the tenant, however, is an open question, since it certainly tended to increase his risks, which under the old system of payment in kind, according to yield, were shared by the landowner.

Thus, with the development, since the Emancipation, of peasant leases, a large part of the area belonging to private landowners, instead of being cultivated by them, passed by means of leases into peasant exploitation. In extending the area cultivated by them, the peasants, however, went farther, and since 1861 had developed considerable activity as buyers on the landed estate market, on which they appeared either individually, or in whole village communities and specially formed partnerships (*tovarishchestvo*). The process, in the course of which, besides their original allotment land (*nadiel*), the peasants have built up a considerable landed property, privately owned and free from the restrictions attaching to the *nadiel*, developed steadily, and its progress may be seen from the table below, showing the increase in the total area of land privately owned by the peasants in forty-seven provinces of European Russia: [1]

	1862	1872	1882	1892	1902
	(In thousands of dessiatins)				
Black-earth belt .	3,276	3,935	5,107	7,921	12,056
Other provinces .	2,469	3,363	5,594	8,331	10,823
Total . .	5,745	7,298	10,701	16,252	22,879

It would be mistaken, however, to consider the whole of this area as an addition to the original holdings of genuine peasant farmers. Russian statistics of landed property were based on the division of proprietors according to the civil estates to which they belonged, and classed them as nobles, merchants, burghers, peasants and "others," irrespective of the sizes of their holdings. Accordingly, the class of peasants included, besides genuine peasant farmers, also a number of persons who, though technically belonging to the civil estate of peasantry, were, in fact, large landowners.

[1] "Russian Farming in the Twentieth Century," *Statistical Handbook*, edited by N. Oganovsky (Rus.), Moscow, 1923, pp. 60–1.

In 1905, it was estimated that, of the 13·2 millions of dessiatins which belonged to individual peasants, as distinguished from whole villages and partnerships, only 32·2% represented genuine peasant holdings, not exceeding 50 dessiatins each, while all the rest was accounted for by holdings varying from 50 to over 1,000 dessiatins. Thus, in 1905, genuine peasant farmers, apart from their original allotment land (*nadiel*), had 4·2 millions of dessiatins owned by them individually, and 11·6 millions of dessiatins bought by whole villages and partnerships, or a total of 15·8 millions of dessiatins.[1] The balance of the total area of 24 millions of dessiatins, or roughly about 8 million dessiatins, belonged to more or less large landowners of peasant origin, who had not legally severed their connection with the class.

The extension by the peasants of the area of land in their possession by purchases, at the expense, in the first instance, of the landed gentry, was a characteristic feature of the agrarian evolution of Russia since the Emancipation. From what has been said above about the respective conditions of the two principal parties concerned on the morrow of the reform of 1861, this development was only natural. Many landowners, after having struggled for a few years, in the vain hope of eventually pulling through, finally despaired of the possibility of making good and saw nothing better than to sell their land, mostly to peasants, who, hard pressed as they were for more land in many cases, were, as a rule, prepared to pay higher prices than anyone else. The growing agrarian overpopulation in certain parts of the country made the extension of peasant farming imperative, and this extension, effected by means of leases or of purchases, tended to raise the rental value and the price of land by leaps and bounds. It will be remembered that, owing to the peculiar constitution of his budget, and particularly of his costs, which permits the peasant to treat a large part of the gross yield of his holding as net income, the peasant farmer is generally better fitted to withstand agricultural depression than a farmer employing hired labour. The same peculiarity of his position enables him to pay, on the whole, higher rents than a capitalistic farmer could afford, as well as prices for land which, for another class of buyers, would be uneconomically high. A peasant, if he wants land, as he mostly does, is usually ready to go to the length of working on it, until it is finally paid for, not only without any profit, but even at a rate of remuneration for his own labour often considerably lower than the wages of an

[1] N. Oganovsky, Article on "Landownership in Russia" in the 1915 edition of the Russian *Encyclopædia* of Brockhaus & Efron.

agricultural labourer for similar work. As a result, in the matter of prices, the peasants, under any agricultural conditions whatever, either on a rising or falling market, can outbid other classes of purchasers. Accordingly, with peasants active on the landed estate market in Russia since the Emancipation, there arose a situation which, to a superficial observer, could only appear as highly anomalous, namely, one of rapidly rising land prices during the worst period of agricultural depression Russia had ever known. The general trend of land prices in European Russia from 1860 to 1889, according to regions, is shown below: [1]

Movement of Land Prices, 1860–89

	Roubles per dessiatin				Percentage of Increase		
Regions	1860	1870	1883	1889	1860–70	1870–83	1883–89
Northern	—	—	—	—	—	—	—
North-Eastern	—	—	—	—	—	—	—
Petrograd	13	18	29	44	38	61	34
Moscow	22	(33)	55	55	50	66	—
Western	13	17	31	38	31	82	23
South-Western	43	66	111	120	53	68	8
Ukraine	34	56	98	99	64	75	1
Central Agricultural	43	78	111	96	81	42	−14
Middle Volga	30	50	78	67	66	56	−14
Eastern	13	18	32	34	38	77	6
New Russia	21	31	48	85	48	55	77
South-Eastern	—	—	—	—	—	—	—
Caspian	—	—	—	—	—	—	—

Of the land which slipped out of the hands of the less efficient or the poorer among the gentry, some passed into the possession of the commercial community, "merchants" coming next to the peasants in the extent of their purchases. In this case, as this has been pointed out by certain writers on the subject, purchases of land on a rising market were largely in the nature of a speculative investment by the moneyed class of the country. This would appear to be borne out by the fact that, with the rise in land prices, the sales of land by merchants increased actually faster than their purchases. To a large extent, however, their purchases involved forests, which they bought with a view to exploitation. Since, until the beginning of the current century, there were, in most parts of Russia, no effective laws in force to prevent the destruction of forests, the purchasers cleared the timber in the course of a few years and then the land, now barren, was put again on the market and often bought by peasants, who turned it into arable or pasture. On the whole, the part played by the business man in this particular sphere of activities could not be described as constructive. Far from appearing in the countryside as pioneers

[1] N. Troinitzky, *Du mouvement des prix des terres en Russie*, 1860–1889, Report to the International Statistical Institute, September, 1891.

of capitalism in agriculture—a rôle attributed to them by several writers with Marxist leanings—the merchants, as landowners, have been rather agents of destruction and spoliation, with no more regard to the forest wealth of Russia, than to the " cherry orchards " in the old homes of the ruined gentry.

The changes in the distribution of landed property in Russia during the period dealt with may be seen from a comparison of the results of the Surveys of Landed Property of 1877–8 and of 1905. The table below shows the distribution of land according to the different classes of owners in forty-nine provinces of European Russia, including the three Lithuanian and the three Baltic provinces and Bessarabia, but excluding the Don, Stavropol, Kuban and Terek.

	Area in millions of dessiatins		Increase or decrease	
			Millions of dessiatins	
	1877–8	1905		Per cent
The State	150·4	145·1	− 5·3	− 3·5
Appanages	7·4	7·8	+ 0·4	+ 0·5
Institutions	8·5	6·9	− 1·4	− 16·5
Peasants' *nadiel*	116·7	128·9	+ 12·2	+ 10·5
Gentry	73·2	46·2	− 27·0	− 37·0
Merchants	9·8	12·9	+ 3·1	+ 31·6
Burghers	1·9	3·8	+ 1·9	+ 100·0
Peasants	5·0	29·0	+ 24·0	+ 480·0
Others	1·7	2·1	+ 0·4	+ 2·4

It may be seen from the above figures that, on the balance for the quarter of a century which elapsed between the two surveys, the peasants have purchased by far the greatest area, adding to their original holdings no less than 24 millions of dessiatins. The merchants, though the actual turnover of their purchases and sales during the period was very large, have not increased the area in their possession by more than 31·6%. The burghers (*meshchane*), whose property had doubled, were a very small group among the landowners. Moreover, except for the difference of legal status, many burghers who lived in small country boroughs, hardly distinguishable from villages, were actually peasants under a different name. The greatest losers, as should have been expected, were the landed gentry, as the principal landowning class of Russia in the past, before the Emancipation. Their total losses amounted to 27 millions of dessiatins, or 37% of the area they owned in 1877–8. Indeed, only the strongest and the most economical among the old landowning class were able to survive the long-drawn depression, and even these did so mostly by drawing heavily on the capital value of their estates.

Such, in the main, was the agrarian evolution of Russia in the second half of the nineteenth century, when, on emerging from serfdom, the country was slowly evolving the elements of a modern economic organization. The old medieval régime of self-sufficient natural economy in the countryside was still struggling against the advancing tide of commercialization, and only gradually, step by step, did money economy, with the help of the railways, gain ground on the stubborn past. The agrarian organization, based on open-field farming and on communal tenure, both survivals of a period when the cultivator lived by tradition and had no need for initiative and rapid adaptation to changing requirements, was still holding the peasant farmer tight in its grip. The economic forces, which were bound eventually to burst these remnants of medieval bonds, still impeding the progress of the countryside, were only gradually being mobilized. Until these forces had become strong enough to cut through the vicious circle formed by the slowness of industrial growth, on the one hand, and the poverty of the countryside, on the other, the agrarian situation could not improve. As I have pointed out in an earlier part of this chapter, the first signs of a breach in this vicious circle began to appear at the close of the last century, just at the time when the agrarian crisis had reached its climax. Probably the first among the factors which had helped Russia in getting out of this deadlock, was the extension of the railway net which, by that time, had already effected a certain economic transformation over large areas throughout the country. Wherever the railways penetrated, they infused new life into the countryside and stimulated the commercialization of farming. The industrial development of Russia, slow as it was, by the close of the century had resulted in the appearance of several important industrial centres, which made the influence of their demand for agricultural produce felt over more or less extensive districts. Thus, the growth of the Donetz coalfields and of the metallurgical industry of the Krivoy Rog, in the South of Russia, dated from the seventies. The same period saw the extension of the textile industries in the Moscow region. By the close of the century, the oil industry began to expand with considerable rapidity, bringing Russia to the front of the oil-producing countries of the world. The manufacturing industries, which, in 1863, on the morrow of the Emancipation, were represented by 16,659 factories and works with 419,600 workers and an aggregate output valued at 351 million roubles, by 1900 had increased to 38,141 establishments employing 2,373,400 workers and producing goods to the aggregate value of 3,438,000,000 roubles.[1] The

[1] Varzar, V., "Factories and Workshops," in *Russia: its Trade and Commerce*, edited by A. Raffalovich, p. 107.

urban population of Russia, between 1867 and 1897, had more than doubled, and its proportion had risen from 10·6 to 13·0%.[1] The customs tariff policy of the Russian Government, which, since the close of the 'seventies, had assumed a decidedly protectionist turn, though it had the effect of checking the general growth of the turnover of foreign trade, which, between 1875 and 1899, had increased only by some 10%, did not prevent the growth of Russia's exports by 24% during the same period.[2] From 1861 to 1900, the turnover of Russia's foreign trade increased as follows:

	Exports	Imports	Turnover
	Millions of Roubles		
1861–1870	222·7	225·9	448·6
1871–1880	454·3	488·2	943·0
1881–1890	622·2	471·8	1,094·0
1891–1900	659·8	535·4	1,195·2

Among the exports from Russia, cereals held easily the first place, and with the industrialization of Western Europe, in the second half of the nineteenth century, they increased gradually, as may be seen from the figures below, showing the exports of the four principal cereals (rye, wheat, oats and barley) in quinquennial averages:

1861–1865	1,226,928 tons
1866–1870	1,965,600 ,,
1871–1875	2,968,875 ,,
1876–1880	4,212,002 ,,
1881–1885	4,408,333 ,,
1886–1890	6,028,333 ,,
1891–1895	6,182,172 ,,
1896–1898	7,943,497 ,,

Thus by the end of the century, the market for Russian agricultural produce, both at home and abroad, had expanded considerably. The rise in spot prices, caused by the extension of the railways and the consequent cheapening of transport, stimulated production for the market even in the more remote districts. Moreover, since the later 'nineties, there came an important change in the condition of the world market. The agricultural market began gradually to improve, and the prices of cereals in the principal exporting harbours of Russia, which had been seized in a headlong fall since the close of the 'seventies, started on an upward move, which had then continued without interruption until the outbreak of the Great War. By the cumulative effect of the economic changes, which developed slowly in the course of four

[1] Pokrovsky, V., "La population," in *La Russie à la fin du XIX siècle*, edited by W. de Kovalevsky, p. 65.

[2] Pokrovsky, B., "Commerce exterieur," *Ibidem*, p. 688.

decades, from the Emancipation to the end of the century, the vicious circle which had been holding Russia in a dead grip, was at last broken. The agricultural industry had been given a chance. The time had come when, at last, the peasant could gain by shaking off the last remnants of medieval organization, which still bound him and hindered his progress, and when the agrarian problem, for so long insoluble owing to the peculiarity of Russia's economic position, could be solved. On the threshold of the twentieth century, the "industrial revolution" in Russia, which had been so slow in coming, had at last dawned; and the economic situation, as it developed since, had dictated to the Russian Government those far-reaching agrarian reforms which were being carried out in Russia during the decade immediately preceding the war and were bringing about the rapid transformation of the countryside.

CHAPTER V

THE AGRARIAN REFORMS: ENCLOSURES (1907–1916)

At the beginning of the twentieth century Russia was faced with an agrarian situation which had to be dealt with and improved at all cost. Both politically and economically, the condition of the peasantry involved a direct danger to the State itself. The countryside was in perpetual ferment, the discontent of the peasants expressing itself in sporadic outbreaks of agrarian disturbances and outrages. The latter, since the close of the last century, increased gradually in scope and importance, until, during and immediately after the revolutionary outbreak of 1905, they spread as wildfire throughout the whole country. Financially, Russia could not exist with the great bulk of her population in such economic conditions as those revealed by the inquiries of the "Commission of the Centre" in 1901. The situation was brought to a crux by the Japanese War and the revolution of 1905, which rendered Russia the same service as that for which she was indebted to the Crimean War of exactly half a century before. As the latter showed conclusively the impossibility for Russia, as a Great Power, to go on living on serf labour, so the former two proved that the very foundation of her economic system, namely, her peasant farming, was in too precarious a condition to support the mighty superstructure of the Empire. Something decisive had to be done, and that at once, if Russia was to be saved from becoming a colossus on feet of clay. The radical remedies offered were many, most of them emanating from the parties of the Left, and all based on the principle of increasing the peasants' holdings by means of the expropriation, either with or without compensation, of the land belonging to large proprietors. Nationalization, socialization and municipalization of land were in the air. Yet, especially considering that, as shown by the figures given in the preceding chapter, the peasants, as a class, were already the largest landowners in Russia and that, of the State lands and large estates, the greater part by far consisted of forests, and only a relatively very small proportion could be transferred to the peasants as arable, these measures could not improve the position of the peasants to any marked extent. Moreover, they did not provide a radical solution, since, obviously, if the organization of the countryside and, in particular, the systems of tenure, were not changed, the crisis would only be staved off for perhaps a generation, after which the agrarian overpopulation would again assume dangerous proportions, without any recourse, this time, to similar measures of extending the area of peasant holdings. The Government, accordingly, set its face firmly against compulsory expropriation of land in any form, going even so far as to dissolve the First Duma, in 1906, on the agrarian issue. The necessity of extending the area of land in the peasants' possession, however, was fully recognized, since, after the exhaustive inquiries of the official Committees and the local Zemstvos, as well as the Survey of Landed Property of 1905, it was perfectly clear that "land-hunger" among a large section of the rural population, far from being a figment of revolutionary imagination, was very real and acute indeed. This extension had to be effected by all means short of the violation of the rights of private property, which the Government set out to defend against the advancing tide of Socialism.

Such extension of peasant holdings, however, was not enough. Russia needed a positive programme of agrarian reforms, which would reorganize her countryside and her agricultural industry on modern lines. She needed a man of great vision and courage to take the helm at this juncture. She found such a man in Stolypin, who assumed the duties of President of the Council of Ministers, or Premier in 1906, and held the post uninterruptedly until his death, in 1911, at the hand of a political assassin. During his five years in office he succeeded in passing all the legislation necessary for the execution of the scheme of far-reaching agrarian reform, generally associated with his name. Though most elements of the agrarian policy of Stolypin were, indeed, present to a more or less marked extent in the previous legislation and activity, at least in embryo, to him was due their welding together in a comprehensive whole, their extension to proportions in which they could meet the situation and the carrying of the plans into execution with unflagging energy and perseverance. Moreover, the scheme of agrarian reorganization, as it presented itself to its author, was far more than a measure of purely agricultural or economic policy. For Stolypin, his agrarian legislation was part and parcel of a definitely thought-out programme of social reconstruction, aimed at the creation in Russia of a strong, independent peasant democracy, a bulwark of law and order in the country and a solid foundation of the economic and political system of the Russian Empire.

The policy of Stolypin involved practically every aspect of the agrarian problem. It set out, in the first instance, to improve the conditions of peasant farming by the substitution of compact enclosed holdings or farms for the existing medley of scattered strips in the open field of the village. It also provided for the improvement of the system of tenure in open-field villages, by helping the peasants to eliminate inconveniences caused by the mixing together of lands belonging to different owners, etc. It included the provision of technical and financial assistance to peasant farmers for the improvement of their holdings and cultivation. Through the State Peasants' Bank, whose statute was revised so as to enable it greatly to extend its operations, mainly for the benefit of the poorer peasantry, it helped the peasants to buy more land on liberal conditions of credit, thus inaugurating a scheme of widespread internal colonization. Last, but not least, it opened up Asiatic Russia to the surplus population of the European part of the Empire, by developing a system of assisted emigration beyond the Urals on a large scale, and organizing the colonization of Siberia.

The principal feature of the new agrarian policy was, certainly, the gradual substitution of enclosed peasant holdings for open fields, involving the complete reorganization of the Russian countryside. It was also the part of the general scheme of the Stolypin legislation which gave rise to most violent criticism, rallying the various currents of opposition round the banners of the "populists" or *narodniki*, who saw in it a direct attack on the sacred institution of the rural commune which should have carried Russia past the evils of modern Capitalism straight into the Socialist millennium of the future. That the policy of Stolypin was essentially based on the recognition of the necessity of doing away with the commune, as a survival of the past which could not be fitted into the scheme of modern economic organization, and hindered the development of the country, was perfectly true. Equally true it was that Stolypin was "betting on the strong" and basing his policy on the more energetic and enterprising, if not, as had often been said by his opponents, on the richest, elements of the peasantry, whose gradual pauperization by the commune he wanted to stop at all costs. Yet, the land-settlement legislation of 1906–1911, revolutionary as it proved to be in its effects on the Russian countryside, involved no direct violence on communal tenure, wherever it was still a living institution, with enough vitality to withstand the shocks caused by the legal permission to withdraw from it, given to its individual members. In fact, the withdrawal from the commune, either by individual peasants, or by whole villages, was purely optional, and if the movement assumed enormous proportions, it was mainly because the change answered the purpose of a large number of peasants. In some cases, it must be admitted, there may have been a certain amount of administrative pressure in favour of separation, exercised by minor local officials who wanted to earn distinction, but no amount of such pressure could account for the enclosure, within less than a decade, of about 10% of the total number of peasant holdings in Russia. Technically, the new legislation did not introduce any entirely new principle, but simply represented an extension of the provisions contained in the legislative acts of 1861, which the changes in the general situation made expedient or even imperative.

Indeed, already the *General Statute of Peasants*, issued at their emancipation in 1861, provided that any village with communal tenure (*mir*) could, by a two-thirds majority, effect a transition to hereditary tenure (*podvornoie vladenie*), thus allocating holdings to its members in perpetuity. The Statute also contained a clause, according to which, in those cases in which the land of the village was clear from redemption payments, individual peasants were allowed to claim separation from the commune and the allocation to them of their share of the common land in the form of a compact holding. The *Statute of Redemption* went even farther, by allowing any individual peasant to pay his share of the redemption debt before the expiration of the term, and then to claim the separation of his holding and its allocation to him in a compact plot. Thus, from the start, communal tenure was treated in the legislation rather as an intermediate stage in the development of peasant landownership, rendered necessary, primarily, by the need of adequate security for the redemption credits. As soon as the peasants cleared their debts under this head, they had the option of separating from the commune and enclosing their holdings. The conditions of the first few decades after the Emancipation, however, were far from favourable to the growth of peasant enclosures. Not only did the necessity of paying the whole amount of the redemption money at once make separation from the commune possible only for the richer peasants, but the general economic conditions of the country were such as to check, rather than encourage, the evolution in this direction. Peasant farming, to enjoy the full advantages of enclosures, had to attain a much higher degree of commercialization, than that which was common until the close of the last century. Moreover, enclosures were practically unknown throughout the greater part of Russia, where the actual experience of the peasantry was limited to open-field farming, with the result that the advantages of self-contained compact holdings could not be gathered from direct observation. Early enclosures of common land by individual peasants or whole villages, of which cases have been registered since the 'seventies, were confined mainly to a relatively restricted area, namely, the Western provinces of Russia, where independent farms (*hutor*) either have existed, occasionally, of old, or have been introduced by alien settlers, and the local peasantry had the opportunity to learn their advantages by observation. The region in which the enclosures movement, prior to the land-settlement legislation of 1906, assumed any considerable proportions, extended from the Western frontiers of Russia to the West of the provinces of Petrograd and Novgorod, the Western districts of the province of Smolensk, the Western part of the Ukraine and New Russia. In this part of the country, about 20,000 enclosed holdings have been carved out of the village allotment lands (*nadiel*), involving a total area of approximately 200,000 dessiatins. In the rest of Russia, cases of enclosures were quite exceptional, and the village commune reigned supreme. By the close of the century, when conditions began to change, and the situation was becoming more favourable to the growth of the movement, a legal check had been put to its extension. Since the 'eighties, after the assassination of Alexander II, the Government's policy had been aimed at the creation of some kind of bulwark against the growth of the proletariat, as the principal bearer of revolutionary ideas. The proletariat was recruited from the landless peasantry; *ergo*, the peasantry was to be kept on the land and its proletarization prevented at all cost; and the strengthening of the communal bond was a means to that end. By the Law of December 14th, 1893, the redemption of their land by individual peasants before the expiry of the term of credit was made conditional on the consent of the village assembly by a two-thirds majority. How far this measure succeeded in diminishing the movement of enclosure, may be seen from the reduction in the amount of lumped redemption payments which declined from 965,000 roubles in 1893 to 44,000 roubles in 1896.[1] The law also introduced important changes in the provisions of the original enactments in so far as they concerned the freedom of the

[1] C. v. Dietze, *Stolypinsche Agrarreform und Feldgemeinschaft*, Leipzig und Berlin, 1920, S. 42.

peasants in disposing of their holdings. Hitherto, peasant holdings, when fully paid for, could be disposed of freely, but the law of December 14th, 1893, permitted the sale of allotment land held in perpetuity only to persons who either were members of the same village community, or joined it on the acquisition of the holding. In other words, no one, except a peasant, could buy land belonging to the *nadiel.* If a whole village, holding land in common, wanted to sell part of that land, it had to obtain special permission from the provincial authorities or the Minister of the Interior. The law also prohibited the mortgaging of peasant holdings absolutely. Thus, the peasants' land was put under a régime of severe restrictions, intended to prevent its transfer to other classes of proprietors. The immediate purpose of these restrictions was that of preventing the proletarization of the peasantry, but, since, on the one hand, the increase of rural population could not be prevented by such measures, while, on the other hand, what could not be done by the sale of holdings, could easily be achieved, in circumvention of the law, by its letting on lease to others, it was not achieved, and the impoverishment, if not the technical proletarization of the peasants, proceeded apace.

While the peasants have thus been tied faster to the commune, another enactment, namely the Law of June 8th, 1893, was aimed at the elimination of some of the features of communal tenure, considered particularly objectionable from the technical view-point. The law fixed a minimum term of twelve years between general redistributions of land, while partial redistributions for the purpose of readjusting the holdings of individual peasants were forbidden. This enactment, indeed, tended to create a condition of greater security of tenure among the peasants, but, since the principal obstacle to agricultural progress was not so much the insecurity of tenure, as the system of open fields, with scattered strips and compulsory cropping, it failed to strike at the root of the evil.

The positive results of the attempt to stem the tide of impoverishment and actual proletarization among the peasantry by means of strengthening the hold of the commune on the individual were practically nil. By the end of the century, as I have tried to show above, the crisis reached its climax, and new avenues of escape from threatened ruin had to be sought for. The "Committee of the Centre," appointed in 1901, was followed by another inquiry, planned on a very large scale, whose principal purpose it was to discover the principal needs of the countryside and to devise suitable measures to cope with the situation. The initiative of this undertaking belonged to Mr., subsequently Count, Witte, former Minister of Finance, who at the time was in a position of honorary retirement from active political work, as Member of the Council of the Empire. *The Special Committee on the Needs of the Agricultural Industry*, appointed by the Emperor on January 22nd, 1902, with Witte as President and local committees throughout the country, conducted inquiries for nearly two years, bringing its investigation to completion in December, 1903.[1] It took the opinion of 13,460 witnesses, and the results of its work were published in fifty-eight volumes of about 25,000 pages. The work on the material collected was continued until March, 1905, when the Committee was dissolved. Much of the material was of great value, and served later as basis for various measures, but, probably owing, in the first instance, to the organization of the Committee and its local bodies, which consisted of people of every shade of opinion, the positive results of this grandiose attempt to work out Russia's salvation by local inquiries were rather meagre. The Committee had not worked out any definite, practicable scheme of improving the conditions of the peasantry, though it pointed out a number of desirable changes and improvements. On the problem of the commune and open fields, the opinions were divided, though the majority inclined against communal tenure, as one of the principal obstacles to agricultural progress. While the Committee deliberated, however, events developed with great rapidity, forcing the Government to take decisions. It was clear that no measures of repression alone, however stern, would meet the situation, and that the demands of the peasants for improvement in their position had a very real foundation of fact. The early years of the twentieth century, accordingly, saw the first tentative attempts on the part of the Government, while sternly suppressing agrarian riots, to improve the legal and economic position of the peasantry. The first measure of economic nature, which was the forerunner of the reforms to come, was the abolition, on March 12th, 1903, of the joint liability of the village community for the payment of taxes and of redemption instalments by its members. This was a very important step forward, since it removed the uncertainty of the individual peasant as to the amount of his payments. The next important step, which did more than anything else to free the way to the subsequent campaign for the substitution of enclosed holdings for communal tenure, was the

[1] The actual "Transactions" of the Committee were not available to the writer, whose account here is based on the *Memorandum on the Peasants Problem*, by S. Witte, summarizing the results of work; and W. D. Preyer's *Die Russische Agrarreform*, Jena, 1914, pp. 146 ff.

cancellation, by the Ukaze of November 3rd, 1905, of the balance of redemption payments, still outstanding to the amount of over one milliard roubles. Combined with the repeal of joint liability for taxes, this measure did finally away with the fiscal reasons for the existence of the village commune. Indeed, had the original Statute of Peasants of 1861 not been amended by the restrictive law of December 14th, 1893, from 1906 onwards any peasant would have been free to claim the allocation to him of his share of the common land in a compact holding, and to dispose of it at will. Owing to the restrictions imposed on peasant allotment land (*nadiel*) in 1893, the position was now different, and the peasant could only regain his former freedom of disposal by special legislation. This waited until the advent to power of Stolypin, who, a year later, pushed the matter to its logical conclusion. In the meantime, the Government passed certain enactments rendered necessary or expedient by the new situation.

Thus, by the close of 1905, the way to the gradual dissolution of the commune had been cleared, but the measures already passed were not yet part of a deliberately thought-out policy; rather, they represented partial concessions on the part of the Government, which eliminated the most obviously harmful and oppressive features of the existing conditions. At this time, it was mainly sought to increase the area of land in the peasants' possession and to help them in raising the standards of cultivation; the problem of emigration beyond the Urals, as a means to relieve rural congestion in European Russia, also began to attract increased attention. As to the problem of communal tenure, the Government did not seem to have finally made up its mind.

That the agrarian problem was to be tackled more seriously in all its aspects, was emphasized, among other things, by the reorganization, by the Ukaze of May 6th, 1905, of the former Ministry of Agriculture and Domains, which became the Central Administration of Land-Settlement and Agriculture, emphasis being laid on its function as an organ of agrarian policy. By the same Ukaze, the colonization of Asiatic Russia which, hitherto, had been controlled by the Ministry of the Interior, was transferred to the re-organized Ministry of Agriculture, to which a special Colonization Department, with wide powers, was added. The organization for dealing with the land-settlement problem was created by the Ukaze of March 4th, 1906, which set up a Central Committee of Land Settlement, attached to the Ministry of Agriculture (as, for the sake of brevity, the Central Administration of Land-Settlement and Agriculture will be called), as well as Provincial and District

Land-Settlement Commissions. Later, in 1911, the organization was extended by the appointment of Regional Plenipotentiaries of the Committee of Land-Settlement, whose duty it was to co-ordinate the activities of local land-settlement commissions with those of local administration, as well as to exercise general supervision and control over the work of land-settlement in their respective regions. The activities of the land-settlement organizations involved practically every aspect of agrarian policy, from internal colonization to the assistance to the peasants in the improvement of conditions of tenure and methods of cultivation, but it was not until the appearance of the enclosure legislation of Stolypin that the main direction of their work had been definitely traced. From the promulgation of the Ukaze of November 9th, 1906, that is the first enactment which inaugurated the enclosure movement, onwards, the Land-Settlement Commissions became the real instruments of transformation of the Russian countryside.

The essential features of the agrarian legislation of Stolypin were embodied in three enactments, namely, the Ukaze of November 9th, 1906, the Law of June 14th, 1910, and the final Law of May 29th, 1911, which, as the *Statute of Land Settlement*, was incorporated in the Code (Vol. X, part 3).

According to clause 1 (Sect. I) of the Ukaze of November 9th, 1906, "every head of a peasant family, holding allotment land (*nadiel*) by right of communal tenure, is entitled at any time to claim the appropriation to him on the title of private property of his due share of the said land."

Clause 2 provided that, in village communities, in which there had been no general redistributions of land in the course of the last twenty-four years, the peasant desiring to withdraw from the commune was entitled, besides his homestead, to all the land that was in his actual possession, as a member of the village commune, at the time when his request was made. In other cases, that is in communities in which redistributions have taken place in the last twenty-four years, clause 3 entitled any individual peasant to his due share of communal land; in case if, at the time of separation, his actual holding exceeded his share, the excess had to be purchased by him from the village community and paid for *at the average price of land fixed for the purpose of redemption* at the Emancipation. Clause 4 provided that the peasant who had his share of common land appropriated to him preserved his right of use over those parts of the village land, which were not divided among the members of the community, but exploited jointly, such as forests, meadows, pastures, etc. Clause 12 provided that the peasant who chose to separate from the commune was entitled to claim the allocation to him of his share of common land, "as far as possible" in a single plot. According to clause 13, if the enclosure of the land claimed proved technically impossible, the community could redeem the land in question at a price fixed by agreement or by a decision of the Court. But if the request for enclosure was made at the time of a general redistribution of land in the commune, the latter was bound to grant it and had no alternative of compensation in money (clause 14). Clause 1 of Section III of the Ukaze introduced an important alteration in the existing law, based on custom, according to which a peasant holding was considered as belonging to the peasant family (*dvor*) as a whole, and not to the head of the household personally, by enacting that "individual peasant holdings . . . are the private property of the heads of the households, in whose possession they are" . . . Section IV of the Ukaze provided that "whole village communities, whether holding land in common or in perpetuity, can effect the enclosure of the holdings of their members by a majority of two-thirds of all the peasants possessing the right of vote in the village assembly."

It may be seen that the enactment just quoted deals with two distinct acts involved in the process of separation from the commune, namely, in the first instance, the appropriation to the claimants of their shares of common land in perpetuity, as their private property, which might or might not lead to its subsequent enclosure, and, secondly, the actual enclosure of the land thus appropriated. It is clear that the first of these two steps was only necessary in villages with communal tenure (*mir*), since in those with open-field holdings in perpetuity the need for it did not arise, every peasant being owner of his strips. Accordingly, in the villages with communal tenure, the peasants desiring to enclose their holdings had, as a preliminary step, to obtain a deed of appropriation for the land they actually held; only after the appropriation (*ukreplenie*), they could proceed to the actual separation and enclosure of their holdings. The provisions of clauses 2 and 3 gave rise to much criticism on the part of the supporters of the commune. The former, it was said, unduly favoured the enclosure of their land by the more fortunate members of the community, to the detriment of those less favourably situated, while the latter, by fixing the payment for any excess of land appropriated by individual peasants at prices of nearly half a century ago, also actually robbed the community for the benefit of the individual. The first of these two objections can hardly be maintained, since the clause applies to village communities in which redistributions have been in abeyance for so long a time as to create a presumption in favour of the complete abandonment of the principle of equalization of holdings. The latter objection, referring to the prices payable for the excess of land appropriated, over and above the share of the given family in the common land, rests on much firmer ground. Indeed, this clause can only be interpreted in the sense of indirect encouragement to enclosures. Considering, however, that in enclosures the Government saw the principal means by which to save Russian peasant farming from its distress, it was only natural that, while avoiding violent measures for the dissolution of the commune, in its legislation it accorded a somewhat preferential treatment to those of the peasants who followed the desired course. Enclosures could be effected either by individual peasants (*vydiel*), or by whole villages (*razverstanie*). The importance of the clause, conferring the ownership of the holding on the head of the household, as distinguished from the peasant family (*dvor*), in connection with the transformation of peasant farming on individualistic lines, is obvious, since the unity of control thus achieved was essential under the new régime. The dangers involved in the disposal of the holding being thus vested completely on an individual who, by his conduct, could leave the family destitute, had to be faced; besides, as a matter of fact, the protection afforded by the old custom to the dependants was more apparent than real, and cases of families being left destitute by a drunken or profligate head were not uncommon.

The Ukaze of November 9th, 1906, was issued between two sessions of the Duma, in virtue of clause 87 of the Constitution, which provided for the right of the Executive to legislate by direct Imperial Ukazes provisionally, such enactments to be submitted to the Legislature for approval subsequently. The original Ukaze, therefore, had to go to the Duma and the Council of the Empire, but, the times being very troubled, it remained in force, without formal legislative sanction which, at the time, it could hardly obtain, owing to the strong opposition of public opinion and of the two first sessions of the Duma to anything emanating from the Government, for nearly four years, during which it had been put to the test and considerable headway had been made in the direction traced by it. The Law, which took its place, was passed only on June 14th, 1910, and followed exactly the same course, with the only difference that, in its attitude, it was in some respects more radical and far-reaching than the Ukaze from which it originated.

The principal alteration introduced by the *Law of June 14th*,

1910, consisted in the provisions of clauses 1 and 2, which distinguished between villages with communal tenure which had carried out redistributions of holdings since the original allocation to them of land after the Emancipation, and those in which redistributions have not been actually practised. Those communes, the allocation of land to which dated from before January 1st, 1887, and which had never had a general redistribution of holdings since, that is for at least twenty-three years, were considered as, by prescription, having adopted the system of tenure in perpetuity. Accordingly, all such villages, of which there were many in various parts of the country, since the publication of the law, changed from communal tenure to tenure in perpetuity (*podvornoie vladenie*), with the result that the process of enclosure, in their case, was simplified, by the first step, i.e. the individual appropriation of the land held, becoming unnecessary. Another stimulus to the enclosure movement, as compared with the Ukaze of November 9th, 1906, was the provision of clause 45, according to which the breaking-up of the whole of a village's land into enclosed holdings, in the case of communes with tenure in perpetuity (*podvornoie vladenie*), was made dependent on a simple majority of votes in the village assembly. In the case of villages with communal tenure (*mir*), such breaking-up (*razverstanie*) was left to the decision of a majority of two-thirds, as before.

The enactments referring to peasant enclosures and other aspects of the land-settlement policy associated with the name of Stolypin were finally completed and codified in the *Law of May 29th*, 1911, known as the *Statute of Land-Settlement*.

Clause 1 of this law must be quoted *in extenso*, since it enumerates systematically all the particular operations included in the general term of "land settlement." The clause reads as follows:

"On the basis of the provisions of the present Statute, the following land-settlement proceedings, initiated in accordance with corresponding applications by the parties concerned, may be carried out:

(1) The enclosure of the lands belonging to separate villages forming part of a single village community;

(2) The enclosure of the lands belonging to separate hamlets or parts of villages;

(3) The enclosure of the holdings of individual members of village communities and of villages holding land separately;

(4) The complete breaking-up of the land of whole village communities or of villages holding their land separately into enclosed holdings belonging to their individual members;

(5) The elimination of the mixing-up of lands, on which land-settlement operations are being carried out in accordance with the provisions of the present Statute, with neighbouring landed properties;

(6) The consolidation of lands, held under different titles, included in the same estate for the purpose of enclosure;

(7) The delimitation of various lands, used jointly by peasants and private landowners;

(8) The delimitation of the lands subject to land-settlement operations, from neighbouring landed properties in those cases in which the establishment of external boundaries is necessary for the carrying out of the said land-settlement operation."

Thus, the law provided not only for individual enclosures of peasant holdings, but also for the improvement of the conditions of tenure of separate villages, whose lands were, not unfrequently, mixed up with those of other villages and hamlets belonging to the same village community with, sometimes, several thousands of members and thousands of dessiatins of land. Here, the inconveniences of open-field tenure, with scattered strips, were reproduced on a large scale and applied not to individuals, but to whole villages. Articles (1) and (2) of clause 1, quoted above, refer precisely to such cases of "group enclosures," that is of the consolidation and delimitation of the land belonging not to individual peasants, but to whole communities. According to the report of the Agrarian Commission of the Duma, accompanying the draft law on its introduction to the Legislature, "the two first articles may prove very important for the improvement of the conditions of tenure in those localities in which the idea of communal tenure is still strong and where, while averse from destroying the commune, the peasants would yet desire to improve the conditions of their farming." The rest of the clause deals with "individual enclosures," involving the consolidation and enclosure of the holdings of separate peasants, or the breaking-up of the land of whole villages into individual enclosed holdings. The latter could be effected in one of the two following forms: either the detached, absolutely self-contained farm, technically known as *hutor*, with the peasant's house and farm buildings situated on the plot itself, or the *otrub*, which could consist of one or more (generally no more than three) separate plots, according to conditions, with the house and farm buildings remaining, as before, in the village. From the purely agricultural view-point, the *hutor* was certainly far preferable to the *otrub*, but the extreme isolation of the farm, as well as the expense of transferring the buildings, the difficulties connected with the sending of children to school, etc., even apart from local conditions, which made the organization of completely self-contained holdings often difficult or even impossible, not unfrequently caused the peasants to prefer the less technically perfect form of settlement on the *otrub*. This was recognized by the Agrarian Commission of the Duma which, in its report, referred to the question in the following terms: "However desirable may be the most perfect form of land-settlement, namely the *hutor*, and whatever advantages it might possess as compared with the less perfect settlement on an *otrub*, any kind of pressure and, particularly, any compulsion exercised in the attempt to develop settlements on *hutors*, even with the support of the majority, is impossible. There are too many conditions both in favour of the measure and against it. Very often the purely individual qualities of the peasant concerned are decisive in this respect" . . . With regard to those of the peasants, who preferred to remain in the commune, the Law on May 29th, 1911, also abstained from undue pressure, by providing that, when the land of the village was being broken up into enclosed holdings, those who desired to continue under communal tenure, were entitled to claim the allocation to them of their due share of the village land in a compact plot for exploitation in common.

The operation of the law, according to clause 2, extended not only to the land belonging to peasants, but to lands forming the property of all classes of small-holders, irrespective of their civil estate. It applied not only to the allotment land of peasants (*nadiel*) but also to the lands owned by them outside the original allotments, in so far as these came into the scope of land-settlement operations, which, in the case of privately owned holdings situated in the immediate vicinity of the village, became a necessity, if the enclosed holding of the peasant in question was to be made completely self-contained. This led to the formation of enclosed holdings consisting of lands held under different titles, part being absolute private property, and the other being subject to certain restrictions affecting the disposal of peasant allotment land. Clause 3 of the *Statute of Land-Settlement* provided that such composite holdings became absolute private property in their entirety, unless the proprietor chose to make them subject to the law referring to the peasant allotment land, which, however, he could only do if the privately owned part of his holding was free from mortgage to any person or institution, except the State Peasants' Bank, to which alone peasant allotment land could be mortgaged, in accordance with the *Ukaze of November 15th*, 1906. The latter enactment provided that peasant holdings in the village *nadiel* could be mortgaged to the State Peasants' Bank, for the purpose either of the purchase of allotments sold by those members of the commune who left the village and emigrated, or for supplying the balance of the purchase price of any land bought, over and above the credit granted on it by the Peasants' Bank, or, finally, with a view to financing agricultural improvements. This Ukaze thus repealed the provisions of the Law of December 14th, 1893, which forbade mortgages of allotment land absolutely, but limited the permission to transactions with the State Peasants' Bank, otherwise leaving the restriction in force. Generally speaking, the greater freedom of disposal of his holding, granted to the peasant, was carefully hedged about with a view, while permitting the transfer of land within the peasant class from one individual to another, to avoiding the loss of land by the peasant class as a whole. The land, which had once belonged to peasants, had to remain peasant land for ever after, though it was permitted and expected to change hands from the less efficient to the more efficient elements among the peasantry. The buying-up of peasant holdings by persons belonging to the peasant class technically, and thus the building-up of large capitalistically-organized estates out of what used to be peasant small holdings, was provided against by the Law of June 14th, 1910, which, in clause 56, prohibited the acquisition by any individual, within the same administrative district (*uiezd*) of an area of peasant land exceeding, in the aggregate, six peasant allotments (*dushevoy nadiel*) of the maximum size fixed for the locality at the Emancipation in Great and Little Russia, and in those parts in which no maxima and minima had been fixed in 1861, in excess of a specified area. Thus, the protection of the peasantry, as a class, against the loss by them of their land was fully maintained by the new legislation. The transfer of land within the class, though it was made free, was not legally allowed to bring about an excessive accumulation of land in the hands of separate individuals, which would put them actually outside the boundaries of the class: no engrosser of peasant land was legally allowed, by the mere concentration of peasant land in his hands, to rise higher than the position of *Grossbauer*, with a total area varying, according to locality, from some 15 to, perhaps, 50 or 60 dessiatins.

In some cases, the enclosure of peasant holdings was complicated by the existence of contracts of lease, concerning the whole or part of the land involved in the proceedings. This difficulty was provided for in clause 17 of the Law of May 29th, 1911. According to it, if the contract of lease concerned some separate part of the land belonging to the peasant whose holding had to be enclosed, a similar plot of land out of the new consolidated holding had to be substituted for the land originally leased. The tenant, however, was entitled to claim the cancellation of the contract, if the arrangement did not suit him. If the contract involved the whole land of a commune or partnership of peasants (*tovarishchestvo*), the latter, in order to make enclosure possible, could break the agreement, in which case the tenant could only claim compensation for that part of the outlay on the land leased, which he had not yet had time to recover from the exploitation of the lease. If parts of the land leased by a commune or association had to be enclosed, the contract could be broken in respect only of the parts involved, but in this case the tenant was entitled to claim not only compensation for his unrecovered outlay, but, if he chose, the complete cancellation of the agreement.

A characteristic advance made by the Law of May 29th, 1911, as compared with the earlier enactments, was that it insisted on the consolidation of individual holdings being carried out simultaneously with the appropriation (*ukreplenie*) of the land, while in the Ukaze of November 9th, 1906, and the Law of June 14th, 1910, the two acts were considered as independent of each other.

Land-settlement, according to clause 6 of the Law of May 29th, 1911, could be carried out either by agreement of the parties concerned, or it could be compulsory with respect to one of the parties. According to clause 7, homesteads, self-contained farms with farm buildings, vegetable gardens, vineyards, as well as other lands of especial value to their owners, specified in the text of the law, could only be made subject to land-settlement operations by the express desire of their owners. Enclosures could be compulsory with regard to one of the parties concerned in the land-settlement operations in the following cases:

(1) At the request of any number of individuals, at the time of a general redistribution of land in the commune, if the applications preceded the decision of the village assembly concerning the redistribution of holdings (clause 35).

(2) Apart from general redistributions, at any time, if the number of persons applying for enclosures amounts to at least one-fifth of the total number of families in the commune, when the total membership of the latter does not exceed 250, or to at least fifty, if the membership of the commune exceeds 250 families (*dvor*). Whenever the enclosure of the holding was technically possible without injury to the rest of the village, the latter has to agree to it, even if the

application was made by only one of the members of the commune (clause 36).

(3) At the request of the simple majority of the peasants of a village who chose to continue in communal tenure, with regard to the portions of arable land which have been appropriated in perpetuity (clause 37).

(4) Without the request of any of the parties concerned, at the enclosure of the lands belonging to separate villages or hamlets, with regard to the parts of these lands appropriated by individual peasants (clauses 23–27).

(5) Without the request of any of the parties concerned, if a peasant who preferred to remain in open fields, refused to exchange some of his strips, which it was necessary to include into a newly-formed enclosed holding, against other strips, with regard to the appropriated strips of the objector (clause 37).

It may be seen that, while the legislation avoided direct violence with regard to the commune, it tried to do its best to smooth the way to its gradual dissolution by carefully eliminating the obstacles from the path of those peasants who chose to enclose their holdings. In the main, reliance was placed entirely on the gradual development of the movement, encouraged by the example of the pioneers of enclosures.

The legal position, to sum up the above account of the outstanding points of the land-settlement legislation, was briefly this. Any member of a village commune (*mir*) was free at any time to claim the appropriation to him of his share of the communal land, for which he obtained a title-deed of ownership, and then to have it consolidated and enclosed (*vydiel*) by the local District Land-Settlement Commission, the whole expense of surveying, etc., being borne by the State, except for the housing of the personnel during the operations and the supply of the necessary labour, which rested with the peasants concerned. Unless the enclosure was found impossible, on account of the unavoidable disturbance it was bound to cause in the tenure of the rest of the village, it was carried out there and then; otherwise, it had to wait for the next general redistribution of land in the commune, when the latter was legally bound to meet the request. The request for enclosure had to be granted by the commune in all cases when it was made by a whole group of its members, amounting to no less than one-fifth of the total, or to at least fifty in communes with a membership of over 250 families. Whole villages with communal tenure could decide to enclose their holdings by a majority of two-thirds of the village assembly. In the case of villages with tenure in perpetuity and those assimilated to them by the Law of June 14th, 1910, individual enclosures were not allowed, since their permission would create, among the rest of the village, an uncertainty of tenure even worse than that which existed in redistributory communes, in which redistributions were at least legally limited. In these villages, enclosures had to involve the whole of the village communities concerned, and could be decided upon by simple majority of the village assembly (*selsky skhod*) ; they took the form, therefore, of the complete breaking-up of the village land into enclosed holdings of either the *hutor* or the *otrub* type (*razverstanie*).

While the legislative work, dealing with enclosures and other forms of land-settlement, of which an outline has been given above, continued from 1906 to 1911, the actual transformation of the Russian countryside, started by the Ukaze of November 9th, 1906, went on uninterruptedly from 1907 onwards.

The activities of Land Settlement Commissions extended gradually. In 1906, District Land Settlement Commissions were established in 171 districts of European Russia. In 1913, in the 468 administrative districts comprised in the area of land-settlement operations, 463 such Commissions were already at work. It was on them, in the first instance, that lay the main burden of the actual work on the spot, since they had to carry out the enclosures and other measures involved in the land-settlement policy of the Government.

The area of land-settlement included a total of 11,864,641 peasant households (*dvor*), with 118,721,024 dessiatins of land. Of these, 23·2% of the households, and 17·3% of the land, were held by the peasants concerned in perpetuity (*podvornoie vladenie*), while 76·8 and 82·7% respectively were in communal tenure (*mir, obshchinnoie vladenie*).[1]

Of these latter, according to the data of the Ministry of Finance for 1912, a total of 3,489,898 households with 30,656,940 dessiatins of land, were affected by the Law of June 14th, 1910, as not having practised redistributions since the allocation to them of land after the Emancipation.[2] Henceforward, they passed into the category of villages with tenure in perpetuity. Thus, in over one-third of the villages with communal tenure, the equalization of holdings by redistributions, which formed the essential characteristic of the commune, had been in abeyance for at least a quarter of a century since 1887, and generally even for a much longer period. This seems to point that, at least in many cases and localities, the com-

[1] *Q.v.* A. Koefoed, *The Russian Land Settlement*, 1914 (Rus.), p. 15.
[2] *Q.v.* C. v. Dietze, op. cit., p. 60.

mune had already been a moribund institution for some time past, if not since the Emancipation.

The actual progress of land-settlement and enclosures throughout the whole area affected by the reforms, from 1907 to 1913, may be seen from the figures given below, in which a distinction is made between the various forms of land-settlement operations, following the order in which the transition from the commune to the final enclosure was accomplished.

The first step was that of the appropriation (*ukreplenie*) by the peasant of his share of the common land, in accordance with the provisions of the Ukaze of November 9th, 1906.

In forty provinces of European Russia (excl. of the Don, Terek, Minsk, Podolia and Volynia), up to January 1st, 1915, the number of peasant households (*dvor*) which have applied for appropriation reached 2,719,074; the appropriation had been completed in the case of 1,979,241, with 13,884,959 dess. of land, or 22% of the members of village communes with 13·9% of their aggregate land.

The reasons which prompted the peasants to apply for the appropriation of their holdings were very different. Among the applicants one could see both the best farmer of the village and the nominal peasant, who earned his living in town or factory and for whom the land was rather a drag than an asset, which he either neglected or let on lease to a neighbour, being unable, under the old law, to dispose of it otherwise. The real peasant farmer appropriated his land, generally, with a view to enclosure; the nominal peasant or one whose holding was too small and who preferred to emigrate, with a view, mainly, to sale. The effect of this combination of heterogeneous elements of the peasantry among those who secured title deeds for their land was that the average sizes of appropriated holdings were considerably smaller than those of holdings generally. Thus, the average holding per family in the village communes, in 1911, was 10 dessiatins, while the appropriated holdings, from 1908 to 1913, averaged, according to periods, from 6·44 to 7·81 dessiatins. This difference in the sizes of holdings may be partly accounted for by the fact that appropriations of land were much more common among the former serfs than among former State peasants, whose holdings were generally much larger, and among whom the communal system of tenure displayed much greater vitality. That vitality, to a considerable extent, was the direct consequence of larger holdings, which enabled the State peasants, for the time being, to continue farming on old lines. Partly, it was due to the fact that the great bulk of State peasants were concentrated in localities outside the black-earth belt, where farming was more or less a part-time occupation, other earnings supplying most of the peasants' income. Accordingly, the stimulus to separate from the commune was stronger among the former serfs, than among the better-off State peasants, and this could not fail to reduce the sizes of appropriated holdings below the average for all groups of the peasantry. Generally speaking, it would appear that the two extreme groups among the members of village communities, namely, those possessing most land, on the one hand, and those with least land, on the other, were most numerous among the peasants who secured title-deeds for their holdings, and that the peasant middle-class was relatively the least numerous. Since, however, no figures for the whole country, containing such particulars are available, all data referring to separate localities and covering only a small part of the area, such general conclusions must be made and accepted with extreme caution, rather as indications of possible tendencies, than as statements of facts.

Apart from the appropriations carried out in virtue of the Ukaze of November 9th, 1906, there was the issue of certificates (*udostoveritelny akt*) to those members of the villages, in which communal tenure was abolished by the Law of June 14th, 1910, who applied for the formal appropriation of their holdings. The number of peasant holdings for which certificates of appropriation were issued from July, 1910, to the end of 1913, was 317,800, involving 1,918,600 dessiatins of land.[1]

By the end of 1913, about 2,500,000 peasant family holdings have been finally appropriated, while a number were still in course of appropriation. Within the short space of some seven years, the rural commune ceased to exist, and was replaced by individual ownership of land, duly confirmed, in over one-quarter of the villages with communal tenure within the area of land-settlement.

The appropriation of holdings in perpetuity by individual members of rural communes, in virtue of the Ukaze of November 9th, 1906, and of the Law of June 14th, 1910, however, was only the first step in the transformation of peasant tenure envisaged by the land-settlement legislation. The next step, which completed the process of transformation, was the actual consolidation and enclosure of the holdings concerned, which, according to the above mentioned two enactments, was optional, but later, by the Law of May 29th, 1911, was made compulsory in all cases of appropriation of land. While, taken in itself, the appropriation of holdings did not involve any radical alteration in the conditions of tenure and cultivation, the consolidation of the holding into an *otrub* or its enclosure in

[1] *Q.v.* Oganovsky, op. cit.

the form of a self-contained *hutor*, brought about a complete transformation of both, and changed the face of the Russian village entirely.

The development of the new system of tenure, or of modern individualistically-organized peasant farming proceeded with great rapidity generally, though the rate of its growth varied considerably according to locality.

On the eve of the Revolution, on January 1st, 1917, the position with peasant enclosures, according to the latest available statistics, was as follows: [1]

Regions	Number of Enclosures	Area in dessiatins (dessiatin = 2·7 acres)	Average size of Holding (dessiatins).	Per cent of Holdings enclosed
Northern	9,627	173,220	17·9	2·7
North-Eastern	17,204	259,996	15·1	1·7
Petrograd	82,868	837,702	10·1	17·3
Moscow	173,519	1,389,466	8·0	8·3
Western	88,173	909,531	10·3	16·0
Ukraine	152,517	858,192	5·6	12·9
South-Western	84,046	470,465	5·5	7·5
Central Agricultural	130,477	877,263	6·1	7·9
Middle Volga	126,995	1,143,262	9·0	10·3
Eastern	106,123	2,272,068	21·4	16·2
New Russia	203,220	1,842,761	9·0	29·2
South-Eastern (exclusive of Kuban)	47,872	753,427	15·7	—
Caspian (Astrakhan only)	5,824	182,569	31·3	8·9
Total	1,228,964	11,336,512	9·2	10·7

By the end of 1916, no less than 10·7% of the total number of peasant families in European Russia have enclosed their holdings and become economically independent of the commune, while the area of land actually enclosed in the form of either *hutor* or *otrub* reached 9·5% of the total area of peasant allotment land (*nadiel*).

Considering that these results have been achieved in the course of a single decade, of which the last quarter passed under war-time conditions, seriously hindering and restricting the movement, they cannot fail to impress the student as striking. Indeed, in themselves, they are proof of the fact that, at the time of the inauguration of the agrarian reforms by Stolypin, the Russian countryside was already ripe for individualization and only awaited the opportunity of breaking away from the bonds imposed by communal tenure and the open-field system. To a greater or lesser extent, this applied to the whole country, though, as mentioned

[1] Based on figures given in an article on "The Forms of Tenure" by P. Pershin, in the collection of articles on land settlement *O Zemlie*, Moscow, 1921.

before, and as shown by the wide range of variations in the percentages of enclosed holdings in separate regions, the new agrarian legislation met with more widespread response in some parts of Russia than in others. Generally speaking, the more or less rapid extension of the enclosure movement depended mainly on two factors, namely, on the degree to which peasant farming in a given locality was commercialized, on the one hand, and on the familiarity of the local peasantry with the advantages of enclosed holdings, as compared with open fields, on the other. The latter consideration was extremely important, since, being by nature diffident of any theory, but possessing a very keen eye for practical improvements, the Russian peasant could not be expected to revolutionize his system of tenure until he had seen for himself the advantages of the alternative offered. It is not surprising, therefore, that certain parts of Russia were more prepared to take advantage of the new legislation, as soon as they were open to them, while other localities lagged behind, gradually getting into the stride, in the wake of a few pioneers who made the superiority of enclosed holdings clear to their neighbours. It will be remembered that, in the Western provinces of Russia, from Petrograd and Novgorod down to White Russia, the Ukraine and the South-East, the enclosure movement, encouraged by the example of the neighbouring Baltic provinces and Poland and of German, Czech and Esthonian settlers, as well as by the surviving *hutors* of the Ukrainian Cossacks, dated from long before the Stolypin legislation. In these localities, the peasants could see for themselves the technical advantages of enclosures on concrete, tangible and thoroughly well-known examples. Moreover, as I have tried to show in the part dealing with the agricultural geography of Russia, the Western regions of the country are among the most commercialized districts, where the peasant farmer is more closely dependent on the market than in the Eastern parts. Here, indeed, the conditions were more favourable to the extension of enclosures than in most other parts of the country. Accordingly, the percentages of the area and of families involved in the enclosure movement in the West were generally above the average for the whole country. Thus, the percentages of the area of peasant allotment land actually enclosed were: 15·6 in the region of Petrograd, 17·6 in the Western region, and 11·1% in the Ukraine. Only the South-Western provinces, owing to special local conditions, which prevented the growth of enclosures, were below the average, with 7·6% of the land enclosed. Here, indeed, many reasons made the development of the movement particularly difficult. The holdings were the smallest in the country, and a complicated system of legal servitudes, unknown in Russia proper, had to be done away with before the enclosures could take place. Some of the village lands were held on lease by sugar factories, which was another complication. Moreover, the relatively high standards of cultivation practised by the peasants of these provinces, also proved a serious obstacle to the consolidation of holdings, necessary for the radical re-organization of the system of tenure.[1] Accordingly, only in the province of Volynia, where conditions with regard to the sizes of holdings and the systems of cultivation were more favourable, did the percentage of enclosures rise above the average, reaching 11·9% of the area; in the province of Kiev it was 7·7, and in that of Podolia it did not exceed 1·9%.

Not less characteristic for the West of Russia, than the high percentages of the area enclosed, is the principal means by which enclosures have been carried out, which, more than anything else, points to the extent to which the country was ripe for the change. In fact, throughout the Western half of Russia, by far the greater part of enclosures have been effected by common consent of whole village communities, the whole of the village land being broken up into compact holdings by the vote of the village assembly. The actual percentages of enclosures carried out in this manner (*razverstanie*) reached 83·9% in the region of Petrograd, 91·5% in the Western region, 72·8% in the Ukraine, and 75·5% in the South-West. The percentages of individual enclosures, mostly against the will of the majority of the peasants in the village concerned, on the other hand, were very low. A study of the figures for the separate provinces is interesting, because it supplies striking proof of the extent to which the popularity of enclosures in the individual provinces of every region of Russia depended on the situation and economic characteristics of the locality. The more the locality was developed economically, and the more commercialized was its agricultural industry, the more common was enclosure of holdings by whole villages, as compared with that effected by individual peasants.

The centre of Russia, round Moscow, while economically possessing the characteristics essential for the successful development of peasant enclosures, was hindered in its evolution towards the individualization of land tenure by the strong communal tradition of the peasantry, on the one hand, and the practical absence of self-contained peasant farms in the provinces concerned, which could teach the rural population the advantages of enclosure, on the other. The enclosed farms on which peasant-buyers were

[1] *Q.v.* A. Koefoed, op. cit., p. 128.

settled by the State Peasants' Bank, mostly since the beginning of the current century, were, indeed, almost the first concrete examples of this kind of settlement in Central Russia. Accordingly, in the region of Moscow, with the exception of its Western and Northern provinces, which adjoined localities with widespread enclosures, and the Central Agricultural region, the percentages of enclosures were relatively low, namely, 8·4% of the area in the region of Moscow, and 6·8% in the Central Agricultural. The same applied to the Middle Volga region, with the exception of the province of Saratov, where the percentage was very high (22·1%). As distinguished from the West, Central Russia generally had a predominance of individual enclosures over those effected by common consent of the whole village communities concerned, a sure sign of the movement not yet having achieved the extent of popularity which it possessed in the Western parts of the country. While in the region of Moscow the percentage of enclosures by common consent was still very high (72·8%) in the Central Agricultural region it was 37·0, and in that of the Middle Volga it did not exceed 28·9. In the two latter regions, 53·2 and 65·1% respectively of the number of enclosures were carried out by the request of individual peasants asserting their individualism against the opposition of the age-old tradition of communal tenure, which was still strong among the bulk of the peasantry. While, in this part of the country, the enclosures were making headway, the movement started somewhat later than in the West and developed more slowly.

In the North and North-East of Russia conditions generally were such that the enclosure movement could not be expected, for generations to come, to assume large proportions. Economically and socially, the need for the transformation of the system of tenure, so acutely felt in the rest of the country, had not yet become a vital problem in these localities, with their sparse population, relatively very large holdings, and generally primitive conditions of isolated economy. Besides, these provinces had a very large proportion of former State peasants, among whom strong communal traditions, coupled with extensive holdings, were a serious obstacle to the individualization of tenure. It is not surprising, therefore, that in the Northern region only 2·2% of the peasants' land was enclosed, while in the North-Eastern enclosures accounted only for 1·6% of the land.

The highest percentage of peasant enclosures, however, was not in the West, but in the Southern and Eastern parts of the black-earth belt, with their extensive production of cereals for the home and foreign market. The commercialization of peasant farming in these regions assumed different forms from those characteristic of the West, with its relatively intensive systems of cultivation and well-developed dairy farming and flax growing, but it certainly contributed to the growth of individualism and initiative among certain groups of the peasantry, and thus created conditions favourable to the development of enclosures. Stimulated by local conditions, enclosures in the black-earth steppes developed with great rapidity. By the end of 1916, in New Russia 27·2% of the peasants' land was enclosed. In the Eastern region and the Southern part of the region of Middle Volga, the peasants of the two adjoining provinces of Samara and Saratov have enclosed 28·1 and 22·1% respectively of their land. The characteristic form of enclosure in the black-earth steppes was the *otrub*, the homesteads of the peasants remaining in the villages, while their holdings were consolidated into compact plots. The *hutor*, or detached self-contained farm, was far less popular in these parts of the country than in the Western provinces and the centre, mainly on account of the shortage of water which, especially in the Eastern part of the belt, caused the concentration of homesteads in the immediate vicinity of the available sources of water supply.

The Eastern and South-Eastern outskirts of European Russia, from the province of Ufa down to Astrakhan, and from there, through Stavropol, to the Don, were relatively little affected by the enclosure movement, for which the necessary conditions were still more or less completely lacking. In the province of Ufa, the enclosures involved 5·5% of the peasants' land; in the province of Astrakhan, 7·7%; in the Don Territory, 1·2%. The percentage of area enclosed in the province of Stavropol cannot be ascertained, since it was not included in the 1905 Survey of Landed Property, but the number of enclosures was considerable, reaching 34,903. Since, however, 93·6% of the enclosures were effected by individual peasants, only 6% being accounted for by the enclosures of whole villages, it would appear that the movement had not yet gained a strong foothold in the masses of the peasantry of Stavropol.

The results achieved by the enclosure movement by the end of 1916, after ten years of development, are shown graphically on map No. 3.

That the movement was really the beginning of a rapid transformation of the Russian countryside, in which modern organization asserted its triumph over the survivals of an old system of tenure which did not fit into the whole scheme of capitalistic society, was generally conceded, even by opponents of the Stolypin reforms. The opposition, as far as it was based on considerations of policy and expediency, as distinguished from the mystical faith in the commune as an instrument of social transformation, characteristic of the *narodniki*, pointed out, in the first instance, to the dangers with which the freedom of disposal of their holdings threatens the peasantry as a class. The critics of the reforms, in their books, in the Press and on the tribune of the Duma, conjured the vision of the Russian peasantry rendered landless by the greed of the *kulak*, who, supported by the Government's "betting on the strong," would get hold of the land of his neighbours and grow rich by their utter ruin. At the beginning of the reforms, when the real trend of events had not yet had time to show itself clearly, it was expected that the first to make use of the new legislation would be the richest peasants or those who live by usury and exploitation. Yet, the actual development of the movement proved that, as a rule, the exaggerated fears of the opponents of the reform have not been justified by events. That the Government, indeed, based its policy on the stronger and more enterprising peasant farmers, was perfectly true, and freely acknowledged by the authors of the legislation themselves. But the progressive, energetic peasant farmer need not necessarily be either a relatively rich man, or a *kulak*, exploiting his neighbours; it may even be said that, in the main, it was the peasants with average-sized holdings that accounted for the majority of pioneers of improved methods. To keep above the poverty line, they were more urgently in need of raising the yield of their holdings than any other group of the peasantry, and it is, probably, the common experience of those intimately acquainted with the Russian countryside that it is among them that improvements found the readiest reception. The experience of the first decade of enclosures in Russia would appear to prove conclusively that, on the whole, it was this group of peasants which responded most readily to the new legislation, since, in comparing the average sizes of enclosed holdings in the several regions and provinces concerned with the average sizes of peasant holdings in the same localities, it may be seen that the former, far from being larger than the latter, and thus pointing to the enclosures mostly involving richer peasants, rather tend to keep below the average of all holdings. The table below shows the average sizes of enclosed holdings, compared with those of all holdings, according to regions:

Regions	Average sizes in dessiatins Of all holdings	Of enclosures
Northern	22·4	17·9
North-Eastern	16·0	15·1
Petrograd	11·2	10·1
Moscow	7·9	8·0
Western	9·2	10·3
Ukraine	6·1	5·6
South-Western	5·5	5·5
Central Agricultural	7·8	6·1
Middle Volga	8·3	9·0
Eastern	19·7	21·4
New Russia	9·7	9·0
South-Eastern	—	—
Caspian	—	—

It may be seen that, except in a few cases, in which the average sizes of enclosures are slightly above those of all holdings, the former are generally a little below the average. A study of the figures for separate provinces would disclose an even more general tendency in this direction, regional excesses being due in most cases to the influence on the average of some especially large deviations in separate provinces. It would appear, therefore, that the fear lest enclosures should be taken advantage of mainly by the richer section of the peasantry, at the expense of the less favoured, have not been justified by facts, the average peasants being, as this should be, in the front rank of the movement.

The Stolypin legislation was, indeed, responsible for a considerable volume of transfers of peasant land as between individual members of the peasant class, though, as I have pointed out above, the law took care to prevent both the loss of land by the peasantry as a class and its concentration in the hands of the few to the detriment of the many. The information concerning the actual turnover of peasant land since 1906 is rather scarce, but some of the available figures throw a good deal of light on the development, as well as on its principal causes.[1]

The extent of the movement may be judged by the fact that, up to the end of 1912, a total of 2,153,400 dessiatins of peasant land has been sold by 581,000 peasants, the average area involved in the transaction being 3·71 dessiatins. From the size of the areas sold it may be concluded that the sellers belonged, as a rule, either to the poorest elements among the peasantry, who earned their living mostly in outside occupations, and were often all the better for being able to get rid of their land at a price, or that the sales involved parts of holdings. Statistics collected by the Government and the Zemstvos in a number of provinces, give some idea of the

[1] For the figures given below the writer is indebted to the article of Prof. N. Oganovsky already quoted.

actual reasons which prompted the peasants to sell the whole or part of their holdings.

According to official data for the same period, referring to five provinces, the distribution of sellers according to reasons responsible for the disposal of the land, was as follows:

	Percentages
Sales by peasants who have been engaged in outside occupations and did not farm their land	39·8
Sales by emigrants to Siberia	14·3
Sales by peasants who settled on land privately bought	5·5
Sales due to failure of crops	1·8
Sales due to poverty	17·0
Sales by drunkards and profligates	3·4
Sales by invalids, peasants too old to work or single	9·0
Sales for other reasons (mainly with a view to improve the cultivation of the rest of the land)	9·2
	100·0

In the five provinces concerned, as well as in the provinces of Simbirsk and Novgorod, for which statistics have been collected by the Zemstvos, the largest group consisted of nominal peasants, who, while working in towns, factories or other occupations, sometimes thousands of miles away from their village, could not, under the old agrarian régime, dispose of their holdings otherwise than by leasing them to some other members of the commune. Now, at the first opportunity, they broke their bond with the land completely, the holding passing, in most cases, to the peasants who actually cultivated it before on lease. The change brought about by such sales was merely formal. If, indeed, it resulted in the final "proletarization" of the seller—the bugbear of so many opponents of the reforms, both "populist" and conservative—it did but give a legal sanction to a fact accomplished long since. The advantage was that it effected the transfer of the land from a nominal to a real peasant, living on his land and from it.

Next came the group of sellers who disposed of their old holdings with a view to settling on new ones, either in Asiatic Russia, or on lands purchased in the same or other provinces, mainly through the State Peasants' Bank. Obviously, no objections could be raised against such sales, seeing that the money thus obtained was essential for the success of the new venture. Here, the Stolypin legislation only achieved one of its principal objects.

In the five provinces concerned, these groups accounted for 59·6% of the sellers. In the provinces of Simbirsk and Novgorod they made up 51 and 39·6% respectively.

Sales due to failure of crops and to poverty may be combined, as they actually are in the statistics for the province of Novgorod, since it is the poorer peasants that are driven to such extreme measures by bad seasons. Here one comes across the actual proletarization of a group of the peasantry, which, in the five provinces for which official statistics are available, amounted to 18·8% of the sellers. In the provinces of Simbirsk and Novgorod, according to Zemstvo statistics, the percentages were 14 and 14·2 respectively. In this case, what was taking place, was the gradual disappearance from the village of those among the peasantry, who were, either by reason of the too small size of their holdings or by their own unfitness for farming, or for some other reason, unable to keep head above water as independent peasant-proprietors. Under no known systems of social and economic organization had it proved possible to avoid this going-by-the-board of the unfortunate and the unfit, but it is far from certain that, tied up to his miserable piece of land in the village, the unsuccessful peasant farmer had a better chance than when, having sold out, he could try his hand on some other job. In farming, besides, no man can hope to succeed, unless his heart is fast in his land, whatever his material position or the area of land at his disposal. A man is either born a farmer, or not, and if not, the best he can do is to look for a living elsewhere. In so far as the agrarian reforms involved the choice by a number of people of the latter alternative, they were all for the good, helping to put the right pegs in the right holes. The more so that, with the industrial progress of Russia in the twentieth century no one, able and willing to work, could fail to find more or less congenial employment.

On the other groups of sellers it is hardly necessary to dwell.

The characteristic of the purchasers, which supplies the other side of the picture, can be gathered from the statistical data referring to the number of purchasers for every 100 sellers. According to official figures, the purchasers, up to the end of 1912, numbered 554,500 which, to 581,000 sellers, makes 95·6 per 100. In other words, on the whole, the statistics contain no indication of any engrossing or concentration of land among a small section of the peasantry, which was so much feared by the opponents of the reforms. If anything, the figures would appear to support the impression made by the agrarian evolution of Russia before the war on those who observed it at close quarters, that it signified the triumph of the genuine peasant farmer over those whose heart, though they lived in the country, was not in the land. Once again, as in the past, the word peasant, with the progress of the reform, would mean farmer, while in the decades preceding the Stolypin legislation it lost its original meaning and became the formal designation of nothing more than a civil estate.

Yet, necessary and even vital for the country as were the agrarian reforms of Stolypin, and the enclosure movement in particular, the latter had one very weak spot, which, though characteristically seldom mentioned by the opponents of the legislation, was a great potential source of danger for the whole social system of Russia.

While, from the economic view-point, the policy of the Government was that of creating a prosperous peasantry out of the available material, politically it aspired at building up a strong class of sturdy, independent peasant farmers, on whom, as a conservative force of law and order, it could in future rely in its struggle against the onslaught of the revolutionary elements. The process would take time and, in the course of its gradual evolution, would be bound unavoidably to produce a considerable amount of social friction in the countryside. Yet, with the ultimate goal looming large and certain in its eyes, the Government, who only just succeeded, with great effort, in putting down an extensive revolutionary outbreak, decided to face the difficulties and dangers of the period of transition. The lot was cast, but some time was bound to elapse before the prospective "strong" peasants became really strong enough not only to act as a bulwark of law and order, but even to feel secure in their own position, so recently occupied. Unavoidably, appropriations and enclosures of their land by individual peasants were accompanied by a certain amount of friction with those still remaining in the commune, especially in those parts of the country in which communal traditions were still strong and where enclosures gained ground slowly, by individual initiative, in opposition to the majority.[1] The countryside, still seething with discontent, with the revolution of 1905, the agrarian disturbances and the measures of repression still alive in its memory, was rent asunder by another conflict, which only awaited an opportunity to express itself in violence. Since it crushed the revolution of 1905, the Government was strong enough to prevent any trouble on this score. With the enclosure movement gaining ground with considerable rapidity, in a generation the peasant owners of enclosed holdings, whatever their original relations with the other members of the village, would have grown strong enough, both economically and numerically, to feel secure against any aggression, if, by that time, any incentive to it were to survive. In the meantime, however, the pioneers of enclosures, faced with numerous enemies,

[1] The position in this respect in various localities is very well analysed by P. Pershin in his article quoted above.

were still in a small minority, barely reaching 10% on the eve of the revolution of 1917. Economically, the first years on their new holdings were a time of hard uphill struggle. All went well until they had the support of the State to fall back upon in case of any hostile acts on the part of the majority, and, had it not been for the Great War, which made the Revolution possible, by now they would probably have doubled in both numbers and resources. But Fate decreed otherwise. The first onslaught of the revolution, by sheer force of numbers, submerged and destroyed the enclosures of the preceding decade. But the great movement, once started, could not be killed by force: there is, indeed, no more striking proof of its essential soundness and inherent vitality, than the fact that, by 1921, quite spontaneously, it started again throughout the country, forcing its legal recognition on the reluctant Communist Government in the Agrarian Code.

CHAPTER VI

INTERNAL COLONIZATION AND EMIGRATION BEYOND THE URALS

THE remodelling of peasant tenures, which was the outstanding feature of the reforms associated with the name of Stolypin, though essential to enable the peasants to raise the standards of cultivation, could not provide a complete solution of the agrarian problem. The "land-hunger" though it was not the only cause of the miserable condition of the Russian peasantry, was real enough for a large section of the rural population of Russia, especially in the more congested parts of the country, such as the Agricultural Centre, the Ukraine and the South-West. In a large number of cases, sometimes involving whole villages, the holdings were too small to enable their owners to exist on their yield under any practicable system of farming. Indeed, as the Department of Agriculture put it in its Memorandum to the Special Committee on the Needs of the Agricultural Industry, in 1902, in some villages "the holdings were so small as to exclude every possibility not only of the adoption of improved systems of cultivation, but even of practising any regular cropping at all." [1] With regard to such cases the Department of Agriculture insisted on the necessity of special measures, of which the principal were assisted emigration to new regions and the development of the activities of the State Peasants' Land Bank in the settlement of peasants on privately purchased holdings, mainly in the less congested localities. These two expedients had to play an important part in any scheme of effective agrarian reform by helping to relieve rural congestion in the more densely populated districts and to bring about a better distribution of population throughout the country. Both had achieved conspicuous development under Stolypin, as parts of his general scheme of agrarian reconstruction.

A. Land Settlement through the State Peasants' Bank

The first efforts to assist peasants in the purchase of land by the grant of long-term credits out of public funds could be traced

[1] V. Rosenberg, *From the Chronicle of the Peasant Problem* in Manuilov's *Essays on the Peasant Problem*, Vol. I, p. 182.

to the 'seventies of the last century. It was then that three Zemstvos, namely those of the provinces of Novgorod, Tver and Yaroslav, established special funds for this purpose. Their example was followed by a number of other Zemstvos, some of which developed their activities in this direction to a very considerable extent. The machinery which helped the peasants to finance their purchases of land included also the joint-stock land banks, whose first appearance in Russia also dated from the early 'seventies. But the problem of increasing the area of land in the peasants' possession assumed such importance very soon after the Emancipation, that the Government found it necessary for the State to take a hand in this movement, with a view to assisting it and, at the same time, controlling its course. This brought about the foundation, in 1882, of the *State Peasants' Land Bank*, which began its activities in 1883.

The original idea of the promoter of the Bank, Bunge, the then Minister of Finance, was to make of it an institution with very far-reaching objects in the sphere of agrarian policy, but the Council of the Empire, when the draft statute came before it for discussion, assumed a more cautious attitude, and made of the Bank a simple land mortgage establishment, distinguished from other institutions of a similar kind only in its being owned and managed by the State, and its custom being limited to peasants. According to clause 1 of its statute, the Bank was established "with the object of helping the peasants in the purchase of land offered for sale." The Bank was authorized to grant advances on approved transactions either for 24½ years at 8½% p.a. inclusive, or for 32½ years at the lower inclusive rate of 7½%. The proportion of the purchase price advanced was generally fixed at 75%, though in exceptional cases, at the discretion of the management, it could be raised to 90%. Credits were available to whole village communities (*selskoie obshchestvo*), to peasant partnerships (*tovarishchestvo*) and to individual peasants. A rigid limit of credit was fixed, amounting to 125 roubles per male soul under communal tenure and to 500 roubles per family (*dvor*) if the land was held in perpetuity. The latter sum applied also to all individual purchasers. The funds necessary for the operations of the Bank were raised by the issue of 5½% Mortgage Bonds of the State Peasants' Land Bank redeemable by drawings.

At the beginning, while launching this new venture, the Government proceeded with great caution and endeavoured to avoid involving itself into too heavy commitments. At the time, indeed, when the Peasants' Bank was being planned and established, Russia was thoroughly overhauling her financial system, which required complete remodelling on modern lines, and the currency problem, aggravated by the financial strain of the Turkish War of 1877–8, was in urgent need of attention. This concentration on financial and monetary reconstruction had the effect of checking the development of measures which, in the nature of things, could not be expected to repay the necessary outlay in the near future. The Peasants' Bank, though designed to improve the economic position of the peasants and thus eventually to benefit the State in every respect, could not be run on strictly commercial lines and, if the situation required, had to be worked even, for a time, at a loss. Accordingly, the Government was not disposed to extend the activities of the institution beyond certain narrow limits. Though, indeed, Bunge, as the actual creator of the Bank, during his tenure of office, emphasized the social aspect of its work, as an institution called upon to perform important public functions, and looked upon its immediate returns as a matter of relatively little moment, his successor, Vyshnegradsky, in his attitude to the Bank, displayed the qualities rather of a financier than of a statesman. While, in spite of the limitations imposed on the Bank by its statute, Bunge had succeeded to a certain extent in actively helping through it the poorer elements of the peasantry, who most needed assistance, in the later 'eighties, under Vyshnegradsky, the institution was put on a purely commercial footing and required, first and foremost, to avoid possible losses and to secure profits on its transactions. Accordingly, the activities of the Bank, which had reached a considerable development during the first few years of its existence, while it was controlled by Bunge, underwent a change in their extent and their character since 1887, under the direction of Vyshnegradsky. The advances which, under Bunge, had averaged 85 to 90% of the purchase price of the land, were greatly reduced, and, in 1889–92, did not exceed some 66 to 75%. This either prevented the poorer peasants from availing themselves of the assistance of the Bank, or forced them to have recourse to other sources of credit for a large part of the necessary amount, mostly at high rates of interest and on generally far less favourable conditions. The Peasants' Bank, which, in 1883–1886, had made considerable progress towards its ultimate goal of becoming *the* peasants' bank and acquiring effective control over the landed estate market, in the course of the following few years lost much of what it had hitherto achieved in that direction. Indeed, while, when, after the first trial year, during which its operations were very small, the Bank got into its stride, in 1884–1886, the peasants bought with its financial assistance 277,000 dessiatins a year on the average, from 1887 to 1895, the average was reduced to 175,000 dessiatins. The greater part of their purchases during this period was financed either by the joint-stock land banks, whose charges were more moderate and whose procedure was simpler, or by various other means such as the special funds of the Zemstvos, emergency capitals of the village communities or private loans. In some cases, peasants had recourse to credit institutions which, by their nature, were not intended for this kind of business, but found it profitable and convenient. Thus, for instance, in 1896, in the province of Chernigov, three Municipal Banks alone made advances to the peasants for the purchase of land to the total amount of 390,000 roubles, while the advances of the Peasants' Bank in the same province in the course of that year did not exceed 190,000 roubles.[1] Some idea of the extent to which the Peasants' Bank, during the first decade of its existence, was still far from playing a controlling part on the land market, may be obtained from the following rough calculation. Between 1882 and 1892, the total area of land privately owned by peasants outside the *nadiel* in 47 provinces of European Russia, had increased by 5·55 millions of dessiatins, from 10·70 millions to 16·25 millions of dessiatins. During practically the same period, from 1883 to 1892, the area purchased by the peasants through the Peasants' Bank amounted to 1·89 millions of dessiatins, or about 34% of the total increase. To attain a position of control on the landed estate market, the Bank had still to extend its activities greatly. During this initial period of its work, the Peasants' Bank, moreover, was precluded from dealing in land on its own account, which alone would have enabled it to intervene on the market by the competitive offer of land at reasonable prices. The institution, which was in later years to play so important a part in the agrarian evolution of Russia, was still groping more or less in the dark, now swayed in one direction, now in another. Though, in the course of this experimental stage of its existence, the Bank had helped the transfer to peasants of over 2·4 millions of dessiatins (about 6·5 millions of acres), to the aggregate value of about 108 millions of roubles, these figures appeared utterly insignificant in comparison with the problems with which it was faced. When at the close of the first decade of its activities, the Council of the Peasants' Bank submitted to the Government a draft amendment of its original statute, in the accompanying Memorandum it frankly admitted that " the results of the first nine years of the

[1] V. Rosenberg, op. cit., p. 164.

Peasants' Bank work must be described as unsatisfactory or, at any rate, practically negligible."

In 1892, the appointment of Witte as Minister of Finance inaugurated a new epoch in the financial history of Russia, and in the course of the next few years, though, until 1895, no radical alterations had been effected in the constitution and scope of activities of the Bank, certain improvements were made with a view to extending its work. In the first instance, in 1894, the Bank was assigned a working capital, which was to be accumulated gradually out of annual grants from the receipts of the Treasury on account of redemption payments, from 1895 onwards. The annual charges of the Bank were reduced from 8½ and 7½% to 6½ and 5½%, the terms of repayment being extended from 24½ and 32½ years to 26½ and 38½ years respectively. But it was not until the publication, on November 27th, 1895, of its new statute, embodying most of the experience gained in the course of the preceding period, that the Peasant Bank was enabled to extend its activities considerably.

The *Statute of November 27th*, 1895, introduced important changes in the Bank's functions. The most important of these was the authorization of dealings in land on the Bank's own account, though the extent of the purchases was not allowed to exceed the amount of the institution's own capital. Although the limits thus set to the initiative of the Bank in the sphere of land-settlement, as well as its possibility of active intervention on the landed estate market, were very narrow, the recognition by the statute of such dealings as part of the regular business of the Peasants' Bank signified a new departure. The Bank was also permitted to grant loans on mortgage of land originally purchased without its assistance. This enabled the purchasers, if their existing financial arrangements were less advantageous to them than those provided by the Bank, to transfer their mortgages to the latter. The advances, which under the old statute were limited to 75% of the purchase price, and only in exceptional cases could be increased to 90%, were raised to the latter figure. Special valuation of land was substituted for the former system of normal prices, which were entirely at variance with the real land values. The rigid norms of advances per male soul or per family, which took no account of local conditions, were replaced by local maxima, based on the principle that purchases through the Bank must not, in any individual case, exceed the limits of an area which can be cultivated by a peasant family without hired labour. These limits were fixed on a fairly liberal scale, from 25 to 40 dessiatins per family in the centre of Russia to 40–60 and in some cases even as much as 75 dessiatins on the outskirts of the country, in regions of extensive farming.

The publication of the new statute was followed by a considerable extension of the activities of the Peasants' Bank, as well as by certain modifications in its policy. While, in the 'nineties, the Government was engaged in efforts to prevent the proletarization of the peasants by strengthening the village commune and its hold on the individual, some inklings of the coming change in the attitude of the authorities to the awakening of the individualist spirit in the countryside were beginning to appear here and there. The need of creating, by means of gradual evolution, a peasant class with a stronger instinct of individual ownership and a general middle-class outlook, was growing. Thus, while discussing the policy of the Peasants' Bank in 1889, the Council of the Empire expressed the view that, while the Bank is not called upon to assume the task of general land-settlement of the peasantry, "its real task, though more modest, is by no means less important. If it were to help the development of a class of small landed proprietors among the peasants, it would not meet with its present difficulties in dealing with defaulters." . . . "By financing the better-to-do among the rural population, the Bank would render a real service to the country by the creation of a class of small proprietors who appear to be a source of prosperity for the country and a strong bulwark of law and order." [1] The same idea found expression in the draft of the new statute prepared by the Council of the Peasants' Bank in 1892. [2] Though the draft submitted by the Council of the Bank had not been passed, and the new statute did not contain any specific provisions concerning the mode of tenure, the view that the Bank was called upon to help building up a class of independent peasant proprietors, mostly from the ranks of the more prosperous peasants, found its way into the actual policy of the institution.

The second period of the history of the Peasants' Bank, inaugurated by the publication of its new statute at the close of 1895, lasted for exactly ten years, from 1896 to 1905, and was marked by a considerable increase in its activities. Indeed, in the course of these ten years the Bank had helped to transfer into the hands of the peasantry a total of 5,864,245 dessiatins of the aggregate value of 536,747,000 roubles: an average of 586,400 dessiatins a year as against 175,000 dessiatins a year in the preceding decade. The average number of transactions, which, in 1883–1893, was

[1] Y. Skalon, *The Peasants' Bank and its Defaulters*, Manuilov's *Essays*, Vol. II, pp. 14–15.
[2] *Ibidem*, pp. 55–8.

1,148 a year, increased to 4,525 in 1896–1905, and the amount of advances rose from 6,338,000 to 40,994,000 roubles a year. The influence of the Peasants' Bank as a controlling factor on the landed estate market also increased considerably. While, in 1883–1892, the Bank was responsible only for about one-third of the increase in the area privately owned by peasants, in 1893–1905 its share in the extension of peasant land-ownership reached approximately 75%. During this period, the peasants increased the area in their possession by 8·49 millions of dessiatins, of which the Bank had contributed to the extent of 6·39 millions of dessiatins. Since the later 'nineties, indeed, the Peasants' Bank became a power of considerable national importance, and its influence became increasingly felt in the life of the Russian countryside.

The policy of the Bank during this period of its work was that of fostering the growth, from among the better-to-do elements of the peasantry, of a class of independent peasant proprietors. This it did by using its discretion in the choice of borrowers and by keeping the advances considerably below the 90% limit set by the new statute. Besides, while, in the course of the preceding period, credits have been granted mostly to whole village communities, now the Bank began to give preference to peasant partnerships, mostly formed by the better-to-do peasants, and to individual purchasers. It may be said that, at this stage, the Bank had definitely decided to follow the views of the Council of the Empire, which I have quoted above, and, turning away from the existing agrarian problem as such, in the sense of improving the economic position of the peasant class generally, to concentrate on the creation of a strong peasant middle-class from the available human material. This policy was strongly condemned by most writers on the agrarian problem for its bias in favour of the more prosperous peasants, and its comparative neglect of the poor, whose needs were most pressing. This wholesale condemnation, however, was hardly deserved. The conduct of the Bank should rather be considered as reflecting one of the stages in the gradual evolution of ideas and policy, which prepared the way for the thorough re-organization of peasant tenures and the whole scheme of agrarian reconstruction which had eventually to be carried out, and which was actually inaugurated by Stolypin. At the close of the last century it was generally recognized that the existing condition of the countryside and of farming was unsatisfactory and required improvement, but how such improvement could be achieved was far from obvious. The village commune and open-field farming, strongly condemned, on technical grounds, by a large body of expert opinion, found enthusiastic support among the more articulate sections of the public. Great statesmen of the conservative school, whose views carried weight in Government circles, with Pobedonostzev as their learned protagonist, warned the Government lest, by any rash step, they might bring about the dissolution of the commune, which would threaten the whole social and political structure of Russia with dire consequences. Faced with these conflicting opinions and an economic situation whose prospects, in the 'nineties, were none too clear, the Government was, not unnaturally, disposed to temporize, and its attitude to the problem was wavering and uncertain. The last decade of the nineteenth century represented a distinct stage in the evolution of Russian agrarian policy; a stage, during which, in a tentative way, the separate elements of a far-reaching programme of agrarian reconstruction were being evolved, which, later on, the genius and the courageous will of Stolypin had welded into a finished comprehensive whole. So it was with the Peasants' Bank; and the same can be said of Siberian colonization, as the reader will see in a later part of this chapter; while in other respects during the last few years of the nineteenth and the first years of the twentieth century, down to 1906, the way had been cleared for the reformer.

It is, however, the third period of the history of the Bank, during which it was called upon to play a prominent part in the scheme of Stolypin's agrarian reforms, that, beyond comparison, was the most brilliant in that institution's career. The urgent necessity of extending the work of the Bank, clearly emphasized in the Memorandum of the Department of Agriculture quoted at the beginning of the present chapter and referring to 1902, was generally recognized at that time. It became obvious that the Bank had now to come forward as an active organ of land-settlement and internal colonization, for the benefit, first and foremost, of those peasants who were most in need of assistance. As a credit institution, its functions, though greatly extended since 1896, were by the nature of things of a passive kind, while the authorization to buy land on the Bank's own account for the settlement of peasants, important as it was in principle, in its practical application was confined within limits so narrow as to reduce it to almost complete insignificance. In fact, at this stage, these ventures into active land-settlement policy could be described as experimenting on a small scale. Now, with the peasantry throughout Russia in a state of revolt, the time for experimenting was over, and the Bank had to be called upon to do its bit in the solution of the agrarian problem to the full extent of its capacity. Accordingly, by the *Ukaze of November 3rd*, 1905, which inaugurated this new phase in the development of the Peasants' Bank, it was authorized to purchase land on its own account, for the settlement on it of peasants, without restrictions as to the expenditure involved, and to meet its outlay under this head by increasing its issues of mortgage bonds. The Bank was also permitted, in the case of peasants with very little or no land, under certain conditions to grant loans up to 100% of the purchase price of the land. With a view to enabling the Bank to extend the reserve of land at its disposal, the *Ukaze of August 26th*, 1906, provided for special facilities for the transfer to it of estates mortgaged in other credit institutions. The reserve was also increased in a direct way by the *Ukaze of August 12th*, 1906, which provided for the transfer to the Bank, for the settlement on them of peasants, of about 1·9 millions of dessiatins of land belonging to the appanages. The *Ukaze of August 27th*, 1906, authorized the sale to peasants, through the Bank, of the available agricultural land belonging to the State in various parts of European Russia, to the total of some 7 millions of dessiatins. The *Ukaze of October 14th*, 1906, reduced the inclusive charges of the Peasants' Bank for its loans to 4½% *per annum*, thus bringing them down to the level of those of the Bank for the Nobility, the more favoured position of whose customers had, hitherto, been the subject of much adverse criticism. The final cancellation of the outstanding balance of redemption payments by the Ukaze of November 15th, 1905, opened the way to an extension of the Peasants' Bank work in another direction, and on November 15th, 1906, it was authorized to grant loans on the security of allotment land (*nadiel*), for the purpose of either buying more land or of improving the methods of cultivation. Indeed, with regard to mortgages of allotment land, the Peasants' Bank was given a monopoly, no other credit institution being permitted to transact business in this particular class of property.

It may be seen that, from 1906 onwards, the Peasants' Bank was placed in a position in which it could really play an important part in the solution of the agrarian problem. This stage of its history was marked by the prominence given to its function as an instrument of active land-settlement policy, in which the needs of the poorer peasant farmers became a matter of special preoccupation. The Bank was called upon to perform a twofold service. In the first instance, it had to extend the area of land in the peasants' possession and to help in relieving the congestion in some districts by the transfer of some of their population to other localities. This, indeed, was its principal function. Its other function consisted in providing concrete examples of enclosed peasant holdings in those districts in which this form of tenure did not exist and its advantages were unknown to the rural population. During this period, individual purchases of land in enclosed holdings were accordingly encouraged in preference to those of whole villages or peasant partnerships, and settlements on compact self-contained farms of the *hutor* type were accorded preferential treatment. Thus, the poorer peasants who could not afford to pay any part of the purchase price in ready money, were granted loans up to the full amount of the price, if the land was to be used as a *hutor*. If the peasant continued to live in the village and did not turn his newly-purchased holding into a self-contained farm, the loan was not permitted to exceed 95% of the price. With a view to enabling individual members of peasant partnerships (*tovarishchestvo*) to enclose their respective shares of the land owned jointly, the statute of the Bank provided, as a condition of collective purchase, that the partnership contract should contain a clause by which the partnership bound itself to allocate to any of its members, whenever requested, his share of the land in a compact plot. The principal direct means by which the Bank encouraged settlement on enclosed holdings, however, was its policy of selling land from its own reserve in conveniently situated compact holdings. During this period, indeed, the Bank has been trading not so much in mere land, as in specific peasant holdings, and the so-called "Bank farms" (*bankovsky hutor*) had a very good reputation and were in great demand. In the central provinces of Russia it was mainly through these holdings that the advantages of enclosures were demonstrated to the peasantry and the enclosure movement stimulated.

The extension of the activities of the Peasants' Bank from 1906 onwards, as well as the great changes in the character and direction of its work, may best be seen from the table below, in which the principal results of the three consecutive periods of its history are compared.

Summary Results of the Peasants' Bank work from 1883 *to* 1912

1. *Area of Land bought through the Bank*

	Village Communes.		Partnerships		Individuals		Total
	1,000 des.	Percentage	1,000 des.	Percentage	1,000 des.	Percentage	1,000 des.
1883–95	935·7	38·8	1,431·8	59·3	44·2	1·9	2,411·7
1896–1905	998·8	17·0	4,727·4	80·6	138·1	2·4	5,864·2
1906–12	1,095·0	14·5	3,536·0	46·7	2,935·0	38·8	7,566·0

2. Amounts of advances to purchasers

	Mil. rbls.	Percentage	Mil. rbls.	Percentage	Mil. rbls.	Percentage	Mil. rbls.
1883–95	32·8	39·8	48·5	58·9	1·1	1·3	82·4
1896–1905	67·7	16·5	332·9	81·2	9·4	2·3	410·0
1906–12	119·7	11·9	354·8	35·5	362·5	52·6	836·9

The most striking feature of this table, apart from the large increase in the turnover of business transacted by the Bank, is the great extension of purchases by individual peasants during the third period. Indeed, from about 2% of the area bought by the peasants through the Bank on the average between 1883 and 1908, the proportion accounted for by individual buyers rose to 38·8% in 1906–12. Here, the definite individualistic turn given to the Bank's work under the Stolypin régime is clearly evident.

The table below shows the growth of the Peasants' Bank work in yearly averages for each of the three consecutive periods of its development.

	1883–95	1896–1905	1906–12
Number of transactions *p.a.*	1,148	4,525	36,651
Area dealt with, *p.a.*, 1,000 des.	185·5	586·4	1,080·0
Value of land bought through the Bank, *p.a.*, mil. rbls.	8·3	53·7	142·9
Amount of advances, *p.a.*, mil. rbls.	6·3	41·0	119·5

It will be remembered that, by the Ukaze of November 15th, 1906, the Peasants' Bank was authorized to grant loans on the security of peasant allotment land (*nadiel*). The business done in virtue of this enactment, which was given legislative sanction as the *Law of July 5th*, 1912, did not, however, assume large proportions, though it increased steadily from 1907 onwards. In the course of six years, from 1907 to 1912, such transactions numbered 4,736 and involved 21,429 dessiatins of land, on which 5,766,000 roubles have been advanced.

A characteristic feature of the activities of the Peasants' Bank during the third period of its work was the development by it of an active policy of land settlement and internal colonization in connection with its extensive purchases of land on its own account on the market, as well as with the transfer to it, for this purpose, of a considerable area of agricultural land belonging to the appanages of the Imperial Family, in virtue of the Ukaze of August 12th, 1906. The land thus acquired by the Bank during this period, along with the outstanding balance of the area acquired between 1896 and 1905 and the relatively insignificant addition of the holdings of defaulters forfeited to the institution, formed a reserve on which, since 1906, the Bank has been able to build up an effective policy of land-settlement.

The movement of the land reserve of the Bank, from 1906 onwards, according to regions, was interesting because it revealed which parts of European Russia still afforded facilities for internal colonization on a more or less considerable scale on the eve of the Great War. Incidentally, the table below shows also how vast was the area dealt with by the Peasants' Bank under this particular head during this stage of its development.

Movement of the Land Reserve of the Peasants' Bank, 1906–1911

Regions	Area available on Dec. 31, 1905	Additions, 1906–11	Disposed of, 1906–11	Available, Jan. 1, 1912
	In dessiatins (1 dessiatin = 2·7 acres)			
Northern	—	3,074	1,654	1,420
North-Eastern	53,607	182,078	97,753	137,932
Petrograd	1,696	105,498	36,791	70,403
Moscow	28,048	324,944	206,363	146,627
Western	27,227	259,452	89,804	196,875
South-Western	9,194	206,915	131,314	84,795
Ukraine	4,951	248,406	190,279	63,078
Central Agricultural	29,424	674,096	450,903	252,617
Middle Volga	42,029	1,183,128	481,341	743,816
Eastern	45,651	1,189,859	388,046	847,464
New Russia	1,659	217,741	174,426	44,974
South-Eastern	—	205,535	69,982	135,553
Caspian	—	130,234	27,585	102,649
Total	243,484	4,930,960	2,346,241	2,828,203

It may be seen that, in European Russia, the East and South-East, from the Middle Volga down to the Sea of Azov, the Caucasus and the Caspian, provided, on the eve of the war, relatively considerable facilities for internal colonization, partly due to the sparsity of their population, as in the case of the Eastern, South-Eastern and Caspian regions, partly accounted for by special causes which placed at the disposal of the Peasants' Bank at that time more or less important areas of land in various localities. The revolution of 1905, which was accompanied by agrarian disturbances on an unprecedented scale, produced a considerable panic among landed proprietors and brought about, in the course of the next few years, a great increase in the sales of landed estate. Indeed, while, in 1903–5, on the eve of the revolution, the landed gentry sold, on the average, about 630,000 dessiatins a year, in 1906 its sales were estimated at 1,342,000 dessiatins, and in 1907 at 1,985,000 dessiatins. From 1908 onwards, the figures diminished again, but in the meantime the Peasants' Bank has been able to build up a considerable reserve of land in the congested agricultural provinces of the central part of Russia which, under normal conditions, could have afforded very little scope for such operations. This explains the relatively large area available for settlement in such localities as the Central Agricultural, the Middle Volga, the Ukraine and the South-Western regions.

With the extension of the land reserve of the Bank, the share of transactions arising out of its policy of active land-settlement in its total turnover of business tended generally to increase.

The progress of the Peasants' Bank work, as a credit institution, financing the purchases of land by peasants, on the one hand, and as a seller of land on its own account for the purpose of land-settlement and internal colonization, on the other, from 1906 to 1913, is shown in the table below:

	Area of Land bought		Value of Land bought	
	From the Bank's Reserve	From Land-owners, with Bank Credits	From the Bank's Reserve	From Land-owners, with Bank Credits
Years	Dessiatins		Roubles	
1906	39,244	482,960	2,755,610	61,430,602
1907	180,148	752,701	20,268,520	100,312,464
1908	324,956	694,058	37,675,070	96,332,508
1909	551,307	675,807	74,882,888	97,572,616
1910	764,771	784,949	94,836,186	109,990,216
1911	679,658	717,979	78,712,040	101,522,580
1912	371,687	545,626	45,073,483	78,845,193
1913	410,118	484,023	45,895,808	73,009,246

During the last two or three years of this period, the activities of the Bank have slackened down to some extent, the decline being especially marked in the volume of purchases direct from landowners, with the Bank's financial assistance. This slackening down, after a period of abnormal activity caused by the fluid state of landed property after the revolution of 1905, was the natural consequence of the return of the country to more settled conditions. By 1913, the turnover of the Peasants' Bank may be considered as having also reached a level which, under existing conditions, should normally be maintained with slight variations until the exhaustion of the available reserve of land. Before the war, the available area in European Russia, suitable for immediate settlement and cultivation, was already limited, and was being rapidly disposed of through the Bank. There remained vast tracts of unreclaimed waste lands, which show well on the map, but cannot be put to any profitable use without large preliminary expenditure. Had, however, land prices continued to rise, and had the general upward trend of the agricultural industry, so pronounced in the years preceding the war, been maintained, extensive areas of such now useless land might have eventually been reclaimed and made available for settlement, as part of the Bank's reserve.

In settling peasants on the land sold to them out of its reserve, the Bank, during this period, did its best to promote enclosures. Accordingly, in 1913, for instance, of the 1,079,328 dessiatins prepared for disposal in the course of that year, only 5·2% were to be sold to whole villages and peasant partnerships, while 38·3 and 43·6% respectively were to be sold as compact enclosed holdings of the *hutor* or *otrub* types; besides, 6·3% of the area went with the enclosed holdings as woodlands and pastures for joint use by a number of neighbouring farms. The balance of 6·6% consisted of the remaining parts of the former estates, which the Bank had broken up for the settlement of peasants, such as country houses, sugar refineries, spirit distilleries, etc., offered for sale to a different class of buyers or temporarily leased by the Bank to tenants. Thus, in 1913, 88·2% of the whole area offered for sale by the Peasants' Bank, was made up into enclosed holdings, and their purchase by the peasants introduced this new and superior form of tenure into many a district in which its advantages have hitherto been unknown.

The development of the individualistic trend in the Bank's work during this period is also evident from the respective shares of the three classes of purchasers, namely village communes, partnerships and individuals, in the purchases of land through the Bank. From 1906 onwards, the evolution in this respect proceeded as follows:

	Communes		Partnerships		Individuals	
	From the Bank	From Land-owners	From the Bank	From Land-owners	From the Bank	From Land-owners
Years	Percentages					
1906	16·8	35·1	80·0	63·4	3·2	1·5
1907	37·7	37·7	59·8	60·7	2·5	1·6
1908	11·1	27·4	50·1	68·7	38·8	3·9
1909	2·7	13·5	18·9	76·1	78·4	10·4
1910	2·0	15·5	5·0	64·9	93·0	19·6
1911	2·3	6·4	4·5	63·3	93·2	30·3
1912	2·4	5·0	4·1	59·8	93·5	35·2
1913	1·6	5·0	3·8	59·7	94·6	35·3

It may be seen that while, at the outset, the share of individual buyers was very small indeed, from 1908 on it began to increase rapidly, the increase being especially marked in the case of purchases from the Bank's reserve, where the institution was in a position to exercise complete control over the character of transactions. In a less direct way, the Bank exercised a strong influence also over the transactions it assisted financially, with the result that in the case of direct purchases from landowners the share of the individual buyers increased also at the expense of those of the two other classes.

Looking back at the history of the State Peasants' Bank, from its foundation, in 1882, to the outbreak of the war, one cannot help being impressed by the enormous work it had succeeded in accomplishing.

The area of land which, from 1883 to 1912, the Peasants' Bank had helped the peasants to buy, reached an aggregate of 15·8 millions of dessiatins or, roughly, 43 millions of acres. The importance of this figure is brought into relief by comparing it to the total area of England and Wales, which it exceeds by some 5 to 6 millions of acres.

It is not surprising, therefore, that one of the foreign students of Russian land-settlement, the well-known German expert on agrarian problems, Professor Max Sering, writing in 1911, expressed the view that "in the domain of internal colonization, by far the greatest activity among all countries is being displayed by Russia." [1] The record of the State Peasants' Bank, whose work had been untimely interrupted by the war and brought to a premature end by the revolution, was, indeed, unparalleled in history.

It would be mistaken, however, to judge of the Peasants' Bank work by the size alone of the figures involved in its transactions. Large as these were, no less important was the nature of the services rendered by the Bank in the solution of the Russian agrarian problem, especially during the last period of its activities. It may be said that, in some cases, those of the transactions which involved very small areas, were among the most important for the progress of peasant farming. This was especially the case with some of the purchases effected by whole villages, in which the land bought formed the whole or part of the area retained by the landowners at the Emancipation. As I have pointed out in a preceding chapter, these "cut-away lands" (*otriezki*), economically and socially, were an unmitigated evil, and the sooner they were incorporated in the *nadiel*, the better it was, ultimately, for all those concerned. Particularly important this was in those numerous cases, in which the land of the village and that of the estate were mixed together, each cutting deep into the other and breaking its continuity. It was most important that this abnormal position should be remedied without recourse to so drastic a measure as compulsory expropriation. The work done by the Bank in remedying this state of affairs, though it was perhaps among the least conspicuous of its services, should also be included among the most important of them.

[1] Max Sering, "Innere Kolonisation" in Elster's *Wörterbuch der Volkswirtschaft*, 2nd ed., 1911.

The Peasants' Bank was also responsible for the building-up in Russia, alongside of the villages, of a class of independent peasant farmers, whose farms showed the advantages of enclosures to the neighbours. The great importance of such examples was fully appreciated when the enclosure movement was set on foot in 1906, and, as I have pointed out, the Peasants' Bank policy, during the third period of its history, was deliberately directed to this purpose. Here, as in the case of the "cut-away lands," the results of the Bank's work could not be measured merely by quantitative standards, and a few well-chosen settlements, involving, perhaps, a relatively negligible area, but spread over localities which needed to be taught the advantages of enclosures, could prove more productive of beneficial results than other transactions involving millions of acres. It was precisely this service that the Peasants' Bank had rendered in those parts of Russia in which enclosed holdings were otherwise practically unknown.

B. Land-Settlement in Asiatic Russia

The modern history of Siberian colonization is closely connected with that of the Great Siberian Railway, whose construction was started in 1892. Though, indeed, a thin stream of settlers had been flowing beyond the Urals since the reign of Ivan the Terrible, by the close of the nineteenth century the Asiatic dominions of Russia, with the exception of some Western districts of Siberia, were still a new country and afforded extensive, if not unlimited, scope for systematic colonization. As late as the 'eighties, there was still enough colonizing space in the outlying parts of European Russia, and the bulk of peasant migration had been directed to the left bank of the Volga, to New Russia, the Northern Caucasus and the South-East. Only when these opportunities began to show signs of exhaustion, the possibilities of Siberia and of other parts of Asiatic Russia emerged from comparative obscurity. The first stimulus to systematic colonization of Asiatic Russia was mostly political in origin, and was due to the necessity of strengthening the Russian population in the Far East. It dated from the 'sixties, that is a period when, with the *laissez faire* doctrine at the very height of fashion, organized and assisted emigration to new regions was generally out of favour with the Russian Government, and the encouragement was accordingly confined to settlement in those localities in whose colonization the need was most pressing from the political point of view. Thus, the Regulations of 1861, dealing with settlement in the Amur Region, provided for the free allocation to settlers of 100 dessiatins (270 acres) per male soul, as well as for the temporary remission of public liabilities in taxes and services. In 1866, similar Regulations referring to settlement in the region of the Lower Ussuri river, besides the free grant of land, gave the settlers subsidies of 100 roubles per family. The Special Regulations of 1865, concerning the colonization of the Altai mining district, and prompted, in the first instance, by the desire of developing that rich locality economically, provided for the free grant to settlers of fifteen dessiatins per male soul and of timber for buildings. Other enactments of the same period, in so far as they dealt with peasant migrations at all, referred mainly to settlement in the East of European Russia, and in the first instance on the State-owned lands of the provinces of Samara, Ufa and Orenburg. Any emigration to Siberia that had been proceeding at the time was entirely spontaneous, unorganized and uncontrolled.

A certain change in the official attitude to emigration became noticeable since the early 'eighties, when the aggravation of the agrarian crisis brought to the forefront the economic, as distinguished from the purely political, aspect of the problem. This aspect has been emphasized, since the 'seventies, by several of the Zemstvos in the provinces most seriously threatened by the growing agrarian overpopulation in a number of representations to the Government, in which they laid stress on the urgent necessity of organizing and assisting emigration on an adequate scale. The change of policy found expression in the *Provisional Regulations of July 10th*, 1881, which permitted emigration, with a view to settlement on unoccupied State-owned lands, to all those peasants whose holdings were below one-third of the maximum fixed for the locality at the Emancipation. The actual decision as to the cases to which the Regulations were to apply, however, was delegated to the local authorities concerned. This enactment, indeed, though it was to some extent symptomatic of the change in the Government's attitude, had little practical effect. More important was the *Law of July 13th*, 1889, which was the first to put the movement on a definite legal footing. The Law of 1889, on which, for the next fifteen years, the whole development of Siberian colonization had been based, dealt with the conditions of emigration to State-owned lands and provided for special facilities with regard to the release of the intending emigrants from their communal liabilities in the old places of settlement, as well as the allocation to them of holdings of sufficient size in the chosen locality. Moreover, the law contained provisions concerning the temporary remission of taxes and the postponement of conscription, as well as the grant to settlers of loans for their initial expenses and their maintenance until the first harvest. From 1889 onwards, the yearly credits for the purpose of assisting emigration and settlement, which, in the preceding decade, did not, as a rule, exceed the paltry sum of 20,000 roubles a year, were increased to 80,000 roubles, of which amount three-fourths were earmarked for loans to settlers. Thus, a beginning has been made in assisted emigration and land-settlement in new regions, but it was still planned on a very small scale.

The development of Siberian colonization entered on its next phase since the early 'nineties, when the construction of the Siberian Railway was started. The colonization of the vast tracts of land in Siberia, opened up by the railway, became one of the principal tasks of the Committee of the Siberian Railway, established in 1892, of which the Grand Duke Nicholas, Heir-Apparent to the Throne, was appointed President. He took great personal interest in the development of Siberia, and even on his accession, as Emperor Nicholas II, did not relinquish this post. In Kulomzin, as Chairman of the Drafting Commission, on which devolved the task of actually studying the problems of the Committee and preparing the materials and drafts for its deliberations, Siberia found an enthusiastic and capable protagonist, to whom it was indebted for much of its progress in the course of the following two decades. From 1893 onwards, the Committee of the Siberian Railway was in charge of the planning and financing of Siberian colonization, while the Ministry of the Interior was responsible for the control of the emigration from European Russia and the actual settlement of the emigrants in the new regions. The financial resources necessary for the purpose were provided by the Treasury, in the form of the so-called " Fund of the Auxiliary Enterprises of the Siberian Railway," which originally amounted to 14 million roubles, but has been subsequently roughly trebled by the allocation to it of supplementary credits. Between 1893 and 1904, when the Committee was dissolved, it prepared for settlement a total area of 11·5 millions of dessiatins for the accommodation of about 1·2 millions of settlers of both sexes. In the course of these twelve years, from 800 to 900,000 peasants from European Russia had been actually settled beyond the Urals. Emigration to Siberia, which, in the course of the preceding decade, averaged roughly 10,000 a year, increased to an average of 70–75,000.

The activities of the Committee of the Siberian Railway coincided in time with the worst period of the Russian agrarian crisis. Siberian colonization, with the facilities it afforded for the settlement there of millions of peasant families and the consequent relief of agrarian overpopulation in European Russia, therefore, assumed considerable importance in the eyes of the Government, as well as in wide circles of the general public. The necessity of developing the movement to its full extent was generally recognized by the close of the last century. In a report submitted to the Emperor by Witte, in 1902, on his return from a tour of Siberia and the Far East, that statesman insisted on the urgent need of development in this direction, saying that the existing extent of emigration to Siberia "neither meets the requirements in land-settlement of those peasants of the internal provinces, whose holdings are insufficient, nor answers the need of the enormous territories of Siberia for agricultural colonization." It will be remembered that, about the same time, the Department of Agriculture, in its Memorandum to the Committee on the Needs of the Agricultural Industry, also emphasized the necessity of developing emigration to new regions, as a means of relieving the agrarian situation at home.

Accordingly, when, in January, 1904, the Committee of the Siberian Railway, on having completed its task, was dissolved, and the control of both emigration and settlement in new regions was entrusted entirely to the Colonization Department of the Ministry of the Interior, the development of the movement was planned on a large scale. The Russo-Japanese War, which broke out in February, 1904, however, prevented the putting of these schemes into immediate execution, and it was not until 1906, when the war was over and the revolution of 1905 had subsided, that the progress of Siberian colonization had been resumed.

The legal foundations of the new policy were laid down in the *Law of June 6th*, 1904, which took the place of that of 1889, but only came into force in 1906. The purpose of this enactment, as stated in the preamble, was, firstly, to improve, by means of emigration, the economic position of the peasants in European Russia and, secondly, to strengthen Russia's hold on the more remote parts of her dominions. This law, modified in details by a large number of subsequent amendments, formed the legal basis of the work of Siberian colonization until the Great War and the Revolution of 1917.

The control over emigration and settlement, vested entirely in the Ministry of the Interior in 1904, did not remain long in its hands. With a view to better co-ordination with other measures of agrarian policy, by the Ukaze of May 6th, 1905, the Colonization

Department was transferred to the re-organized Ministry of Agriculture. The Ministry of the Interior, through its local officials, was left responsible only for the organization of actual emigration of peasants from European Russia and the issue to intending emigrants of licenses entitling them to reduced railway fares. Asiatic Russia was divided into a number of Colonization Districts, each in charge of a responsible official of the Colonization Department, with a staff of administrative officers, surveyors, agricultural experts, statisticians, etc., who were entrusted with the task of preparing land for settlement, its allocation to applicants, the financial and technical assistance to settlers and all the rest of the work on the spot. With a view to preparing the ground for settlement, moreover, especially from 1907 on, the Colonization Department used to send to various parts of Asiatic Russia, which were supposed to afford more or less extensive facilities for settlement and development, special scientific and technical expeditions for the investigation and survey of fresh areas. Every spring, expeditions of this kind, consisting of geologists, botanists, hydrologists, agricultural experts and statisticians, were equipped and sent to survey some particular districts of Asiatic Russia, to which it was intended eventually to extend the colonizing activities. Other expeditions, purely statistical, were organized, as need arose, with a view to investigating the conditions and the available facilities of some of the localities of old settlement, which could still accommodate a more or less considerable number of immigrants. The reports of these expeditions and statistical surveys, published by the Colonization Department from time to time during the last few years preceding the war, formed a valuable contribution not only to the study of the immediate problems of colonization but to the general study of Russia's vast and still little known Asiatic dominions.[1] On the eve of the war, as a result of these investigations, the difficult work involved in the planning of the colonization campaign for each successive field season, was becoming gradually more rational, and the results less dependent on chance. The older regions, their population, its economic conditions and the principal elements of their husbandry were, on the whole, made sufficiently clear as a result of statistical surveys, while geological, botanical, meteorological and hydrological expeditions supplied the necessary information concerning the agricultural

[1] It is much to be regretted that, owing to some oversight on the part of the Colonization Department, none of its numerous publications, not even the annual Reports on "Emigration and Land-Settlement beyond the Urals," had been sent to the British Museum: a matter which made the writing of this section often very difficult.

prospects of those localities to which settlement was being gradually extended. Indeed, from 1906, when the business of colonization was definitely incorporated into the scheme of agrarian reconstruction under Stolypin, no efforts have been spared, within the limits of available financial and technical resources, in developing the movement and improving its organization. The Japanese War, which proved so conclusively the weakness of Russia in the Far East and the necessity of developing the background of the exposed frontier regions, had strongly emphasized the importance of Siberian colonization from the political point of view as well. In Krivoshein, as Minister of Agriculture, Stolypin found an able and energetic lieutenant in this particular sphere, who, moreover, before his appointment to the Ministry, had been in charge of the Colonization Department and possessed first-hand knowledge of the problem, its possibilities and its difficulties.

These difficulties were not inconsiderable. Indeed, it must be admitted that, though the Russians, from times immemorial, have been a nation of colonizers, the organization of systematic colonization on a large scale presented many a highly complex problem.

The greatest among these difficulties was that of securing the adjustment of the volume of immigration to Siberia to the available facilities for immediate accommodation. Theoretically, the problem would appear to be relatively easy of solution, and so, undoubtedly, it would be, had not Siberia formed a continuous stretch of territory with European Russia and had not the Russian peasant possessed an almost unlimited capacity for bearing hardships for the sake of some object, either imaginary or real, which he set himself to achieve. As a matter of fact, however, the road to Siberia lay open to anyone who was not deterred by the hardships of the journey, and whose worldly possessions were not so large as to keep him fast to the soil of the old country. Formal difficulties in connection with passports, communal liabilities or other legal bonds were relatively easily obviated or surmounted. The sale of the live and dead stock and, sometimes, the lease of his land in the village to a neighbour for a lump sum, payable at once, provided at least some of the money required for the journey. Once in Siberia, the prospects of the emigrants depended not so much on the money they brought, as on the working capacity of themselves and their families. In most parts of Siberia, even at the beginning of the current century, a large family with several workers had seldom to go far in search of land in the more remote districts, since, after having worked for wages for some time in a village of old settlers, they were in most cases enrolled in the commune and given their share of the communal land. To some extent, this practical certitude of getting land and improving their position, once in Siberia, encouraged unlicensed emigration (*samovolnoie pereselenie*) beyond the Urals. Moreover, since the peasants in European Russia remained naturally somewhat behind the times as to their current information about conditions in various parts of Siberia, the more the colonization advanced, the more serious became the complications involved in the annual influx of unauthorized immigration. The older, more densely populated districts of Siberia tended to get more than their due share of intending settlers, with the result that their accommodation there was being made increasingly difficult. Year after year, large and growing numbers of new arrivals failed to settle and had either to return home, utterly ruined, or to stay in Siberia as labourers, until, at last, they could assemble the wherewithal to move further afield and settle ultimately on their own. With the increase of population in the older districts of Siberia, however, the conditions of the labour market were becoming less favourable to newcomers, and their prospects of making good and settling on their own tended to be postponed. Every year, while in certain localities the influx of immigrants far exceeded the available facilities for settlement, in other parts vast tracts of agricultural land, ready for allocation to settlers, remained unoccupied for indefinite periods.

Various measures have been tried with a view to ensuring at least some measure of control over the influx and the distribution of immigrants, but short of the absolute refusal to permit unauthorized newcomers to settle none of them appeared to be effective. This only possible solution, however, was so drastic and, if recurred to, would have resulted in such hardships for those concerned, that, for reasons of humanity as well as of policy, the Government could not adopt this course and, accepting this spontaneous movement as an unavoidable complication, did its best to mitigate its effects.

The system of licenses for emigration from European Russia dated from the 'eighties. Issued by the local authorities in the old place of settlement, these licenses entitled the intending immigrants to the privileges accorded to settlers in new regions, such as reduced fares, the grant of loans for travelling expenses and temporary remission of taxes and liabilities. Since the Law of July 13th, 1889, licensed emigrants were also granted certain facilities in their discharge from their former communes. The Law of July 6th, 1904, moreover, enacted that licensed emigrants could dispose of their land in the villages either to the village commune or to individual members of the latter, against remuneration, instead of simply abandoning it, as hitherto. Until, by the Ukaze of November 9th, 1906, peasants were permitted to appropriate and to sell their respective shares of the communal land, this provision placed the licensed emigrants in a position of considerable advantage as compared with the unlicensed. It was expected at first that the system of licenses would keep the flow of immigration to Asiatic Russia under control, but in this it failed signally. When, after a few years of increased immigration, following the Law of 1889, and the period of bad crops in European Russia in the early 'nineties, the available facilities for immediate settlement in Siberia, in 1893, were practically exhausted and the issue of licenses was temporarily suspended, the influx continued. The only difference was that, in 1894, the proportion of unlicensed emigrants among the new arrivals to Siberia rose to 78% of the total. The spontaneous popular movement, once set on foot, could not be kept in check so easily. In the course of the next two decades, though it had never again reached this record, the proportion of unlicensed emigrants fluctuated approximately between 30 and 40%.

Even apart from the technical problems this rush to Siberia involved for those immediately responsible for the colonization of Asiatic Russia, the interests of the intending settlers, who had either been misinformed or were ignorant of the difficulties and risks of their venture, had to be considered. Accordingly, from the earliest days of the development of Siberian colonization, the Committee of the Siberian Railway and the Ministry of the Interior endeavoured, as far as possible, to help the situation by spreading reliable information concerning the conditions of settlement in various parts of Asiatic Russia and ensuring that those who contemplated emigration should be able, before actually starting, to judge of the new regions for themselves and even to prepare, provisionally, the ground for their future settlement over there. This gave rise to two characteristic developments, namely to the extensive publishing activities of the Colonization Department, on the one hand, and to the encouragement of preliminary reconnoitring of the places of intended settlement by those peasant families which wanted to emigrate through special scouts (*khodok*), on the other.

The publication by the Colonization Department of popular pamphlets or leaflets containing the necessary information concerning the conditions of emigration and settlement in Asiatic Russia was started in 1894 by the issue of a pamphlet on *Emigration to Siberia*, which, at the cost of 1½ copecks, or less than a halfpenny in English money, supplied the reader with what he most needed to know about conditions in various parts of Siberia and the claims they made on the settlers. Since then, the publication of such booklets has been a regular feature of the Department's work. Far from in any way concealing or minimizing the risks involved in emigration for unsuitable families, such as those with few adult and able-bodied members or too little money to start with, these publications, which could be obtained by peasants everywhere, usually laid special stress on the conditions of success. As all other branches of the activities of the Colonization Department, this particular branch of its work had been greatly extended since 1906, so that, on the whole, anyone who was contemplating emigration was enabled to obtain a reasonably good idea of his prospects before he finally decided to start.

While such publications were intended to broadcast information among the masses, the encouragement of *khodachestvo*, that is, of preliminary reconnoitring of the ground by intending emigrants, aimed, as far as possible, at reducing the risks of the individual families which embarked on the venture. Here also, the first attempts at organization dated from the period of the Committee of the Siberian Railway. Already in 1894, a year of unprecedented difficulties with the enormous influx of unauthorized immigration to Siberia, the Ministry of the Interior instructed the Provincial Governors in European Russia to encourage the preliminary dispatch by intending emigrants of responsible scouts (*khodok*). The next step was made by the *Law of December 7th*, 1896, which enacted that every peasant family contemplating emigration, was entitled, before applying for a license, to send a scout, who was accorded the benefit of reduced fares and was authorized, provisionally, to book a holding for the family concerned and to obtain, on their behalf, the advances granted to settlers for their travelling expenses. By the close of 1896, it became clear that, owing to a large influx of immigrants, the favourite districts in Siberia were again running short of land ready for immediate occupation, and in January, 1897, the Ministry of the Interior was compelled to go a step farther, by instructing the Governors, as a general rule, to issue licenses for emigration only to those families which had already prepared the ground for their settlement in Siberia through scouts. At the same time, the Circular of January 20th, 1897, which contained these instructions, restricted the sending of scouts by whole villages or peasant partnerships, which practice was fairly common, by making it conditional in every case on the permission of the provincial authorities: a measure directed at preventing emigration on a wholesale scale. The Circular laid much stress on the necessity of keeping emigration to Siberia within bounds, partly with a view to relieving the excessive influx of immigrants to Asiatic Russia, partly in the interests of the emigrants themselves. These efforts availed little, however, in bringing more order into the movement, and in the Law of June 6th, 1904, the provisions of the earlier enactments with regard to the scouts have been modified, in the sense that the sending of scouts instead of being, as hitherto, an essential condition of the issue of licenses, was made compulsory in the case of all those who had already obtained a license. Any family which had not yet obtained a license, if it chose to send a scout, had to do so at its own expense entirely, without assistance in the form of reduced fares. This enactment, which remained in force to the last, besides reducing the considerable strain on the already heavily taxed Siberian railway, aimed at making sure, at least, that the licensed emigration should be made as safe as possible.

Thus, the Government endeavoured, as far as possible, to bring the volume of emigration from European to Asiatic Russia under control, but only to meet with persistent failure. To the very last, indeed, unlicensed emigration contributed a very large proportion of settlers in Asiatic Russia. In the five years from 1909 to 1913, in spite of a vast improvement in the colonizing work of the Government, the proportion of unlicensed immigrants to Asiatic Russia fluctuated between 31·2 and 47·1% of the total. The movement was influenced by a great variety of causes, some of them entirely accidental and short-lived. Thus, a year of bad crops in European Russia would generally swell the volume of emigration from the districts affected by the failure. Prosperous times in the old country would reduce the movement considerably, by keeping those who would have emigrated at home, at least temporarily. On the whole, the emigrants to Siberia, being recruited mostly from that part of the rural population whose bonds with the old places of settlement were of the loosest, responded very readily to the influence of a variety of factors. Since licensed emigration, with Government assistance, was a matter of considerable deliberation and some delay, involved in the necessary formalities and the dispatch of scouts, which made it less subject to the effects of sudden changes in conditions, the latter naturally found expression mainly in the volume of unlicensed emigration, carrying away the floating elements of the population.

The distribution of the annual influx of immigrants over the area available for settlement was another, and by no means less formidable, difficulty. The flow of intending settlers kept on pouring mainly into the Western districts of Siberia, in which room for newcomers was getting relatively scarce, while the rest of the country developed very slowly, the population keeping only to the immediate vicinity of the railway. New arrivals tended to congregate in the older districts, whose colonization dated in many cases from the seventeenth or eighteenth century, and land in these localities was growing relatively scarce. Farther afield, in the provinces of Yenissey and Irkutsk, not to speak of Transbaikalia, vast tracts of agricultural land, ready for occupation, and situated within relatively easy reach of the railways and navigable rivers, remained vacant for lack of settlers.

The problem of the means by which to divert the flow of immigration from the more populated districts to the remoter and less populated parts of Asiatic Russia had exercised the Government since the earliest days of the systematic colonization of Siberia, the Steppes and the Far East. The advantages of settlement in the more developed parts of the country were so obvious that some incentive had to be created, which would counter-balance them and entice the intending settlers farther afield. In the early days of the Committee of the Siberian Railway, Kulomzin's Drafting Commission, in one of its Memoranda, suggested that "the suitable adjustment of the amounts of loans is, apparently, one of the very few means available to the Government, by which the effective control of Siberian colonization in the general interests of the State can be achieved." Since the close of the last century, indeed, the variation of the rates of advances granted to settlers, according to districts, became the principal means by which it has been sought to control the direction of immigration to Siberia, and the advances were ever more nicely adjusted to local conditions. Prior to the Committee, there had been no general system in force, advances up to 200 roubles per family being granted on application direct to the Ministry of the Interior, on the merits of each individual case. Since 1894, the limits were fixed at 150 roubles per family in the Amur region and at 100 roubles in the rest of Siberia. The same limits were left in force by the Provisional Regulations of June 25th, 1903. The amounts of the loans, however, were so small, and the range of possible variations so narrow, that it proved impossible by their adjustment to achieve any measure of effective control over the progress of settlement. Accordingly, from 1904 onwards, along with the general increase in the amounts of loans, there has been gradually evolved a system of grants which, by being more flexible and corresponding more closely to local requirements, was expected to be more effective. Thus, according to the system in force on the eve of the war the amounts of loans to be granted to settlers in various districts of Asiatic Russia were revised by the Government every three years, "account being taken of the difficulties of settlement in the given locality, as well as of the special importance for the State of the rapid colonization of the district concerned." [1] The amounts of loans were not permitted to exceed 400 roubles per family in the wooded districts of Transbaikalia and 250 roubles in the rest of Asiatic Russia. In order to encourage settlement in those parts whose colonization was considered especially important, settlers in these localities were allowed to retain part of the advance, up to one-half of the total amount, as a subsidy. On the other hand, loans to settlers who established themselves in existing village communities in Asiatic Russia were restricted to one-half of the normal amount. With the increase in the amount of loans and the possibility of their finer adjustment to local requirements, this financial expedient became more effective in ensuring a certain measure of control over the progress of Siberian colonization; though it failed in achieving a complete solution of the problem. The colonization of Siberia, though assisted and, to some extent, controlled by the Government, still preserved, on the whole, its original character of an elemental process of Russian national expansion.

This aspect of the movement was fully recognized by the two statesmen responsible for the agrarian reforms in Russia on the eve of the war—Stolypin and Krivoschein. In their joint Memorandum [2] on their tour of Siberia for the purpose of personally investigating the conditions and problems of Siberian colonization, in 1910, the two Ministers emphasized the spontaneous character of the movement and the necessity, instead of theorizing, of taking into account the simple consideration that, after all, those who emigrated to Siberia, did so not with a view to serving the interests of the State, but, first and foremost, with the purpose of improving their own economic condition. Accordingly, the only effective means of control was that of creating economic stimuli for settlement in certain regions in preference to other localities. This could be achieved, on the one hand, by the provision of really important special privileges for those settling in the remoter parts of Asiatic Russia, which would be tempting enough to attract there at least certain groups of immigrants; on the other hand, the same purpose would be served by discontinuing the free allo-

[1] Clause 46 of the "Regulations on Migration," *Code of Laws*, Vol. IX, 1912 edition.

[2] P. Stolypin und A. Krivoschein, *Die Kolonisation Sibiriens*, Berlin, 1912.

cation of land to settlers in the more densely populated parts of Siberia. It must be noted, in this connection, that, throughout Siberia, the whole of the land, with the only exception of some 1·2 millions of dessiatins (about 3 million acres) granted to different individuals at various times, was the property of the State and could not be sold to anyone, since even the Ukaze of August 27th, 1906, which authorized the sale to peasants of the unoccupied lands belonging to the State, did not apply to Asiatic Russia. The latter, indeed, presented to the world the greatest known example of land nationalization. The attempt made by the *Law of July 8th*, 1901, to depart from the principle of national ownership of land in Siberia, by authorizing its sale by auction to private individuals and its lease to members of the gentry with a view to subsequent purchase, proved abortive, no land having been either sold or leased in virtue of that much-criticized enactment. Hence, as before, Siberia remained wholly democratic, a country of peasant farmers settled on State-owned land, of which they enjoyed the use in perpetuity against the payment of a small rent (*obrochnaia podat*), fixed periodically for a number of years ahead. The entire freedom in disposing of the whole unoccupied area beyond the Urals, thus possessed by the State, combined with the control of transportation by State-owned railways, greatly facilitated the task of colonization. Now, in their joint Memorandum, the two Ministers suggested that, to divert the stream of immigration from Western to Eastern Siberia, the provisions of the Ukaze of August 27th, 1906, concerning the sale to peasants of agricultural land belonging to the State, should be extended to Asiatic Russia. This would enable the Government, while continuing the free allocation of holdings to settlers in those districts to which it wanted to attract immigration, to apply the provisions of the Ukaze in the more densely populated parts, in which land already possessed considerable value and could, therefore, be sold to applicants. A bill to this effect was introduced by the Ministry of Agriculture into the Duma in 1913, but its passing through the Legislature has been delayed by the war, with the result that it had never been actually passed. Pending its coming into force, however, from 1913 onwards, the system has been adopted in those localities to which it was proposed to apply the new law, to let the land to settlers provisionally on lease, on the understanding that, when the law is passed, they would be prepared to buy their holdings. Thus, by means of a serious departure from the existing agrarian régime of Siberia, the Government expected to achieve more or less effective control over the progress of the colonization of Siberia. How far this attempt could be considered as successful, it is impossible to say, for lack of experience of the actual working of the system, though certainly, on the face of it, it would appear to have at last provided sufficiently strong incentives, especially in the case of the poorer immigrants, to proceed farther East, to localities in which land was still available free, and where special privileges and more generous financial assistance were granted.

Meanwhile, with the demand for accommodation in Western Siberia and in other more easily accessible localities constantly exceeding the available facilities of immediate settlement, the Government had recourse to a measure of far more doubtful expediency. This was the so-called "regulation of holdings of the old settlers," which involved the reduction of the holdings of the old village communes down to the legal limit of 15 dessiatins per male soul. Though, indeed, the Government was legally justified in taking possession of the unauthorized surplus, occupied by old settlers in the past, when land was still so plentiful that no one paid any attention to the area actually taken by the immigrants, the reduction often involved considerable disturbance in the layout of the land of the old communes and affected the peasants' husbandry very unfavourably. Since the reductions thus effected averaged about one-quarter of the area of land in the possession of the old settlers, they were sometimes liable to be badly felt by the villages concerned, which, as a rule, farmed their land on an extensive system and were dependent, therefore, above all on the area at their disposal. As a result, though, indeed, a considerable extension of the area available for the accommodation of immigrants in the most densely populated parts of Siberia had been achieved by this expedient, the old settlers have been adversely affected by the curtailment of their holdings, and their attitude to newcomers was embittered. The same could be said of the withdrawal of the "surpluses" of land in the possession of roaming Kirghiz tribes in the Asiatic Steppes, which led to considerable friction with the aborigines.

Every year, the Colonization Department prepared for settlement, in Asiatic Russia, an extensive area of land for the accommodation of 350,000 male souls annually, so that, had it been possible to control the distribution of immigrants over the available territory, there could be no difficulty about their accommodation in Siberia.

The organization of the whole movement of emigration and settlement had been vastly extended and improved in the course of the last decade preceding the war. This general development of the work of the Colonization Department may be seen from the rapid increase in its Estimates, which, in 1903, did not exceed 3,144,000 roubles. From 1906 onwards, they grew with great rapidity, reaching 24,488,000 roubles in 1910 and 30,200,000 roubles in 1914. The loans to settlers, which amounted to 1·3 millions a year in 1901–3, rose to 7,600,000 roubles (over a quarter of the total Estimates) in 1913. The Colonization Department also assisted the settlers through a staff of agricultural experts and veterinary personnel, as well as by running in Siberia a number of agricultural stores, in which the peasants could buy at little more than cost price practically all they needed in the way of agricultural implements, from a spade or an axe to a binder or threshing machine, as well as seeds, fertilizers, etc. In 1914, these agricultural stores numbered 300, with an aggregate turnover of 8 million roubles and a net profit of 600,000 roubles. The Department also maintained, for the benefit of the settlers and the prevention of epidemics, which could easily be engendered by the congregation at certain points along the route of large numbers of immigrants, a medical staff of its own. On the eve of the war, in 1914, in various parts of Asiatic Russia, the medical personnel of the Colonization Department, posted in 475 "medical stations," consisted of 150 doctors and 650 trained assistants; an organization obviously too small for the needs of the enormous country, with its scattered population, but which, as time went on, was being gradually increased. Thus, of the 475 "medical stations," thirty-two were opened in 1914. Considerable sums were expended on the special expeditions organized every year for the purpose of investigating the natural and economic conditions of the areas proposed for development; in 1913, the cost of these expeditions, including the salaries of the scientific and technical personnel, equipment and other expenses in connection with their work, amounted to 728,800 roubles.

The visit to Siberia, in 1910, of Stolypin and Krivoschein, made for the special purpose of studying the problems of colonization on the spot, in immediate contact with the settlers and the local officials directly responsible for the work, marked the beginning of a new era in the history of the movement. The Memorandum of the two Ministers treated the whole problem with great candour and suggested many important improvements. Some of them, as the finer discrimination in the extent of Government assistance to settlers in various regions, have, at least partly, been put into effect before the war. Other measures, including the modification of the whole land system in Asiatic Russia, have not yet received legislative sanction, when the outbreak of the war brought the development of Siberian colonization to a temporary close.

The few years immediately preceding the war were also marked by an increase in the interest taken in the problems of Siberian colonization by the Zemstvos of those provinces of European Russia which supplied the greatest numbers of emigrants. It has been mentioned before that, as early as the 'seventies of the last century, these organs of local self-government urged the Government to a more active encouragement of emigration to Asiatic Russia or other under-populated parts of the Empire. In the early part of the twentieth century, the Zemstvos began tentatively to take a hand in assisting peasant emigration from their respective provinces, as well as the actual settlement of such emigrants in the new places. This development, though, on the eve of the war, it was still in its infancy, was very important potentially. The outstanding event in this particular sphere, in the early years of the current century, was the endeavour to co-ordinate the scattered efforts of the several Zemstvos of a number of provinces from which emigration proceeded on a large scale. This assumed the form of the establishment of the *Regional Emigration Organization of the Zemstvos of Southern Russia*, which started its work in 1908. The Organization included the Zemstvos of the provinces of Poltava, Kharkov, Chernigov, Kherson, Volynia, Kiev, Voronezh and Ekaterineslav. The latter, however, soon seceded and was replaced by that of Saratov, which joined in 1913. In all these provinces, as well as several others, mostly in the South, South-West and West of Russia, attempts at organizing assistance to emigrants had been made for a number of years. Now, those of them which had joined the Regional Organization, began to extend their activities. Besides assisting emigrants financially, supplying them with the necessary information and organizing parties of scouts to Siberia, now they extended their work over there, on the places of settlement, through special representatives of the Zemstvos who helped the settlers on the spot. In 1913, the Organization had thirty-two representatives in various parts of Asiatic Russia, whose function it was to assist immigrants belonging to the provinces which had joined that body. Besides, eleven inquiry agents were maintained either at important points along the route (Syzran, Cheliabinsk, Irkutsk), or in the provinces of origin, to whom the peasants could apply for information and directions. The Budget of the Regional Organization, made up of the grants of the several constituent Zemstvos, amounted to 94,550 roubles in 1909 and increased to 141,000 roubles in 1913.

Other Zemstvos were feeling their way in the same direction, that of Vitebsk being one of the most active outside the Organization. The extent to which the Regional Organization succeeded in achieving its purpose may be judged by the fact that, since its establishment, in 1908, the percentage of unlicensed emigrants and of returning failures in the associated provinces has been consistently lower than in the rest of Russia, while the numbers of scouts and of emigrants had increased.[1]

The actual progress of the movement of emigration in 1901–14 may be seen from the figures below:

	Emigrants	Scouts	Total
1901	89,088	31,161	120,249
1902	81,921	29,009	109,930
1903	85,824	29,012	114,836
1904	40,001	6,731	46,732
1905	38,750	5,269	44,019
1906	141,294	77,584	218,878
1907	427,339	145,240	572,579
1908	664,777	94,035	758,812
1909	619,320	88,143	707,463
1910	316,163	36,787	352,950
1911	189,791	36,271	226,062
1912	207,027	52,558	259,585
1913	234,877	92,553	327,430
1914	241,874	94,535	336,409

It may be seen that, at the beginning of the twentieth century, the movement was fairly steady, the total being over 100,000. It declined heavily in 1904–5, owing to the Japanese War, only to rise again to unprecedented figures from 1906 onwards, on the termination of the war and the coming into force of the new Emigration Law of 1904. The crest of the wave of emigration was reached between 1906 and 1909, during which period, with the new facilities provided by the legislation and the support of the Government, all the hitherto pent-up potentialities in this respect were released. From 1910 to the outbreak of the war, though the movement continued on a large scale, involving, on the average, over 200,000 actual settlers *per annum*, it had slowed down considerably. This decline was heartily welcomed by the responsible authorities, which, during the preceding few years, have been literally overwhelmed by the human torrent pouring into Asiatic Russia in search of accommodation. From 1910 to the outbreak of the war, the movement had been proceeding at a pace which could, probably, be considered normal under existing conditions and could be expected to be maintained on the whole,

[1] A. Grigorovsky, *The Progress of Emigration in 1912–13*, in the "Agronomichesky Zhurnal," No. 9–10, 1913, p. 118.

with fluctuations from one year to the next under the influence, on the one hand, of either good or bad seasons at home, in European Russia, and, on the other, of the policy of the Peasants' Bank and the reserve of land available at its disposal, which provided an alternative outlet for the peasants seeking accommodation on new holdings outside their villages.

With regard to the distribution of immigrants among the various parts of Asiatic Russia in 1906–14, no detailed figures concerning the numbers of settlers in separate provinces are, unfortunately, available to the writer. As to the total number of immigrants, which, during these nine years, amounted to 3,618,447, over one-half, namely 2,083,560, or 57·5%, had settled in the four provinces of Tobolsk, Tomsk, Yenissey and Irkutsk, while 1,162,754, or 32·2%, had chosen the regions of Akmolinsk, Semipalatinsk, Semirechie and Turgay. Only 372,133, or 10·3%, found their way to the East, beyond the lake Baikal.

Between the Census of 1897 and January, 1914, the total population of Asiatic Russia was estimated to have increased from 13·5 to 21·1 millions: an increase of 56%. The Census of 1916, which did not cover the whole of Asiatic Russia, but included all the provinces of Siberia and the Steppes Region, that is the principal districts of colonization, showed the actual increase according to provinces. The figures are given below:

	1897 (In thousands)	1916 (In thousands)	Increase or decrease per cent.
Siberian Provinces:			
Tobolsk	1,433·1	2,032·9	42·0
Tomsk	1,927·8	4,269·8	121·4
Yenissey	570·2	1,018·4	78·6
Irkutsk	616·5	950·7	54·2
Yakutsk	269·9	232·2	–14·0
	4,817·5	8,504·0	76·5
Steppes Region:			
Akmolinsk	687·6	1,553·7	125·9
Semipalatinsk	684·6	970·5	41·9
Turgay	453·4	881·1	94·3
	1,825·6	3,405·3	86·5
Total for Siberia and the Steppes	6,643·1	11,909·2	78·9

The agricultural development of the region in the course of colonization in the early part of the twentieth century may be judged by the following figures, showing the extension of the area under crops according to provinces:

Increase in the Area under Crops in Siberia and the Steppes between 1901–5 and 1916

	Average for 1901–5	1916	Increase per cent
	(In thousands of dessiatins)		
Provinces:			
Tobolsk	1,165·9	1,702·1	45·9
Tomsk	1,455·0	3,307·3	127·3
Yenissey	411·3	625·9	52·2
Irkutsk	360·5	425·0	17·9
Yakutsk	13·3	33·8 (1917)	—
Akmolinsk	266·8	1,324·1	396·0
Semipalatinsk	87·1	439·0	404·0
Turgay	289·0	1,003·4	247·0
Total	4,048·9	8,860·6	118·9

Thus, in about fifteen years the area under crops in Siberia and the Steppes had more than doubled, the increase being especially marked in the Steppes and in the province of Tomsk, that is in those localities to which the immigration during that period had been greatest. The Western districts of Siberia, and in the first instance the province of Tobolsk, though they still continued to absorb rather more than their fair share of new settlers, were already too densely populated and too developed agriculturally to allow of so rapid an extension of arable. In fact, certain districts of the province of Tobolsk were already in course of transition from the old unregulated cropping to the three-course system, thus making up for the reduction in the available reserve of virgin soil by economy in the use of the land. Farther East, in the provinces of Yenissey and Irkutsk, in the midst of the Siberian forests (*taiga*), the agricultural development proceeded more slowly, owing partly to the greater difficulty of cultivation, partly to the preference of the immigrants, in spite of all measures of encouragement, for the less remote districts.

It would, however, be entirely beyond the scope of the present study to engage into a more detailed account of the agricultural development of Asiatic Russia. In a work dealing with the agricultural evolution of European Russia, Siberian colonization and farming have only to be considered in so far as their progress affects the solution of some of the problems arising out of the principal subject. Among such problems, the most important is that of improving, by means of land-settlement beyond the Urals, the economic conditions of the peasantry in European Russia; and it was this particular aspect of the colonization of Siberia that has been generally foremost in the minds of the legislators and of the public.

In this connection, the prospects of Siberian colonization have been the subject of considerable discussion and not a little controversy. It may be said that, on the whole, expert opinion, by the close of the last century, while it recognized the great possibilities of the agricultural development of Asiatic Russia, was inclined to take a cautious, if not actually a pessimistic view, of the part land-settlement in Siberia could be expected to play in the solution of the agrarian problem in European Russia.[1] A closer study of the actual conditions of Siberian colonization, as well as the practical experience gained in the 'nineties in connection with the activities of the Committee of the Siberian Railway, served to disappoint any exaggerated expectations in this respect and to show that, though, indeed, Siberia was an enormous country, with immense natural wealth, its capacity of immediate absorption of agricultural immigration fell far short of the measure necessary for the effective relief of agrarian overpopulation in European Russia. Moreover, for purely technical reasons, even with improved means of transport and greatly increased financial resources, the actual transfer of population beyond the Urals at a rate which would keep pace with its growth in the old country, would certainly have been impossible. Accordingly, in their Memorandum quoted above, Stolypin and Krivoschein expressed the view that the part land-settlement in Siberia could legitimately be expected to play in the solution of the agrarian problem in Russia proper should not be overestimated, because, unless the yearly emigration should rise to millions, instead of hundreds of thousands, its effect on rural congestion in the home provinces would be relatively very small. Enormously important as it was in itself, land-settlement in Asiatic Russia had to be considered only as one of a series of concerted measures intended to improve the economic condition of the peasantry of European Russia; and it was as such that the colonization of Siberia was incorporated in the comprehensive scheme of Stolypin's agrarian reforms, in which it figured along with enclosures, land-settlement through the Peasants' Bank and a whole system of other measures designed to assist the peasants in raising the standards and the yield of their farming.

Dealing with the agrarian aspect of Siberian colonization, it may be interesting, in conclusion, to examine some numerical data bearing on this subject, summarized below according to regions. The table shows the distribution of emigrants who left for Siberia during the twenty years from 1896 to 1915, according to their locality of origin, along with the average sizes of peasant holdings

[1] A. Kaufmann, *Emigration and Colonization*, p. 157 *and passim*.

in the region, the percentages of small holdings below 5 dessiatins (13·5 acres) and the relation the number of emigrants bears to the population of the region, as recorded by the Census of 1897. The combination of these figures is intended to throw some light on the extent of influence exercised on the volume of emigration by the "land-hunger," as well as to give an idea of the part played by it in relieving the pressure of population in the congested agricultural districts.

Regions	Average Size of Holdings in 1905 in dessiatins	Per cent of Holdings under 5 dessiatins	Number of Emigrants, 1896–1915	Emigrants relatively to Rural Population, Percentage
Northern (Vologda only)	15·5	12·5	16,510	1·3
North-Eastern	16·0	10·0	—	—
Petrograd	11·2	5·5	32,438	1·7
Moscow	7·9	18·8	192,260	1·6
Western	9·2	8·9	457,212	11·8
South-Western	5·5	57·6	346,261	4·9
Ukraine	6·1	45·6	864,077	12·8
Central Agricultural	7·8	22·5	848,319	8·4
Middle Volga	8·3	18·7	189,396	2·7
Eastern	19·7	7·6	205,875	4·4
New Russia	9·7	20·4	523,823	10·5
South-Eastern	—	—	—	—
Caspian	—	—	—	—

It may be seen that, indeed, the proportion of emigrants, as a rule, was far too small to relieve the pressure of population on the land in European Russia. It was highest in the two Western provinces of Vitebsk (17·2%) and Mogilev (14·4%), but even this did not exceed, on the average, 0·86% and 0·72% *per annum* relatively to the population at the beginning of the twenty-years period. The third province of the Western region, namely that of Minsk, with only 6·1% of emigrants, however, reduced the average for the whole region, which came, accordingly, second in Russia, with 11·8% of emigrants during the period dealt with. As a region, the Ukraine took the first place, with 12·8%. In this part of the country, with the low average sizes of holdings and the high proportion of holdings below the economic size, as well as with practically no local outlets for the surplus of hands outside agriculture, emigration beyond the Urals could be looked upon as due primarily to the effects of agrarian overpopulation. The percentage of emigrants was high in all the three provinces of the Ukraine (Chernigov, 13·1; Kharkov, 10·1; and Poltava, 10·9%). In the South-Western region, with holdings still smaller than in the Ukraine, and a larger proportion of small-holders, the percentage of emigrants was relatively very low, namely 4·9% in the course of twenty years. Though the great demand for agricultural labour on the spot, due to the high development in that region of large capitalistically-organized farming and to intensive cultivation, with sugar-beet playing an important part in the rotation, may have been responsible, to a certain extent, for keeping the peasants from emigrating, the principal cause of the surprisingly low proportion of emigrants must, probably, be looked for elsewhere. Indeed, this region, situated close to the Western frontiers, supplied large numbers of emigrants to other parts of the world, and, in the first instance, to the American continent. The Western and South-Western provinces, besides supplying large numbers of seasonal agricultural labourers to Germany and Austria, whence many of them proceeded to America, were also a favourite recruiting ground for German shipping companies engaged in the transport of emigrants to the New World. A large proportion of the annual emigration, accordingly, instead of going to Siberia, left Russia altogether. Some of them settled and made good in the United States or in Canada, but others were less fortunate in being tempted to try South America, where they mostly found themselves stranded in appalling conditions, at the mercy of the owners of large estates or farms, who never cared what happened to such imported foreign labour during the slack season, so long as they were sure of a regular supply of fresh hands. The traffic, reminiscent of the slave trade of the past, and involving every year a considerable number of peasants from the Western provinces of Russia and from Poland, in the course of the years immediately preceding the war was a matter of serious concern to the Russian Government and has been a subject of detailed investigations.[1] In the South-Western provinces, emigration abroad must have accounted for a very large proportion of the total number of persons who left their old homes to try their luck elsewhere.

On the whole, it would appear from a close examination of the distribution of emigrants according to localities that, though the sizes of holdings and the degree of "land-hunger" undoubtedly exercised considerable influence on the extent of the movement, they were far from being the only, or even the principal, factor affecting the phenomenon. No less, if not actually more, important was the general economic organization of the localities concerned. Thus, purely agricultural districts, other conditions being equal,

[1] In the spring of 1914, as a member of the staff of the Colonization Department, the author has been privileged to hear the official report of the person, to whom the investigation of this problem on the spot, through all its stages from Germany to America, had been entrusted, and the impression is still fresh in his mind.

would appear to supply a higher percentage of emigrants than localities with more developed outside earnings. The influence of the economic organization of the locality is, indeed, well marked in the case of the region of Moscow, on the one hand, and of the Central Agricultural, on the other. The Moscow region, with its well-developed industries, its numerous outlets for the surplus of hands, its good communications and its relatively varied and intensive systems of farming, in spite of the considerable percentage of small holdings, and the small sizes of holdings generally, as well as of indifferent soil, was responsible for no more than 1·6% of emigrants. The industrial provinces of the region had only small fractions of 1% (Moscow 0·04 and Vladimir 0·09%). The Central Agricultural region, on the other hand, with practically the same sizes of holdings and a slightly higher percentage of holdings below 5 dessiatins, but with better soil, had 8·4% of emigrants. The Eastern provinces of Samara and Ufa, purely agricultural and engaged entirely in extensive cereal farming, combined very large average holdings and a low percentage of small ones with a relatively large percentage of emigrants, amounting to 4·4%. The purely agricultural Western region which, it will be remembered, had the high proportion of 11·8% of emigrants, is another case in point. New Russia, with its extensive cereal farming, without local industries or any sources of outside earnings for its peasants, except in the metallurgical district of the Krivoy Rog, in the province of Ekaterinoslav, in spite of excellent black-earth, average holdings larger than in the Moscow region, and a proportion of small ones only slightly higher, registered 10·5% of emigrants. In certain cases, as the Southern provinces and the province of Vitebsk, the proportion of emigrants may have been somewhat raised by the assistance afforded to its development by the Zemstvos of these provinces, which, as I have had the opportunity to point out, have actively encouraged and helped the movement.

This digression, with which I conclude my account of the development of emigration beyond the Urals and land-settlement in Asiatic Russia, would appear to suggest that the movement, far from being accounted for entirely by the relative degree of "land-hunger" in various localities, as had often been contended, was the complex result of a variety of factors. Here, as elsewhere, it is difficult to generalize in dealing with a country of the size and variety of Russia, and conclusions have to be reached by the careful analysis of local conditions and peculiarities. If any tentative conclusion would appear possible, it is rather that, while in purely agricultural districts it was the degree of "land-hunger" that was the decisive factor, with regard to the country as a whole the general economic organization of every given locality played the decisive part. The higher the economic organization of the region, the more developed the local markets for agricultural produce and for the surplus of labour on the land, the less need there was for the peasants, even with small holdings, to emigrate, and the lower the proportion of emigrants.

CHAPTER VII

AGRICULTURAL RUSSIA ON THE EVE OF THE GREAT WAR

By the close of the nineteenth century the industrial development of Russia, extremely slow in the course of the first few decades after the Emancipation, had made a certain progress. Russia entered the twentieth century a different country from that which, forty years earlier, had emerged from the medieval social and economic regime based on serfdom. The closing decade of the last century, with Witte in control of the financial and economic development of the country, laid the foundations of further industrial expansion. Russia's principal need, then as always, has been capital for the development of her great natural wealth, and the financial policy of Witte has been directed, first and foremost, to the purpose of helping its accumulation at home, as well as attracting investments from abroad. The currency reform of 1897, which introduced the gold standard in Russia, crowned the policy of stabilization of the rouble, on which the Government had embarked since the 'eighties under Bunge and Vyshnegradsky, as Witte's predecessors in the Ministry of Finance. A carefully devised elaborate customs tariff encouraged the growth in Russia of large industrial concerns and insured high rates of profit to foreign capitalists on their Russian investments. On the foreign capital markets money was plentiful and cheap. The wealth created by the expansion which followed on the Industrial Revolution in Western Europe was on the look-out for profitable employment, and in Russia it saw an extensive field open for investments. Thus, at the beginning of the twentieth century conditions were available in which the hitherto slow rate of Russia's industrial development could be greatly accelerated, and the process of growth which had actually set in was, indeed, such as to justify the application to it of the much-abused name of Industrial Revolution. As an Industrial Revolution in the true sense of the word, the change involved far more than a purely industrial transformation, since it left no sphere of economic, social and political life unaffected. Economically and socially, in spite of the briefness of the time allowed to its development, it transformed the countryside no less than the towns, by the rapid substitution of commercialism for the old self-sufficient natural economy, and of the individualism of enclosed peasant farms for the traditional bonds of the village commune. Politically, the transformation found expression on the substitution of representative government for the autocracy of the Tzar. The Duma, indeed, while it had, for its mother, the political idealism of the Russian *intelligentzia*, could certainly claim the rising industrial *bourgeoisie* for its father.

A few figures will be sufficient to show the rapidity with which the capitalistic development of Russia had been proceeding on the eve of the Great War.[1]

In the first instance, it is best to refer to the growth of the resources of the Russian money market from 1900 to 1913, as characteristic of the general economic expansion. The amounts of deposits and current accounts in the credit establishments of Russia increased as follows: [2]

	1900	1913
	Millions of Roubles	
State Bank	168	263
Commercial banks	536	2,539
Mutual credit societies	178	595
Municipal banks	97	198
Savings banks	662	1,685
Small credit institutions	?	466

The total resources of the money market, including the credit establishments' own capital, increased from 1,388 millions of roubles in 1900 to 4,697 millions of roubles in 1913.

The increase in the capacity of the Russian capital market during this period may be seen from the growth of investments in securities. The total value of Government bonds and other securities placed in Russia, which amounted to 5,115 millions of roubles in 1900, increased to 9,380 millions in 1913.

Between 1900 and 1911, the aggregate capital and debentures of joint-stock companies, other than railways, had nearly doubled: from 1,941 millions it rose to 3,642 millions of roubles.

In the manufacturing industries there has been a noticeable tendency towards an increase in the size of concerns, at the expense of their numbers, pointing to a higher organization of production.

[1] For a fuller account of the period the reader is referred to Miss M. S. Miller, Ph.D., *Economic Development of Russia*, 1905–14; also to *Russia: its Trade and Commerce*, edited by A. Raffalovich, and containing figures bearing on every aspect of the development.

[2] V. A. Mukoseiev, "Money and Credit," in Raffalovich's *Russia: its Trade and Commerce*.

The number of factories and works, the number of hands, and the aggregate value of their output, according to the Statistics of the Ministry of Finance, were a follows: [1]

	Number of Concerns	Number of Workers, in thousands	Value of Output in millions of roubles
1900	38,141	2,373·4	3,438·9
1912	29,965	2,931·3	5,738·1

The average number of workers employed per factory, accordingly, had increased from 62 to 98.

The heavy industries—coal, mining and metallurgy—have made great progress.

The output of coal, which amounted to 555·5 millions of poods in 1895, rose to 1,140·9 millions of poods in 1905 and reached 2,213·8 millions of poods in 1913. The production of iron ore increased from 152·5 millions of poods in 1895 to 282·5 millions of poods in 1905 and reached 532·4 millions of poods in 1913.[2]

The cotton industry, the most important branch of Russian manufactures, had increased the number of spindles from 6·6 millions in 1900 to 8·8 millions in 1912; that of weaving looms—from 151,000 to 224,000, and the weight of cotton used—from 16·0 to 25·7 millions of poods.[3]

The increase in the production of all branches of industry was accompanied by a simultaneous expansion of trade, both home and foreign.

The growth of trade is shown by the increase on the traffic on the railways, which rose from 3,958·4 millions of poods in 1900 to 7,981·0 millions of poods in 1913. The volume of freights handled per verst of track increased from 82,900 to 127,700 poods. Besides, the traffic in manufactured goods and raw materials for industries had increased considerably more than that involving cereals, with the result that the share of the latter in the aggregate freights of the Russian railways declined ftom 22·2% in 1903 to 16·3% in 1913.[4]

Russia's foreign trade, whose average annual turnover, in 1891–1900, amounted to 1,195·2 million roubles, increased to 1,960·5 million roubles in 1901–10 and to 2,779·3 million roubles in 1911–13.[5]

The railway net had been increased, between 1902 and 1912, by a total of 9,855 versts.[6]

[1] V. Varzar, *Factories and Workshops, ibidem*, p. 107.
[2] Miss M. Miller, op. cit., p. 291.
[3] V. Varzar, op. cit., p. 132.
[4] J. Bookshpan, *The Internal Trade of Russia*, Raffalovich, pp. 271–3.
[5] MM. Sobolev, *The Foreign Trade of Russia, ibidem*, p. 301.
[6] K. Zagorsky & E. Geidanov, *Internal Transport, ibidem*, p. 232.

As a result of the rapid development of industry and trade, the national income of Russia had also increased very considerably during the decade preceding the war. According to an estimate made by Professor Prokopovich, the income per head of population in European Russia (50 provinces, i.e. exclusive of the Kingdom of Poland and the Grand Duchy of Finland), had risen from about 66 roubles in 1900 to 101 roubles in 1913, or roughly 50%.[1]

The distribution of population as between town and country also underwent a not unimportant modification, the proportion of urban population having increased from 13% in 1897 to 18% in 1914. While, from 1867 to 1897, in the course of 30 years, it had risen only from 10·6 to 13·0%, that is by only 2·4 points, now, in the little more than half that time, it had advanced by 5 points.

The industrial growth of Russia, even taken alone, could not have left the agriculture unaffected and was bound to exercise a stimulating effect on the progress of farming. As a matter of fact, however, in the course of the years immediately preceding the Great War, Russia was not alone riding on the rising tide of trade and prosperity. The industrial revival following on the protracted depression of the last quarter of the nineteenth century, was world-wide, and had lasted until, in 1914, it was shattered by the war. The prices of agricultural produce on the world market were rising, as may be seen from the table below, showing their movement in those countries of Western Europe, in which Russia was particularly interested:[2]

Prices of Wheat, in Marks per 100 kg.

	England	France	Prussia
1900	133	154	152
1901	125	162	164
1902	132	174	163
1903	126	182	161
1904	138	170	174
1905	145	182	175
1906	138	182	176
1907	147	186	205
1908	155	184	209
1909	186	198	234
1910	158	213	211
1911	155	212	204

In Russia, the trend of prices for agricultural products, which had been declining since the close of the 'seventies, changed by

[1] S. Prokopovich, *Estimate of the National Income of 50 Provinces of European Russia* (Rus.), Moscow, 1918.

[2] Dr. Carl Ballod, *Grundriss der Statistik enthältend Bevölkerungs-, Wirtschafts-, Finanz-, und Handels-Statistik*, Berlin, 1913, p. 108.

the end of the century, and the steady rise continued until, in 1912 and 1913, a slight setback became noticeable in the case of cereals. Though, from one year to the next, there have been more or less marked fluctuations in the prices of the separate cereals, the general trend was well maintained throughout.

The movement of prices of the principal cereals in European Russia may be seen from the table below, giving the average spot autumn prices in copecks per pood (36 lb.):

	Rye	Winter Wheat	Spring Wheat	Oats	Barley
1901–5	68	82	82	62	62
1906–10	90	110	108	69	77
1911–13	87	107	107	75	83

The movement of prices in the exporting harbours is shown in the following table, which gives the average prices of the same cereals at Odessa, this particular port being chosen because, in its case, unlike that of other harbours, quotations are available for all the four principal cereals:

	1901–5	1906–10	1911–13
	Copecks per pood		
Rye	70·4	90·6	84·2
Wheat (Arnautka)	91·9	113·8	113·1
Oats	70·3	80·3	83·2
Barley	63·3	77·7	87·3

The prices of flax and hemp fibre moved as follows:

	1901–5	1906–10	1911–13
	Roubles per pood		
Flax, Livonian, Riga	5·20	4·62	5·52
Flax, combed, Petrograd	4·83	4·54	5·41
Hemp, Oserrh, Riga	4·58	4·92	5·84

Prices of certain other important agricultural products in Moscow are given below:

	1901–5	1906–10	1911–13
	Roubles per pood		
Russian beef, plain	3·69	4·62	4·96
Mutton	3·77	5·04	5·77
Pork	5·22	6·77	7·00
Fat, beef	4·62	5·41	5·59
Butter, best fresh	13·19	15·41	16·25
Eggs, per 1,000	26·84	30·23	32·67

It may be seen that, in so far as the market for agricultural products was concerned, the beginning of the twentieth century was a period very favourable to the farmer. Not only in Russia, but practically throughout the world, the agricultural industry, which, for decades, had been lingering under the effects of continually falling prices, was enabled to breathe again. What agricultural Russia was like during this stage of its evolution, and how the various groups of the Russian agricultural population reacted to their rapidly changing environment, I shall endeavour to show in the present chapter.

A. Large Farming on the Eve of the War

From what has been said of the agricultural conditions of Russia after the Emancipation, it will be remembered that they were generally unfavourable to the development of large farming. By the condition of the agricultural market; by the shortage of regular farm labour; most of all, by the lack of capital and credit facilities, the farming landowner was handicapped at every turn. Moreover, the difficult task of reorganization, necessitated by the abolition of serfdom, was not made easier by the fact that the personnel of the land-owning class, whose whole idea of estate management was based on the existence of serf labour, could not easily adapt itself to new conditions. Accordingly, even apart from a general improvement in agricultural conditions, some time had to elapse, which, in the natural course of events, would substitute a new generation for the old, on whom the traditional ways have grown too strong to be easily discarded. Meanwhile, the land-owning gentry have been painfully adjusting themselves to the change, at the cost of heavy losses in land and money.

To the superficial observer, indeed, the last four decades of the nineteenth century, in so far as large farming was concerned, presented a picture of unmitigated gloom and rapid dissolution. The landed property of the gentry was in course of almost feverish liquidation, most of it finding its way into the hands of the peasants. During the 27 years which elapsed between the Surveys of 1877–8 and 1905, the land-owning gentry had lost about 37% of their aggregate holdings. Those who still stuck to their land, and more particularly, to the exploitation of their estates on their own account, were loud in their complaints on hard times, which were bound to bring them eventually to ruin. Farming landowners, however, were relatively few; and most of them, especially in the central provinces of Russia, used methods of cultivation which, though they certainly reduced the capital outlay to a minimum, held out little hope of progress and prosperity. The cultivation and harvesting was done by peasants, hired to do the work with their own live and dead stock, or, not unfrequently, the land was cultivated on a system of share-tenancy, the peasants, remuneration consisting of a fixed share in the crops, which varied according to the current rates of wages in the locality.

A large proportion of the agricultural land belonging to private proprietors was leased to peasants either direct or through middlemen: a practice on whose many evils I have dwelt in an earlier chapter.

Only a small minority of landowners, in the second half of the last century, cultivated their land scientifically, using improved methods and materials; and among these probably the majority have been really running their farms at a loss and meeting their deficit under this head either from capital, or from the purely rental yield of other parts of their estates—forests, meadows, leases, etc.

An idea of the proportion of their land cultivated by landowners on their own account at the close of the nineteenth century, may be obtained from the following figures, referring to estates mortgaged with the Bank for the Nobility in 1896–1900. Of the total area of these estates, 29% were cultivated by their owners, with their own live and dead stock; 51% were leased, and 20% were partly cultivated by their owners, partly let on lease. Characteristically, along with the aggravation of the agricultural depression in the last quarter of the century, the proportion of land farmed by the proprietors on their own account gradually diminished (40% in 1886–90 and 38% in 1891–95), while leases increased (39% in 1886–90 and 42% in 1891–95).[1]

The generally depressed condition of the agricultural industry; the absenteeism of landowners; the wholesale liquidation of the landed property of the gentry; scientific large farming unable to pay its way: all this seemed to prove beyond doubt that both large landownership and large farming in Russia were on their way to rapid dissolution. So, at least, it was bound to appear to the contemporary. Now, at a distance of several decades, one is better able to discern that, amid the general gloom and the decay of what had outlived its usefulness and could not adapt itself to modern conditions, constructive forces were also at work. With reference to English farming it has been said that it was, as a rule, to the periods of its worst plight that, at least within the last hundred years, it owed most of its progress. *Mutatis mutandis*, the same can be said of Russian farming in the course of the fifty years preceding the war and the revolution. Losses were unavoidable, and the weaker and least efficient members of the landowning class

[1] A. Tumenev, *From Revolution to Revolution* (Rus.), Leningrad, 1925; p. 159.

had to be weeded out, before what still remained of large estates and large farming could be enabled to develop under healthy conditions. The old generation was gradually passing away, many of them after having survived the last acre of their former property. Their successors, in most cases, were bred under different conditions. They had little left to squander, and unless they were prepared to do away with what still remained to them of the land of their forebears, they had to economize and to work to make their property pay. If they were wise, during the agricultural depression they could realize that the least they spent on their land, the best it was in the long run, since thus they left whatever they had available in cash or credit for future use, when the conditions would mend and investment in farming would begin to pay. This, considering the circumstances of the time, justified to a certain extent the antiquated systems of exploitation in use, especially in the central provinces of Russia, to which I have referred above. Under the economic conditions of nineteenth-century Russia, with no capital and no credit, with inadequate transport facilities and an undeveloped agricultural market, the landowner, if he did not want to sell out, could hardly do anything but wait until the slow progress of Russia's capitalistic evolution would eventually bring about her economic transformation and infuse new life into the lingering countryside. During this period of transition and waiting, the antiquated methods of exploitation, inherited from the old economic regime and enabling the farming landowner to reduce his outlay of capital to a minimum, while they may not have been profitable business, were at least a sound policy of self-preservation. In the Northern half of Russia, in so far as the losses suffered since the Emancipation did not actually result in the sale of the estate or of part of the land, they have mostly been met by the felling-down of forests, and the younger generation, on succeeding to the land, found it often denuded of its most valuable asset. Before the estate could be expected to pay, there were two, three or even four decades of waiting to be endured, during which the strictest economy was imperative. Those who could or would not face it, sold out; others stuck to the land and set to the task of repairing the damage and restoring their property if not for their own, then at least for their children's benefit. It must, certainly, be admitted, that the number of failures was great, and that it was only a minority that had survived the ordeal and pulled through. Quite apart from personal characteristics, there were numerous other causes which made the struggle of the Russian large landed proprietor very difficult. Among these, the most important by far was the custom of dividing the estate among the heirs at every succession, which could break up a large, well-balanced and well-farmed estate into small and unmanageable fragments within the lifetime of a single generation. This custom, which even the despotic will of Peter the Great had been unable to alter by his Ukaze of 1714, aimed at the establishment of the law of primogeniture, made, indeed, the survival of large estates and, incidentally, of large farming, extremely difficult and, in any individual case, to a large extent dependent on chance. The attempt, in a modified form, was revived in 1898 by the permission given to members of the land-owning gentry to entail their estates, with a view to preventing their division and to keeping the land in the family; but the custom proved too strong, and the permission had remained a dead letter. At every succession, the land continued to be divided among all the heirs, and unless they chose to succeed as joint owners and to manage the estate as a whole, on their joint account, which practice had its very obvious drawbacks, its exploitation had to be entirely re-organized. As a result, the average estate was relatively small, especially in the older parts of the country, and the great bulk of the land-owning gentry were people with small means, mostly dependent, at least to some extent, on employment as Civil Servants, officers in the Army or Navy, or the practice of some professions. This, while, to some extent, it could have helped some of the landowners to pull through the worst times, could not, on the whole, be favourable to the development of farming on their own account, since it necessitated the absence of the proprietors and the management of the estate by a steward, under the master's occasional control. Yet, in spite of all these obstacles, there existed certain groups of the landowning class, within which, in the midst of the dissolution that was going on, certain constructive tendencies were at work leading to the revival of large farming conducted on scientific lines, which, though small in extent, could not be denied an important rôle in the economic development of modern Russia. Along with the process of decay, there was noticeable a growth of fresh productive agents both personal and material.

While, during the first three decades following the Emancipation, the rapid liquidation of the landed property of the gentry was considered to be an unmistakable sign of its utter and unavoidable doom, by the close of the century voices began to be heard which contended that, though much of the land of the gentry was being thrown on the market, within that class itself there existed groups which, far from appearing as sellers, actually increased their landed property by fresh purchases. Viewed from this standpoint, the process, while it involved the dissolution of one section of the landowning class, pointed also to the consolidation of another section within it, which appeared as if promising to evolve into the agrarian class of the future.[1] To the present writer, looking back to this development from a distance, this view, mostly contested at the time, would appear to have been entirely borne out by the evolution of Russian farming on the eve of the war, under rapidly changing agricultural conditions.

In certain other directions, there has also been some progress. Thus, the gradual evolution of a system of long-term credits for the landowners, of which they stood in so urgent a need on the morrow of the Emancipation, served to ease the financial position of large farming. The movement originated in the midst of the landowning class itself in the 'sixties, when the *Zemsky Bank of Kherson*, the first mutual credit institution of the kind, and the *Mutual Land Credit Society* in St. Petersburg have been established. In the early 'seventies, the following eleven joint-stock land banks were founded: the Land Mortgage Banks of *Kharkov, Poltava, Petersburg-Tula, Moscow, Bessarabia and Taurida, Nizhny-Novgorod and Samara, Kiev, Vilna, Don, Yaroslav-Kostroma,* and *Saratov-Simbirsk.*

Of particular importance to the landowning gentry, however, was the establishment, in 1885, of the *State Land Bank for the Nobility*, which started its operations in 1886. The Bank was founded for the purpose of providing cheap credit to the landowning gentry on the mortgage of their estates, and its activities extended to the whole of Russia, with the exception of Finland, Poland and the Baltic Provinces. The terms of the Bank were very liberal, the inclusive annual charge for accommodation, since 1889, having been limited to 4½%, which was considerably below that of other credit establishments, not excluding the Peasants' Bank, whose rate was reduced to the same figure only in 1906. The Bank, which was designed rather to serve a political, than a commercial purpose, by preventing the utter dissolution of the hard-hit landowning class, which for reasons of State it was considered essential to prevent from breaking its bonds with the countryside, has been working at a loss, and its financial history makes dismal reading. Year after year, arrears in interests were

[1] The first to call attention to this phenomenon was Ionov, at the close of the last century. He was followed by Yasnopolsky, by whom his views were developed. V. Sviatlovsky, in his book on the *Mobilization of Landed Property in Russia* (1907), says that, though it still lacks sufficient statistical corroboration, this view has certain probabilities in its favour (pp. 77 and 97).

being piled up, and in spite of a very liberal attitude with regard to defaulters, numerous estates were forfeited and sold by auction. The Bank, therefore, has mostly been held up to ridicule as a failure, which, from the commercial point of view, it most certainly was. Yet, it must be admitted that though, through it, millions of public money have been squandered in loans to spendthrifts, who thus frittered away the last that was left of their former estates, it had also helped many of those genuine farming landowners, whose uphill struggle would have been hopeless without this much-needed source of cheap credit.

By the close of the last century, accordingly, the Russian landowners were already fairly well provided with facilities of long-term credit. Any landed proprietor, from noble down to peasant, could obtain credit, on the mortgage of his land, in one of the joint-stock land banks, while members of the landowning gentry had special facilities provided them by the Bank for the Nobility. The indebtedness of landed estates developed rapidly. In 1867, with only two mutual credit banks, just opened, in operation, in European Russia proper it did not exceed 99 millions of roubles. In 1886, it reached 485 million roubles, which included the loans of the Peasants' Bank. On July 1st, 1899, the indebtedness of landowners to the Bank for the Nobility, including the Special Branch of the latter, charged with the winding-up of the business of the Mutual Land Credit Society, and to the other eleven land banks, amounted to 1,235 million roubles. On January 1st, 1913, it reached 2,077 million roubles: an increase of about 40% in 12½ years.[1]

There are, unfortunately, no means to ascertain the extent to which, at various times, the available credit facilities have been used for productive purposes. The land banks, as a rule, were not very particular about the uses to which the money was put, so long as their loans were adequately secured on the property. That much of the money thus obtained had been spent unproductively, was perfectly certain. During the first few decades after the Emancipation, probably no more than a small fraction of the loans found productive employment. In the twentieth century, there has probably been some change in this respect, and both the more business-like attitude of the landowners with regard to their estates and to their exploitation, and the fact that numerous estates were being reorganized and entirely re-equipped on modern lines, at more or less considerable expense, would appear to suggest

[1] *Q.v.* M. Hertzenstein, Article on *Agricultural Credit* in Brockhaus; also *Land Credit* in the 1915 edition of the same *Encyclopædia.*

the conclusion that a larger part of the available credits was invested in agricultural production.

The position was decidedly worse with regard to short-term credit facilities. With a very variable demand for cash according to season, which is inherent in the nature of the agricultural industry, the farming landowner, unless he was in possession of considerable cash balances at all times, often found himself faced with difficult problems. The commercial banks, though, within the last ten or twenty years preceding the war, they began to develop this side of their business and to pay more attention to the countryside, were generally far too remote from the average provincial landowner, who held no considerable balances on deposit with them, and their services, as a rule, were available only to the richest landed proprietors, mostly combining agriculture with such branches of industrial production as sugar refineries, distilleries, etc. Moreover, the scattered character of this kind of custom made the task of organizing such credit extremely difficult, if not actually impossible, for the banks, even with branches and agencies. Accordingly, in so far as they intervened on a large scale, the big commercial banks did so in a limited sphere, confining themselves to the financing of such staple produce as grain, and this indirectly, through the medium of railways, which, besides acting as collectors and warefingers, undertook to act for the banks in providing credits on grain delivered to the stations. This, while it eased the position partially, fell naturally far short of meeting all the credit requirements of the farmer throughout the ebb and flow of the agricultural year. As a result, the average farming landowner, having no established connections with any of the banks and, in most cases, being in actual need of a few hundreds or thousands of roubles at a time, could find himself in a quandary even over so limited an amount, in the face of immediate payments to make and with many times the required amount standing in the field.

Some attempts to improve the situation in this respect were made by the State Bank, but the difficulties which prevented the development of this potentially profitable business by the commercial banks stood on its way also, practically frustrating its efforts. When, in 1884, the State Bank was authorized to grant short-term credit to landowners on the discount of their promissory notes, it was intended by this means to meet the seasonal demands for cash in the agricultural industry. Yet, the actual form of accommodation devised was a sort of clumsy compromise between a bill and a mortgage, devoid of the advantages of either of these two varieties of credits. Since the promissory note could not be secured on goods, which remained in the possession of their owners and were scattered all over the width and breadth of the Russian countryside, it had to be secured on the land of the borrower. Accordingly, the loans granted on promissory notes, originally for nine months, were secured on the estate and were limited to 60% of the unincumbered balance of its value or to two-thirds of the amount of working capital required for its exploitation. In 1892 the term was extended to twelve months and the limit raised to 75% of the balance. It may be seen that the amount of credit was determined in accordance not with the actual needs of the business, but with the value of the land. Extremely clumsy and devoid of any connection with any special business transaction which would liquidate it, the promissory note, through renewals, tended to degenerate into a standing liability encumbering the estate on top of any possible mortgages. Since 1898, joint-stock commercial banks were permitted to transact this kind of business to the extent of one-fifth of their capital and reserve, the State Bank accepting for rediscount such promissory notes, provided they had no more than six months to run. The business of the State Bank in connection with these promissory notes, either discounted or re-discounted, had never achieved any marked development. On January 1st, 1913, the balance-sheet of the Bank under this head showed only 13 million roubles. The statute of 1894, along with other modifications in the scope of the Bank's activities, provided also for the granting of credits to landowners for the purchase of live stock and equipment, the construction of sugar factories, spirit distilleries or other works engaged in the treatment of agricultural produce, etc. These credits, however, also failed to meet the need of the farming landowners for genuine short-term accommodation. Though, indeed, with the general development of the money market in the years preceding the war, the large farmer, here also, was in a better position than before, of organized regular short credit facilities he had still none. The regular organization of short-term credit for landowners, indeed, would under all conditions have been a difficult problem, unless it was tackled on co-operative lines. Here, however, the relatively small numbers of farming landowners and the fact that they were scattered all over the country, made organization extremely difficult, if not altogether impossible. The best method, accordingly, would have been the participation of local landowners in rural co-operative societies, which, if developed on a somewhat larger scale, as this would have generally resulted from the inclusion of local large farmers, would have been in a position to meet most of the current demands of the average farming landowner for short-term credit accommodation. In some cases, as a matter of fact, this practice had been adopted, at least on a tentative scale, and its further extension would have probably been to the advantage of all those concerned, had large farming continued to exist in Russia. As it was, however, though obtaining the necessary cash was much easier in the twentieth than in the nineteenth century, the farming landowner had, as a rule, still to depend for his current requirements in credit accommodation on casual sources.

To the agricultural labour problem, as it developed after the Emancipation, I have had the occasion to refer briefly, while dealing with the agrarian evolution of Russia on the morrow of the Emancipation. Though, by the close of the century, conditions in this respect had changed considerably, viewed from either the employer's or the workers' standpoint, the labour problem in Russian farming had never been solved satisfactorily and, up to the last, had remained in a state of chaos, a real jumble of survivals and modern developments. The reason for this, even apart from the natural differences existing between various localities of so vast a country, was, to a large extent, to be looked for in the economic development of Russia since the Emancipation, in the course of which the extreme shortage of capital and of credit and the depressed condition of the agricultural market forced the farming landowners to stick to those forms of labour contract, which, though they may have been avowedly obsolete and utterly inefficient, had at least the advantage of reducing the capital outlay to a minimum. In those parts of the country, whose economic progress was more rapid, and in which the agricultural industry, by the close of the last century, had become more thoroughly commercialized, the survivals of old forms of labour contract, which represented but a slight modification of the conditions which had existed under serfdom, except for the fact that the labourer was now a free man and, in one way or another, had to be paid specifically for any work he was doing, have been disappearing ; in other parts they exhibited greater vitality. The process of transformation in this respect was greatly accelerated in the early part of the twentieth century ; but, though on the way of being modernized, the forms assumed by the relation between the employer and the worker in the agricultural industry still varied greatly and presented many peculiarities and problems.

From the outset, it is necessary to point out that, in Russia, the familiar figure of the English agricultural labourer, professionally engaged in supplying his labour on a certain farm or a number of farms, was non-existent. The farming landowner depended for his labour on the peasants of the neighbouring villages, who, by thus working, eked out the insufficient incomes they derived from their own holdings. His farm servants, such as ploughmen, cowherds, etc., he hired by the year or the season, from among either the single or landless peasants of the neighbourhood and the sons and daughters of peasant families, who went to work out until they got married, or from among wandering professional farm hands, the latter being most common in the steppes, to which there was an annual influx of seasonal migrant labour from other localities. The seasonal demand for extra labour in the regions of extensive cereal farming, where the seasonal variation in employment were particularly pronounced, were met by yearly migrations involving hundreds of thousands, and sometimes over a million, of workers of both sexes.

Considering the importance of the part played in the labour supply of Russian large farming by the employment, as casual or seasonal labour, of local peasants, it is best to start with this particular group of workers.

Though there are no exact statistics available to show how many peasants in various parts of Russia were engaged in providing, from time to time, occasional work as agricultural labourers, since even the Census of 1897 left this question unanswered, the existing approximate estimates, as well as the actual observation of the life of the Russian countryside, tend to show that the proportion was very considerable, especially in the purely agricultural districts. According to an estimate made in 1894, by probably the best authority on the subject, in the 'eighties and early 'nineties of the last century, 25% of the male peasant population of working age in the black-earth belt, and 10% outside it, were, to a greater or lesser extent, working as agricultural labourers.[1] In fifty provinces of European Russia, at that time, the total supply of male labour available for employment in agriculture was estimated, accordingly, at about 2·4 million men: a figure amply sufficient to meet the demands of all classes of farmers depending on hired labour either generally or at certain seasons. From them were recruited farm servants, casual labour and seasonal migrant labourers. Besides, there was a considerable, though statistically unascertainable,

[1] S. F. Rudnev, Statistician to the Zemstvo of Saratov, *Occupations of the Peasants of European Russia*, published by the Zemstvo of Saratov in 1894. His estimate is based on Zemstvo censuses, involving 81 administrative districts in 13 provinces. Quoted here according to L. Kirillov's article on *Agricultural Labour* in the Russian *Encyclopædia* of Brockhaus & Efron.

supply of female agricultural labour, used on an especially large scale during the hay-making and harvesting season.

The relations between the landed proprietor and the labour recruited from the neighbouring villages presented a great variety of forms. As I have mentioned before, in the central provinces of Russia as late as the beginning of the current century it was still relatively common for the farming landowner to have no farm stock of his own and to have all the work of his estate, from ploughing to the carrying of the harvested grain to the railway station or market, performed by peasants, with their live and dead stock, just as this had been formerly done by them under serfdom, with only that difference that now they were paid for their work in money or in kind. The labour was supplied either as time work, and paid at so much a day per team, usually consisting of man and horse, but sometimes, according to the nature of the work done, also of a lad or a woman, or on a piece-work basis, at so much per dessiatin, according to whether it was ploughing, sowing, harvesting, etc. The practice of payment in kind, very common in Central Russia until the close of the last century, had been rapidly going out of fashion since. Under this system, the peasants usually contracted to perform the whole round of work either on the whole, or on a certain specified part of the landowner's fields or meadows, in exchange for a fixed part of the crops, which varied according to the cost of labour in the locality. In some cases, under such arrangement, the peasants even supplied their own seeds, and the labour contract thus became hardly distinguishable from one of share-tenancy.

While, as applied to the cultivation of fields, the system of handing over the whole work to peasants for a fixed payment in money or in kind was gradually dying out, it survived and even was actually spreading in certain parts of the country in hay-making. Thus, in the Petrograd region, it has been making decided progress, at the expense of day-work, within the lifetime of the present generation. The system, indeed, was characteristic of relatively highly commercialized districts with sparse population and a shortage of agricultural labour, most painfully felt at the height of the all-important hay-making season, the principal moment of the agricultural year in the estates of the Petrograd region. The local peasantry were busy with their own hay-making, and the arrangement, which left them more freedom in disposing of their time and their available hands, suited them better than time-work. Indeed, it was practically the only means by which a landowner, in such districts, could secure the necessary labour, since journeymen, at this season, were extremely difficult to obtain, and their actual attendance utterly unreliable. The system actually involved an agreement, by which a certain peasant family or gang (*artel*) assumed complete responsibility for the mowing-down, drying and stacking of the grass on a specified area, at so much per dessiatin, according to the quality of the crop and the nature of the ground. The owner had only to see that the hay, before stacking, had been thoroughly dried, and to settle with the workers after having measured, together with them, their respective plots.

Among the survivals of labour contracts, originally entered into at the Emancipation, the most frequent practically all over Russia, and especially in the Great-Russian provinces, was that involving certain specified labour services on the part of the peasants of some particular village for the use of some fields, meadows or woodlands which had been in their possession before the Emancipation, but, being in excess of the local norm, have been retained by the landowner.[1] Here, the peasants contracted, year after year, so long as the land was left in their possession, either to perform certain specified tasks for the landowner, such as the harvesting of a certain area of crops, etc., or to provide a certain number of hands, at a specified time every year, for some seasonal work. Similar agreements, involving the payment of rent for the lease of an area of land in labour instead of money, even apart from the case of "cut-away land," have been entered into fairly often, especially in the 'seventies and 'eighties, and many of them survived into the current century; but they were gradually dying out since the 'nineties with the growth on money economy, and on the eve of the war they could seldom be met with. Especially in the case of "cut-away land," that vexed question of the Russian countryside, they had little to recommend them to either of the parties concerned, and were forced on both by the dearth of money in the undeveloped national economy of nineteenth-century Russia. With the possible exception of advance hirings, they were probably the worst of all form of securing labour and have been responsible for an infinite amount of friction and mutual recrimination between the workers and the employers.

Advance hirings were also the direct result of the great shortage of capital, which forced the Russian farming landowner to have recourse to every possible means of reducing his outlay, on the one hand, and of the pressing need for cash on the part of the peasants, on the other. The peasants were mostly in urgent need of money

[1] *Vide* about "cut-away land" in the chapter on the Agrarian Problem.

about the end of the year or, in the Northern provinces, during the early spring, and this made it possible for landowners to secure their work for the next agricultural season in advance, by lending them the whole or part of their wages months before their services were actually needed. Since, for the landowner, the winter was usually a period of financial high tide, with money coming in and the need for working resources at the lowest, he could conveniently advance some cash to prospective workers, thus formally binding them by contract for the season. Moreover, as advance hiring, being a credit transaction, was mostly effected at rates of wages considerably lower than those payable during the season, a direct economy on the wages bill could be achieved by this expedient. To what extent this was so, may be seen from the fact that, according to statistics collected in the 'nineties, in the Tambov district, for instance, the cultivation and harvesting of one dessiatin of cereals, which cost 8 to 12 roubles during the season, if carried out by means of labour hired in advance, cost only 4·50 to 5 roubles. In the Elatomsky district, the difference in the cost of harvesting grain crops was no less marked, the cost working out, with seasonal hirings, at 5 to 7 roubles per dessiatin, and with labour hired in the preceding autumn—at only 3 to 4, and sometimes even 2·50 roubles.[1] The system applied to all kinds of labour and all classes of seasonal workers, including hands hired for the whole summer, from St. George's Day to the Fast of St. Philip, in November. The extremely high rate of discount of the money thus advanced made the transaction, in some cases at least, little better than usury. The peasant, moreover, when the season actually arrived, was greatly tempted to evade the liability incurred and to sell his labour to the highest bidder, instead of going to pay off old debts. On the whole, neither party to the contract found satisfaction in this arrangement, and most of the legislation which, from 1863 on, had been passed in connection with the agricultural labour problem, had been aimed at the evils engendered by this anomalous system. Indeed, as the Smolensk Provincial Conference on Agricultural Labour put it in 1897, "nowhere are the relations between employer and worker so strained as in the agricultural industry: the one is forced to hire, and the other is compelled to serve, and each suffers the other as an unavoidable evil." [2] According to the opinion expressed on the same subject, in connection with the revision of the existing legislation on agricultural labour, by the Director of the Fiscal

[1] L. Kirillov, op. cit.

[2] Quoted according to V. Rosenberg, *Chronicle of the Peasant Problem* in Manuilov's *Essays*, Vol. I, p. 52.

Office of the province of Vologda, the contracts of advance hiring were "actually entered into under duress, the part of compulsory factor being played by the poverty of those peasant families which have little land or few bread-winners." [1] That the workers often broke the contracts and let down the landowners at the height of the season, was not unnatural. Indeed, according to the view of an official Committee of the Ministry of the Interior, "no fear of criminal prosecution would prevent the commission by workers of acts to which they are driven by the force of irresistible need." [2] The legislation which, from the Provisional Regulations on Agricultural Labour of 1863 onwards, had attempted to compel the observance of labour contracts, under pain of imprisonment of offenders and their compulsory return to their place of employment through the police, proved a failure. Life alone, in the course of time, could cure an evil produced by the unhealthy economic conditions of Russia during the period of transition from medievalism to modernity, by gradually eliminating its causes. And, indeed, since the beginning of the twentieth century, along with the economic progress of Russia, the system of advance hirings, together with other survivals, had been disappearing with great rapidity.

In many localities, especially during the agricultural season, the supply of labour, however, was quite insufficient to meet the demand. While, in the densely populated provinces of the agricultural centre, the Ukraine and the South-West, the local supply was in excess of the demand at all times of the agricultural year, in the steppes of New Russia and of the South-East, with their sparse population and extensive cereal farming, which tended to make the seasonal fluctuations in the demands for labour very pronounced, the requirements could never be met out of local sources. At the beginning of the current century, the deficiency in seasonal labour of the twelve provinces of the Russian wheat belt, from Bessarabia to the Urals, was estimated approximately at one million workers of both sexes, and it was there that the excess of labour available in the Centre, the Ukraine and the South-West proceeded every spring in search of employment. The great bulk of this migrant labour came from the provinces of Poltava, Chernigov, Kiev, Podolia, Orel, Kursk, Voronezh, Tambov, Riazan and Tula; some from those of Smolensk, Penza, Kazan and Nizhny-Novgorod. They started mostly in April or May, and even as late as the beginning of the twentieth century the majority made the journey on foot, thus saving the cost of a railway fare. In the South, they proceeded to local centres, railway stations or large

[1] V. Rosenberg, op. cit., p. 54.

[2] *Ibidem*, p. 58.

villages, the best known of these markets of migrant labour being Kakhovka and Jankoy in the province of Taurida, through which scores of thousands of hands passed every summer. This migrant labour, numbering, year in, year out, about a million people of both sexes, did not, as a whole, consist of professional labourers, and its composition changed continually. In their bulk, apart from some peasant couples who had nothing to tie them to their villages, the migration involved mostly young people, sons and daughters of peasant families, who thus helped to fill the gaps in the budgets of their respective households. It has been, accordingly, estimated that every season about 30% of the total number of migrant workers were beginners, who had never been "to the steppes" before. Up to 40% of the men and about 87% of the women were unmarried.[1] In years of bumper crops in the South, the seasonal earnings of agricultural labour rose to relatively very high figures, though of later years the practically general adoption of improved harvesting machinery throughout the steppes had enabled the farmers to manage with fewer hands and, to some extent, equalized the demand for labour which, formerly, had been liable to eratic fluctuations. At the beginning of the current century, equipped with modern harvesters, the Southern farmer, though he had to pay higher wages than before, was at least more or less assured that, in a year of good crops, he would not have to see his wheat perishing in the fields for want of hands. From the workers' viewpoint, the position, though it was now devoid to a large extent of some chances of unexpectedly high earnings in case of good crops and a run on the labour market, had also improved, since, while the diminution of the demand for hands due to machinery had been compensated by an extension of the cultivated area, and the wages have accordingly risen, the greater uniformity of demand assured them steadier employment throughout the season. This was particularly important, because the agricultural labour market was entirely unorganized, and the adjustment of supply and demand throughout the vast area of the steppes was generally very imperfect.

The organization of migrant labour and of its distribution, indeed, presented enormous difficulties. Necessary as this yearly influx of seasonal labour was for the agricultural industry of the Russian wheat belt, its evils, social, economic and hygienic, were manifold. The labour conditions of these migrants did not easily

[1] Figures given by Prince Shakhovskoy in his *Outside Earnings of Peasants in Agriculture* (1903), quoted here according to Prof. N. Kablukov, *Conditions of the Development of Peasant Farming in Russia*, pp. 76–7.

lend themselves to control, and their general position invited all sorts of abuse by anyone unscrupulous enough to profit by the ignorance of these strangers, often mere lads and young girls, and their circumstances. Any organized supervision over the working conditions, accommodation and other aspects of the employment of migrant seasonal labour, presents enormous difficulties. When, in the Russian steppes, it involved close upon a million of men and women, scattered over thousands of square miles of country, the conditions of employment escaped all attempts at control, and the physical, social and moral consequences of this enforced *laissez faire* could easily be imagined. Only from the sanitary point of view attempts have been made, since the close of the last century, following the example set by the Zemstvo of Kherson, at organizing sanitary supervision over the places of concentration of migrant labour, which otherwise were easily liable to generate epidemics.

With regard to farm servants, such as ploughmen, cowherds, dairymen and women, etc., the position of the farming landowner, by the close of the last century, had improved considerably, as compared with the preceding decades. The increase in rural population in search of earnings outside the village had produced a type of professional agricultural workers, who earned their living by employment as farm servants (*batrak*) in large estates or with rich peasants. The *batrak* was mostly employed on a yearly contract. Besides, there were workers hired for certain seasons, from the same class of rural proletariat. Such were the summer workers (*poletchik*), whose contracts usually extended, in most parts of Russia, from St. George's Day to Michælmas or the Fast of St. Philip (November 14th). Farm servants were mostly recruited from among the local peasantry, but in the Steppes many of them came from the ranks of migrant labour. Moreover, all over Russia there has been a certain amount of migration of agricultural labour between neighbouring districts and provinces, according to changing conditions and requirements, the contingent of such labourers being generally fluid and fairly sensitive to changes in the agricultural situation.

Considered as a whole, the position of the farming landowner with regard to his labour supply at the beginning of the current century had improved considerably. This improvement, besides, was not wholly due to the growth of agrarian overpopulation, though the latter naturally had the effect of increasing the supply of hands in the countryside. Had the increase in rural population and the impoverishment of the peasantry been the actual cause of the more favourable position of large farming in this respect, as most writers on the agrarian problem suggest, agricultural wages would certainly have been depressed, while, as a matter of fact, in spite of the aggravation of the agrarian situation, the increase in the number of hands was accompanied by a considerable and sustained rise in the pay of the workers and their standard of life. While the large farmer's position improved, in spite of the rise in the cost of labour, owing to the conditions of the agricultural market, the worker has not been the loser in any sense, especially in those localities in which the extension of cultivation was most rapid, such as the South and South-East. Next came the industrial provinces of the Moscow region and the districts adjoining large centres of population, where farming had to compete for hands with other occupations. In the agricultural centre, with its dense population and few facilities of outside earnings for the peasants, the increase was much less marked. Least of all it was in the Western and South-Western provinces, with their purely agricultural production and a large proportion of proletarian and semi-proletarian elements among the peasantry.

The increase in the general level of agricultural wages from the 'eighties to the outbreak of the Great War may be seen from the figures below, showing the average wages of a journeyman in European Russia: [1]

1882–1891	55·3 copecks a day
1892–1901	56·3 ,, ,,
1902–1910	71·7 ,, ,,
1911–1914	91·3 ,, ,,

It may be seen that while, from 1882 to 1901, the pay of the agricultural worker had remained practically stationary, since the beginning of the twentieth century, with the general economic progress of the country, it increased with considerable rapidity. The average wages of labourers hired on annual contracts had increased from 59·7 roubles a year in 1890 to 86·1 roubles in 1910.[2]

Thus, the economic development of Russia since the close of the last century, while benefiting the large farmer, also raised the wages and the standards of life of agricultural labour, in spite of the pressure exercised on the labour market by the growing agrarian overpopulation.

According to locality, the wages of agricultural labour differed considerably, their variations reflecting the economic and agrarian characteristics of the several regions of Russia. The table below, showing the average rates of daily wages of male and female labour

[1] Prof. A. N. Chelintzev, *Agricultural Geography of Russia*, p. 70.
[2] *Ibidem*, p. 71.

at three critical points of the agricultural year, namely the spring, the hay-making season and the harvest, for the period 1906–10, reveals the outstanding characteristics of the various regions very clearly:

Regions	Male Workers			Female Workers		
	Spring	Hay	Harvest	Spring	Hay	Harvest
			Copecks per day			
Northern	80	93	84	46	57	54
North-Eastern	61	75	68	40	52	53
Petrograd	54	71	64	35	43	46
Moscow	67	90	81	41	49	53
Western	56	71	66	33	41	43
South-Western	41	56	63	31	41	45
Ukraine	53	77	86	37	46	56
Central Agricultural	50	66	75	30	38	48
Middle Volga	57	78	72	43	41	48
Eastern	57	74	81	35	43	53
New Russia	70	95	126	49	61	90
South-Eastern	81	95	148	56	69	98
Caspian (Terek only)	82	117	131	56	71	79

The highest rates of wages prevailed in the South-East and in New Russia, to which, it will be remembered, there has been an annual migration of seasonal labour from other parts of Russia. Next came the Northern region, where high wages were mainly due to the great shortage of hands. In the Moscow region, in spite of its relatively dense population and the small average sizes of holdings, the wages were raised by the competition of industry and farming for the available supply of hands. In the Petrograd and the Western regions the level of wages was much the same, and seasonal fluctuations in them were less marked than in localities with a more pronounced cereal-growing bias. In the Ukraine and the Central Agricultural region, wages, low in spring, rose 50 to 60% at the height of the harvest: a fact which, when average wages are given, tends to produce a misleading impression about the labour conditions of these localities, generally overstocked with agricultural labour. Here, indeed, in the spring, when the annual exodus to the steppes was only just beginning, hands were plentiful and very cheap, but as the season advanced, the supply was being reduced by migration to the South, which continued till June, and wages rose accordingly. The lowest rates of wages prevailed, throughout the season, in the South-West, though with regard to this particular region it must be said that their low level was to a certain extent compensated for by the greater steadiness of employment, due to relatively intensive cultivation.

An interesting feature of the table is the way in which it reveals the relative importance in various regions of either cereal or grass farming. Indeed, it may be noticed that, throughout the Northern half of Russia, in the Northern, North-Eastern, Petrograd, Moscow, Western and Middle Volga regions (the latter mostly on account of the provinces of Nizhny-Novgorod and Kazan, that is its Northern part), the highest wages were paid during the hay-making season, while in the rest of the country they reached their highest point during the harvest, which was the climax of the whole agricultural year.

In dealing with the agricultural conditions at the beginning of the current century, in so far as they affected large farming, it may be said that, on the whole, they have been very favourable to its progress. The farming landowner, after several decades of depression, saw his estate, at long last, beginning to pay its way and even becoming an asset of growing value, as a going concern. The principal difficulty which did really hinder the rapid progress of large farming in many localities, during that period, was the agrarian situation. The agrarian unrest among the peasants, which had become so marked a feature of the life of the Russian countryside since the close of the last century, and had only begun to subside after 1906, could not have failed to exercise a most unfavourable influence of the evolution of large farming, not only direct, by acts of violence, strikes, arson and intimidation, but also by undermining that sense of security which is essential for the very possibility of development.

The subject of large farming in Russia should not be dismissed without an attempt being made at estimating the part played by this form of agricultural production in the economic life of the Russian countryside on the eve of the war and the revolution.

One of the greatest difficulties met with by the student of Russian agricultural economics is the extreme scarcity of information bearing on large, as distinguished from peasant, farming.

The available statistics, while they were relatively complete and fairly reliable as to the actual distribution of land between the various classes of proprietors, when it came to agricultural production, failed to distinguish between large and peasant farming. Even when a distinction had been made, as in the case of the statistics of yield of various crops, the basis of classification was not the manner of exploitation, but the ownership of the land concerned, and the crops were distinguished according to whether they were grown on land belonging to private proprietors, or on the allotment land (*nadiel*) of the peasants. Since the former, along with the large farmers' crops, included also all those grown by the peasants on land leased by them from landowners or purchased by private treaty, these statistics failed to convey any information as to the position and yield of crops on the large estates' home farms. The information concerning large farming in Russia was, accordingly, so scarce, scattered and fragmentary, that no general picture either of its position at any given moment, or of its evolution in the course of time, could be obtained.

Fortunately for the historian and the student of Russian agricultural economics this gap had been filled, on the very eve of the revolution and of the end of large farming in Russia, by the Agricultural Census of 1916. This Census, taken during the Great War, in conditions both economically abnormal and technically unfavourable to the satisfactory execution of the task, was forced upon the Government by the urgent necessity of taking stock of the available supplies of foodstuffs and of their output in various parts of the country, with a view to a better organization of the provisioning of the Army and of the centres of population. The figures obtained cannot, therefore, be trusted unreservedly, the more so that, as a rule, there has been a certain tendency on the part of the informants to minimise the area of crops. The possibility of mistakes in the actual registration must also be allowed for more generously than if the Census had been taken under normal conditions. Moreover, the figures are generally below the pre-war standards, on account of the withdrawal of millions of hands, the requisitions of live stock and the disorganization of marketing facilities. With particular reference to large farming, it is necessary to lay stress on the fact that, as a rule, the effects of the war have been far more painfully felt by the large farmers, than by the peasants, and that these effects have been the more far-reaching, the higher the standards of cultivation and the greater the dependence of the producer on the market. Thus, the withdrawal of men, mobilized into the Army, had not affected the area of peasant crops to anything like the same extent, as it did affect those in large estates. In the villages, though the cultivation may have suffered in quality, it has been mostly maintained on very nearly its pre-war level by the women and those groups of the male population, unaffected by the conscription, which remained on the land. Peasant crops, grown to a considerable extent for immediate consumption on the farm or disposed of in the neighbourhood, have not felt the discontinuance of exports and the disorganization of internal transport to nearly the same extent, as did the commercial production of the large estates. Owing to the difference in the class of horses used in large estates, on the one hand, and by the peasants, on the other, the former have also lost relatively more by the requisition of horses for the Army, than the latter. Generally speaking, it may be assumed that the war had reduced the extent and production of large farming considerably more than those of the peasants, and that, therefore, the figures referring to large estates, in the Census of 1916, are distinctly minimized relatively to those of peasant husbandry, though not, on the whole, to such an extent as seriously to affect their comparative value. This value, indeed, can hardly be overestimated, since the Census of 1916 is the first, as well as the last, statistical record of Russian farming as a whole, which distinguishes between capitalistically-organized or, more loosely, large farming, depending on the regular employment of hired labour, and small family farming, practically synonymous with peasant husbandry.

Though, in the present chapter, I generally use the term "large farming," it must be clearly realized that this expression is somewhat loose. In fact, what matters is not so much the size of the arable area, as the distinctive organization of the whole farming concern, which places it into either the one or the other of the two alternative classes. This point is made clear by a consideration of the number of "large" or capitalistically-organized farms in Russia, in 1916, and the average sizes of their arable areas, according to regions.

Numbers of "Large Farms" and average Sizes of Arable

Regions	Number of Farms	Average Area of Crops in dessiatins per Farm (1 dess. = 2·7 acres)
Northern	438	10·3
North-Eastern	3,183	6·2
Petrograd	4,810	13·8
Moscow	29,474	18·0
Western	7,356	44·0
South-Western	7,868	170·4
Ukraine	11,260	74·3
Central Agricultural	12,601	90·6
Middle Volga	3,699	109·5
Eastern	5,479	81·6
New Russia	14,520	106·6
South-Eastern	6,381	?
Caspian	875	52·4
European Russia, exclusive of South-East	101,563	66·7

It may be seen that, throughout the Northern half of European Russia, outside the black-earth belt, the average arable area in capitalistically-organized farms was very small. Indeed, barring exceptional cases, even in large estates the arable branch had always been kept down to the barest minimum necessary for the harmonious working of the system of farming in force and the supply of the household and the farm. The paying branches in this region were meadows, clover, cattle and timber, and cereal cultivation played only an auxiliary part. With still greater force this applied to large farming in the Northern and North-Eastern regions, in both of which it scarcely existed at all. Arable gained somewhat in importance in the Moscow region, and more particularly in its Southern and Western parts. In the Western region the average size of arable was considerably higher than in that of Moscow. But really "large" farming was generally typical of the black-earth zone, in which arable was more profitable than other branches. Here, especially in the steppes, practically the whole of the available area could be turned into arable, and the proportion of land under plough was raised very nearly to the physical limit. The private landowner, in this respect, went about as far as the peasant, and in these parts, in direct opposition to the North, large estates were mostly ploughed up to their full capacity. Arable farming, in the black-earth belt, was practised on a really large scale, with the result that, here, the application to it of this term was not a misnomer.

With regard to the actual extent of capitalistically-organized farming in European Russia, figures are given in the table below, which shows the distribution of the total area under crops between large and peasant farming in 1916:

Regions	Large Farms Area in 1,000 des.	Large Farms Per cent	Peasants Area in 1,000 des.	Peasants Per cent	Total Area in 1,000 des.
Northern	4·5	0·6	794·7	99·4	799·3
North-Eastern	19·8	0·3	5,315·2	99·7	5,335·0
Petrograd	66·4	4·0	1,584·1	96·0	1,650·5
Moscow	531·4	7·8	6,310·1	92·2	6,841·5
Western	323·7	9·9	2,951·8	90·1	3,275·5
South-Western	1,340·5	27·5	3,527·9	72·5	4,868·4
Ukraine	836·3	14·3	5,011·8	85·7	5,848·1
Central Agricultural	1,142·3	12·0	8,365·5	88·0	9,507·8
Middle Volga	405·1	6·4	5,984·2	93·6	6,389·3
Eastern	447·1	6·8	6,097·3	93·2	6,544·3
New Russia	1,547·5	16·5	7,799·9	83·5	9,347·4
South-Eastern, excluding Stavropol	455·1	6·3	6,804·0	93·7	7,259·1
Caspian	45·9	3·3	1,327·9	96·7	1,373·8
European Russia	7,620·5	10·0	68,678·6	90·0	76,299·1

Thus, of the total area under crops in European Russia, exclusive of the province of Stavropol, for which no figures were available,

10% were accounted for in 1916 by farmers regularly dependent on hired labour. According to the separate regions, the proportions varied characteristically, pointing to differences in the relative development of large farming. Roughly, it may be said that its relative importance tended generally to diminish in the direction from the West and South-West towards the North and East. Indeed, the relative importance of large farming was greatest in the South-Western region (27·5%) and in New Russia (16·5%), where a high percentage of crops in large farms marched hand-in-hand with a high proportion of land in the possession of private landowners (40·7 and 37·5 respectively).[1] In other respects, however, the conditions in these two regions were different, the South-West being a district of high farming, abundant labour and low wages, and New Russia—a country of extensive cereal cultivation, high wages and an annual immigration of seasonal agricultural labour from other localities. In the South-West, large estates and large farming were of long standing and had developed, along with the whole agrarian régime of these provinces, during the centuries of Polish rule; in New Russia, they were of quite recent origin, the product of agricultural colonization since the clofe of the eighteenth century, when the region was annexed to Russia as a result of the victorious wars with Turkey under Catherine II. Next came the Ukraine with 14·3% of crops in large farms, and the Central Agricultural region, with 12·0% The two regions, both belonging, in the main, to the places of earliest settlement, were densely populated, and in each in 1905, private estates accounted for 27·3% of the land. Agriculturally, the Ukraine had much in common with the South-West, while the Central Agricultural region was struggling to emerge from a condition in which its cultivation was still almost wholly dominated by cereal-growing: a tendency which, while it had, as yet, scarcely affected peasant farming, could already be clearly discerned among the farming landowners. In both regions the cost of labour was low and conditions favourable to large farming. Farther East and South-East, as well as outside the black-earth belt, the part played by large farming declined considerably. In the Middle Volga region, whose Northern districts lie outside the black-earth zone, with 20·4% of the land in the possession of private landowners, large farming accounted for only 6·4% of the crops. Here, especially in the provinces of Nizhny-Novgorod and Kazan, a considerable proportion of the landed property consisted of forests, while natural meadows were also relatively extensive throughout the region. The cost of labour was higher than in the Central

[1] *Vide* Distribution of Landed Property in 1905, Appendix V.

Agricultural region, and the conditions generally less favourable to the development of large farming. The Eastern region, in which private estates accounted for 18·1% of the land in 1905, had 6·8% of the area under crops to the credit of large farming. With relatively high wages and low prices of agricultural products, the Eastern region was not a favourable ground for the growth of intensive cultivation on a large scale, but extensive cereal-growing in large farms, with the use of labour-saving machinery, could be conducted with advantage, and it was in this form that large farming was practised there. To the South-East, the farther from the centre, the smaller was the share of large farming. In the South-Eastern region, the Don had 6·7% of its area under crops in large farms; the Kuban—5·5%. The Caspian region had only 3·3%. North of the black-earth belt, the same diminution of the share of large farming could be observed. Outside the black-earth, the highest percentage of crops in large farms was accounted for by the Western region, where it reached 9·9%, while the proportion of land in the hands of private proprietors, in 1905, was as high as 53·0%. Labour was relatively cheap, and markets for the produce near and easily accessible. Large farming, which, in the Western districts of the region, which, like the South-West, had for centuries belonged to Poland, was of long standing, found itself, accordingly, placed in relatively favourable conditions. The Moscow region came next, with 30·6% of the land in the hands of private proprietors and 7·8% of the crops in large farms; though it will be remembered that here, as, to a still greater extent, in all the localities North of Moscow, the "large" arable farm was of very small size. In spite of the high cost of labour, in this region large farming found considerable encouragement in the immediate vicinity of the largest Russian consuming centres. In the region of Petrograd, with 33·5% of the land belonging to private landowners, large farming accounted for only 4·0% of the crops: a phenomenon due to the natural conditions of the locality, where, especially in the provinces of Petrograd and Novgorod, by far the greater part of the area is covered with forests and natural meadows, and arable farming was by no means encouraged. In the Northern region, both private landownership and large farming, for obvious reasons, played a negligible part, private estates accounting for 1·3% of the area, and large farms for 0·6% of the crops. Only in the province of Vologda, in the South of the region, had large farming been of any real account; even there, however, its share in the crops did not exceed 0·6%. In the North-Eastern region, though private landowners, in 1905, possessed 22·1% of the land, large farming played a part even smaller, relatively, than in the North, its share in the crops being only 0·3%.

The Agricultural Census of 1916, besides providing considerable material bearing on the position of large and small farming generally, also made it possible, by the comparison of the figures referring to each of these two alternative types of agricultural production, to obtain a certain idea of their distinctive characteristics, the difference between the respective standards of farming in its two forms and the peculiarities in this respect of various localities. An analysis of arable farming, in which the distribution of crops in large farms, on the one hand, and on land cultivated by the peasant small farmers, on the other, is compared, is attempted on page 215.

The table given opposite brings out in considerable relief the outstanding characteristics of each of the two alternative forms of arable farming, as well as, on the whole, the superiority of the rotation of crops in large farms. This superiority was, beyond doubt, to a large extent due to the fact that, while large farming was essentially commercial, the bulk of its produce being marketed, peasant farming was largely controlled by the exigencies of the farmer's own household. Partly, the higher technical standards of large farming should be explained by its being controlled by men of better education and wider outlook, with at least some capital at their disposal and no fetters, such as open fields, which still impeded the progress in this respect of the great bulk of the peasantry. Whatever the reasons for it, it may be seen from the table that, as a rule, the large farmer in Russia concentrated on the cultivation, in the first instance, of the more valuable crops, while in the case of peasant farming the staple articles of general consumption tended to predominate in the rotation. This was especially pronounced in the case of the principal cereals. Here, though the aggregate percentages of rye, wheat, oats and barley, taking the country as a whole, were very nearly the same in both classes, namely 75·2% in large farms and 78·2% in the case of the peasants, their distribution was very different. It would, indeed, appear that while, in the case of the peasants, the staple food of man and beast in the given locality was the crop most commonly grown, in large farms, allowance being made for the influence of natural conditions, commercial considerations played a decisive part in the apportionment of their respective shares to the various crops. Even in peasant farming, however, the influence of the needs of immediate consumption would appear, on the whole, to have been far less pronounced in the more thoroughly commercialized districts

Large and Peasant Farming: Distribution of Crops, 1916
(Percentages of total area under Crops)

Regions	Rye	Wheat	Oats	Barley	Potatoes	Sugar beet	Fod. roots	Flax	Grass	Other Crops
Northern:										
Large farms	31·1	—	33·3	0·9	0·2	—	—	—	22·2	2·3
Peasants	38·3	2·1	33·1	15·9	3·1	—	—	5·2	0·5	1·8
North-East:										
Large farms	25·7	20·1	34·7	2·5	4·5	—	—	0·5	7·6	4·6
Peasants	34·4	12·6	36·5	6·5	0·9	—	—	3·7	1·8	3·6
Petrograd:										
Large farms	23·9	0·7	23·2	6·9	5·7	—	0·9	2·7	35·0	0·9
Peasants	37·2	0·6	24·9	8·9	6·8	—	—	10·7	8·7	2·2
Moscow:										
Large farms	31·8	2·9	32·7	1·0	6·5	—	0·4	1·6	21·1	2·0
Peasants	32·9	0·5	20·5	0·3	6·4	—	—	5·6	3·3	30·5
Western:										
Large farms	34·1	1·6	31·8	5·8	8·2	—	—	0·9	14·4	3·2
Peasants	41·7	1·2	20·2	12·0	8·5	—	—	4·8	3·4	8·2
South-West:										
Large farms	7·4	28·5	20·5	5·5	1·5	14·9	0·2	0·1	7·8	13·6
Peasants	28·6	16·9	16·0	11·6	5·6	1·1	—	0·1	1·7	17·6
Ukraine:										
Large farms	14·4	29·3	17·7	6·6	1·9	11·6	0·6	0·2	10·5	7·2
Peasants	29·5	22·3	11·2	12·8	3·9	0·4	0·1	0·5	3·0	16·3
Central Agricultural:										
Large farms	29·9	10·7	28·9	0·9	2·9	6·6	0·4	0·2	9·6	9·9
Peasants	42·1	7·4	23·9	2·5	3·9	0·1	—	0·2	0·8	19·1
Mid. Volga:										
Large farms	41·1	12·2	26·8	0·8	1·9	—	—	0·3	7·4	8·5
Peasants	43·8	17·8	19·2	1·4	2·1	—	—	0·8	0·2	14·7
Eastern:										
Large farms	24·1	48·1	11·3	1·8	0·8	—	—	0·5	8·3	5·1
Peasants	29·9	40·5	13·1	2·7	1·0	—	—	0·2	0·9	11·7
New Russia:										
Large farms	11·2	38·3	11·4	29·9	0·8	0·8	—	0·7	2·0	4·7
Peasants	11·6	36·1	3·9	36·5	1·3	—	—	0·6	0·8	9·2
South-East: [1]										
Large farms	7·7	45·4	7·9	23·1	0·6	0·5	—	0·9	4·2	9·7
Peasants	9·2	48·8	3·6	24·2	0·9	—	—	0·4	0·7	9·2
Caspian:										
Large farms	11·7	64·9	5·4	6·8	0·2	—	—	—	0·6	10·4
Peasants	15·1	39·5	4·9	6·1	1·6	—	—	0·6	0·8	30·5
European Russia:										
Large farms	18·8	26·1	19·9	10·4	2·2	5·4	0·2	0·5	8·2	9·2
Peasants	29·5	20·8	16·5	11·4	2·9	0·1	—	1·9	1·7	15·1

[1] Exclusive of the province of Stavropol.

of extensive cereal cultivation in the black-earth belt. Thus, in the wheat-growing zone, including the Eastern region, in which rye was cultivated on a large scale and of the finest quality, and where it formed the staple food of the peasantry, the first place, even in peasant farming, was held by wheat, as a commercial crop.

A closer study of the table shows that, throughout the country, peasant farming, in its distribution of crops, pointed to the prevalence of the three-course, or even more extensive, systems of cropping, with a heavy predominance of cereals, both principal and minor, which latter formed the bulk of "other crops." This predominance was consistently and considerably more pronounced in peasant husbandry than in large farming in all parts of the country, while only very slight signs were visible of a transition to improved rotation. Of these signs, in all cases of transition from the three-course cropping to multiple rotation, the most characteristic is the proportion of grass crops, which are the most potent factor of agricultural progress at this stage of development, since they exercise a decisive influence on both arable and stock farming. In this respect the difference between large and small farming was striking. In the whole of European Russia, the percentage of grass among the crops of large farms was 8·2%, while in the case of peasant farming it did not exceed 1·7%. The maxima and the minima respectively were: for large farming 35·0% of grass crops in the region of Petrograd and 0·6% in the Caspian; for peasant farming—8·7% in the region of Petrograd also, and 0·2% on the Middle Volga. Among root crops, the most common one—potatoes—was represented by a slightly higher percentage in peasant farming than in large farms, the figures being respectively 2·9 and 2·2%. Characteristically, as a crop grown primarily for immediate consumption, with a generally low marketing ratio, potatoes have become one of the standard crops in peasant husbandry. Sugar-beet, on the other hand, while it accounted for 5·4% of the crops of large farms, did scarcely reach 0·1% of those of the peasants. The little that there existed in the way of fodder roots, though, indeed, not confined exclusively to large estates, represented any appreciable percentage of the crops only in large farms. One important non-cereal crop, which was cultivated by the peasants to a relatively far greater extent than by large farmers, was flax; but this, grown in most cases as part of the spring crops under the three-field system, owing to its effect on the soil, was far from always being a welcome element in peasant farming, and appeared as such only in those very few cases, in which it was included in a well-devised rotation, along with clover or other grass crops. Accordingly, on the whole, an analysis of the distribution of crops tends to show that the systems of cropping which prevailed on large farms were generally more scientific than those in use among the peasants. It tends also to show that the difference in the standards of farming between the large farms and the peasants was far more pronounced in those localities in which the agricultural industry was more highly developed generally, while in the regions of extensive cereal cultivation, such as the steppes of the South and the East, their technique, at least in so far as the rotation of crops was concerned, was practically the same.

As a result of more scientific rotation and higher standards of farming, the yield of crops on the estates of farming landowners was generally much higher than that of peasant land. Though, owing to the classification adopted in Russian crop statistics, which, instead of distinguishing between forms of production, made a distinction between the crops grown on the land belonging to private landowners, irrespective of the mode of exploitation, and those grown on peasant allotment land (*nadiel*), no real comparison of the yield of large and small farming can be made, some idea of the difference may be obtained even from the available data. It must be borne in mind, however, that in the figures given below and showing the average yield of the four principal cereals in 1901–10

Average Yield of the principal Cereal Crops on Private Estates (a) and on the Peasants' Nadiel (b), in 1901–10

Regions	In poods (36 lbs.) per dessiatin (2·7 acres)									
	Rye		Winter Wheat		Spring Wheat		Oats		Barley	
	(a)	(b)	(a)	(b)	(a)	(b)	(a)	(b)	(a)	(b)
Northern	52	47	—	—	54	48	54	50	57	54
North-East	56	52	—	—	56	52	58	55	57	55
Petrograd	58	47	55	37	49	43	58	50	58	50
Moscow	58	47	—	—	52	44	59	50	54	48
Western	48	39	53	45	43	40	49	44	49	44
South-West	79	70	84	74	60	51	77	65	72	63
Ukraine	68	55	74	59	60	53	66	54	67	57
Central Agricultural	69	52	67	54	55	46	62	47	57	47
Middle Volga	58	49	50	57	43	37	49	41	44	41
Eastern	60	50	—	—	51	42	51	42	45	41
New Russia	56	44	64	48	50	40	59	50	62	52
South-East	52	48	57	47	53	47	61	54	64	59
Caspian (Terek only)	56	51	65	58	53	46	55	58	58	58
European Russia	60	50	64	53	52	45	59	51	58	51

on privately owned land, on the one hand, and on the peasants *nadiel*, on the other, the former is considerably minimized. The reason is that it includes, besides the landowners' own crops, also all the crops raised by peasants either on land leased from private proprietors, or on that owned by them outside the *nadiel*, most of which, especially when it belonged to whole village communities, was farmed exactly in the same manner as the *nadiel*, and yielded accordingly. Failing better information, these statistics have to be used.

It may be seen that, on the average for the whole country in 1901–10, the difference in favour of private estates was 20·0% in rye, 20·8% in winter wheat, 15·5% in spring wheat, 15·7% in oats and 13·7% in barley. Had the landowners' own crops been separated from the rest, the difference would probably have been much greater, 50% being probably a fair estimate of the excess over the yield of peasant land.[1] From the actual yield of crops on private estates in various parts of the country, this figure would appear to represent a good approach to the real position, as an average. For a number of private estates, records of the yield of crops over a number of years have been collected and published by the Department of Agriculture. For one of such estates, in the province of Tula, belonging to the Shatilov family, known for generations as leading farmers, data were available for nearly a century, from 1806 to 1895, and the average yield of their crops, over the whole period of ninety years, was 85·2 poods for rye, 88·8 poods for wheat, and 102·5 poods for oats, the latter being the pride of the estate, famous throughout Russia (*shatilovsky oves*). Now, these averages, compared with these for the province of Tula in 1901–10, show the following differences:

	Rye	Wheat	Oats	Barley
		Poods per dessiatin		
Large estates, 1901–10 . .	60	57	61	—
Peasants *nadiel*, 1901–10 . .	49	—	50	—
Shatilov's estate, 1806–95 . .	85·2	88·8	102·5	—

For another estate, in the province of Orel, figures were available for five years, 1890 to 1894, which included two consecutive seasons (1891 and 1892) of disastrously bad crops. The average yield for the five years, compared to that recorded for the province of Orel in 1901–10, was, however, as follows:

[1] Prof. D. I. Pestrzhetzky (*About the Land*, p. 39) refers to the results of the Census of 1916, according to which the excess of the yield of crops in large farms over those of peasants worked out at 51%—a figure which would appear to confirm the estimate given above entirely.

	Rye	Wheat	Oats
	Poods per dessiatin		
Large estates, 1901–10	61	72	55
Peasants' *nadiel*, 1901–10	45	—	40
Romer's estate, 1890–94	92	76	88

On the whole, for efficiently managed private estates, these yields of crops are by no means excessively large, and similar examples could probably be multiplied into thousands, had sufficient records been available. Here, I refer only to two out of a considerable number of estates described by the Department of Agriculture in its publication.[1] If these may be looked upon as to some extent models of good farming, the present writer can refer to several estates personally known to him, on whose behalf such a claim could hardly be made, but whose yields, year in, year out, were not much lower. In one estate, in the province of Novgorod, which was most intimately known to him, a year in which the rye crop was under 75 poods per dessiatin was considered, for the last two generations at least, distinctly bad. The average for the province of Novgorod, in 1901-10, amounted to 53 poods of rye per dessiatin on private estates and 41 poods on peasant land. In the neighbouring two estates the standard was approximately the same as in the one referred to.

With the available statistics, as well as the general scarcity of information concerning large farming, one can hardly go any farther in trying to bring out its outstanding characteristics, as one of the factors of Russia's agricultural evolution on the eve of the Great War. A statistical comparison of the live-stock branch of large and peasant farming respectively would be of little real value in throwing any light on the relative position of these two forms of agricultural production. On the one hand, statistics, while they give the numbers of animals, say nothing of their quality and value; and it is precisely in the latter that large farming differed greatly from peasant husbandry. On the other hand, if in the case of arable farming, the essential distinction between commercial production and maintenance farming exercises a marked influence on the crops of large and small farms, this influence is far more pronounced in the live-stock branch, and especially with regard to horned cattle. Accordingly, any attempt at finding out from the constitution of herds, for instance, the bias of peasant husbandry, as compared with large farming, towards either dairy production, or the raising of cattle for meat, is most likely to be frustrated by the nature of peasant farming, in which the direct consumption of

[1] Figures quoted according to Pestrzhetzky, op. cit., p. 36, as the original publication of the Department of Agriculture was not available.

milk plays a decisive part, and the herds, therefore, always contain a high proportion of cows. Since dairy production, as a general rule, is a more intensive form of stock farming, than the raising of meat cattle, a comparison of the constitution of herds in large and peasant farms is most likely to suggest the conclusion that peasant farming, in the case of this particular branch, is generally higher than that of the farming landowners: a conclusion which would be rejected as absurd by anyone familiar with Russian agricultural conditions before the war.[1] This is why, in this section, I prefer to refrain from entering into a detailed examination of the stock branch of large and small farms. Moreover, in dealing later with the production of Russian farming, I shall have to refer to the special services rendered in various branches of stock farming by each of the two alternative types, as well as to their characteristics in this particular sphere.

To sum up the position of Russian large farming on the eve of the war, it may be said that, at that time, it was just beginning, economically, to find its feet and to become a considerable factor in the economic development of the country. Though, as the reader could have observed from the figures given above, large or, to be more exact, capitalistically-organized farming accounted only for about 10% of the crops, its share in the advancement of Russia's economic progress could not be measured by quantitative standards alone. Rather it should be said that the farming landowner in Russia stood for quality, just as the vast expanse of peasant production, by which he was surrounded, represented quantity. The large farm, though its aggregate contribution to the area under crops was but one-tenth, produced, per unit of that area, about 50% more, on a conservative estimate, than did the peasants. The large farmer kept fewer cattle per unit of area under crops, than did the peasants, but of far finer quality and higher value. Not only by its superior production did large farming in Russia justify its existence, but also by its cultural rôle in the countryside, in which a real farming landowner, interested in agricultural science, was a not unimportant factor of progress. Much work that could not possibly have been done, unless at enormous expense, by the

[1] This is precisely the mistake which, in the view of the present writer, Professor Chelintzev commits in his work on "Large Farming in Russia on the Eve of the Revolution," published in 1925 in the *Transactions of the Institute for the Study of Russia*, in Prague. From a comparison of the constitution of herds in 1916, he derives the conclusion that "the standards of stock farming in its various branches in large estates were lower than among the peasants," entirely omitting to take into account the essential peculiarities of the two forms of production, which preclude effective comparison.

Government or other public bodies alone, in the scattered mass of Russian peasantry, was being done imperceptibly, day after day and year after year, by the presence in the countryside of men and women of education and culture, working on their land.

As a source of cash earnings for the peasants, large farming, small as was its extent on the eve of the war, also played a not altogether negligible part in the economic life of the Russian countryside.

Last, but not least, from the national point of view of a country dependent for its financial stability and general progress on the existence of a large excess of exports over imports, large farming was responsible for the production, year after year, of an exportable surplus of high-grade wheat and other cereals. With the inherent tendency of the peasants to concentrate, in the first instance, on the cereals supplying the needs of immediate consumption, to which I have pointed above, the disappearance of large farming, working primarily for the market, was bound to produce in Russia the same effects as in Rumania, where the agrarian reforms, by the expropriation of large estates, broke the trade balance beyond repair.[1] Though, in the cross-currents of various forces now playing havoc with the economics of revolutionary Russia, it is difficult to disentangle the separate influences, it appears highly probable that, among the causes of the failure to restore cereal exports, the same factor is operative.

Now, looking back at the part played by large farming in the economic development of Russia in the years preceding the Great War, one cannot help thinking that the country would have gained, had those still numerous landowners, who preferred to let their land to peasants and thus enjoy sometimes large, but always parasitic incomes, put their hands to the plough. Such leases, under Russian conditions, were always open to abuse and to exploitation of tenants, since no law could effectively control their numerous varieties throughout the country. They did, in the long run, no good to anyone, while they ruined the land and embittered the agrarian relations, leaving the genuine farming landowner to pay the price in hatred, in material losses, and ultimately, during the Revolution, in the land on which he had worked.

B. Conditions and Evolution of Peasant Farming

The beginning of the twentieth century was a period of fundamental transformation of peasant farming in Russia. The century

[1] *Q.v. The Agrarian Revolution in Eastern Europe*, edited by Prof. Max Sering, with an Introduction by him; Berlin, 1925 (Rus.).

opened with an extreme aggravation of the agrarian crisis and a growing unrest in the countryside, culminating in the revolutionary upheaval of 1905. This was followed by the agrarian legislation of Stolypin; and since 1907 the country had been gradually settling down to the work of peaceful reorganization and rapidly casting away those survivals of the old agrarian regime which hindered its agricultural progress. The Russian village which, hitherto, had appeared so dismally inarticulate and inert, was rapidly changing, having entered on the most eventful decade in its history since the Emancipation.

The peasantry had, indeed, been affected by a number of influences, partly coming from outside, partly generated in its own midst. The old agrarian organization, inherited from the days of self-supplying natural economy, was bursting asunder under a combination of strains. The countryside was stirred by the surging tide of the industrial revolution, with its concomitant commercialization of farming. At the same time, the process of its internal transformation, involving the growth of individualism in the village, produced and released a variety of forces which, in their turn, tended to bring about important changes in the life of the peasantry.

Owing to the heterogeneous composition of the Russian peasant class, the improvement in the conditions of the agricultural market affected various groups of the peasantry in very different ways. In the under-producing provinces of the Northern half of Russia, the peasants' own crops, as a rule, could only support them till Christmas or, at the very best, till the spring. Accordingly, in this part of the country, as far at least as cereals were concerned, the peasants, for three to nine months in the year, appeared on the market not as sellers, but as buyers of the principal agricultural products, and the rise in grain prices affected them, on the whole, unfavourably. Indeed, part of the loss under this head could be set off by gains on the sale of dairy produce, flax, etc., but the development of those branches of farming, on which this possibility of compensation depended, was greatly restricted by the prevailing communal system of tenure and open fields. The latter, by enforcing uniformity of cultivation and restricting the supply of fodder, tended generally to keep down the yield of all branches of farming, with especially unfavourable consequences for live stock. In order to profit by the improvement in agricultural conditions, the peasant farmer of the under-producing provinces of Russia had to reduce, as far as possible, his dependence on the market for the principal cereals, by raising their yield on his own holding, and at the same time to develop the commercial branches of his production, mostly dependent on live stock. The solution lay, obviously, in the change over from the traditional three-course system to multiple rotation, including grass crops: a solution which, though extremely difficult under communal tenure, in open fields, was made relatively easy by enclosure. Thus, in so far as, in the under-producing provinces, the peasants, though partly dependent on outside earnings, were yet vitally interested in the yield of their land, the change in agricultural conditions decidedly favoured the growth of peasant enclosures. Indeed, it was a characteristic phenomenon that enclosures, while they made little progress in those localities where the peasants had abundant outside earnings, besides which the yield of their holdings dwindled sometimes to insignificance, developed with considerable rapidity in the more purely agricultural districts of the under-producing provinces. In the former, the peasants found sufficient compensation for the rise in cereal prices in the increase of their earnings from other sources than their land; in the latter, they sought to compensate themselves by increasing the yield of their holdings, and were thus unavoidably led to the recognition of the necessity of higher farming, and of enclosure as a means to that end.

In the purely agricultural, mostly cereal-growing, provinces of the black-earth belt, the large peasant farmer, being a regular producer for the market, certainly gained all-round from the improvement in agricultural conditions. As far as the average peasant was concerned, however, the position was not so simple. Indeed, even those peasants, whose crops were barely sufficient for their own maintenance, mostly appeared on the grain market, in autumn, as sellers, only to reappear on it, in the spring, in the capacity of buyers, having oversold under the pressure of their need for cash. On them, the effects of rising grain prices could tell in a variety of ways, according to the extent to which they had actually encroached on their own supplies, as well as to the difference in the prices of the two seasons. With the strong cereal-growing bias of peasant farming, especially pronounced in the black-earth belt, and the under-development of other branches which might compensate him for any losses under this head, the average peasant of the agricultural provinces of Russia could hardly be said to have benefited to any marked extent by the rise in the prices of agricultural products. At the same time, the agricultural situation could not fail in bringing home to the average peasant farmer of the black-earth belt what he stood to gain by increasing his own output, and thus to create, for the bulk of the peasantry, such stimuli to enclosure and intensification of farming, as could not have existed, had agricultural conditions remained depressed. The peasant small-holder, unless he specialized in some particular form of intensive production, such as market-gardening, tobacco-, vine- or fruit-growing, etc., which he was only able to do under especially favourable conditions and in some particular localities, was dependent on outside earnings and, being definitely a loser from rising cereal prices, could only find compensation for them in an increased remuneration of his labour.

Characteristically, on the whole, the effects of the improvement in agricultural conditions generally, and the influence of rising prices in particular, tended to encourage the individualist movement in the Russian village, as expressed in enclosures, on the one hand, and to bring about a considerable change in the distribution of land among the various groups of peasants, on the other. Indeed, now that farming was beginning to pay, the difference between the genuine peasant farmer, vitally interested in the yield of his holding, and the nominal peasant, who earned his living outside farming, was greatly accentuated. Among the former, the change in agricultural conditions naturally developed a keener interest in the improvement of cultivation. The efficient peasant farmer welcomed enclosure as a means of escape from the incompetent tutelage of the commune, and the movement developed with great rapidity throughout the country, and especially in the steppes, with their commercialized cereal farming.

While the genuine peasant farmer's attitude to enclosures was thus favourably affected by the improvement in agricultural conditions, the general economic development of the country encouraged the nominal peasant to break his bonds with the land and try his luck as a landless wage-earner in the towns and industrial centres. Prior to the Ukaze of November 9th, 1906, such nominal peasants could not dispose of their holdings in the village outright, and mostly had to let them to some other members of their communes, while they were absent, or to leave their families to manage the land, while they were employed elsewhere. The Stolypin legislation introduced a drastic change in the position of such peasants, by allowing them to sell their land, and the industrial growth of Russia on the eve of the war decided many of them to exchange their amphibious existence for that of real industrial workers.[1] Thus, while, on the one hand, the economic evolution of twentieth-century Russia encouraged the enclosure movement among genuine peasant farmers, on the other hand it contributed to the final breaking of all bonds with the land by those elements of the

[1] *Vide*, Ch. V, statistics concerning sales of allotment land.

peasantry, whose connection with farming was only slight or even nominal. In other words, the industrial revolution, coupled with the improvement in agricultural conditions and a rise in agricultural prices, while it accelerated the progress of agrarian reorganization in Russia, was also bringing about a rapid decomposition of the formerly inarticulate mass of Russian rural population into two distinct classes: the genuine peasants and the proletarians. A similar evolution had accompanied the industrial progress of Western Europe as well, where by the middle of the nineteenth century, the social transformation may be said to have been completed, and the structure of modern capitalistic society evolved. In Russia, it had been delayed until the twentieth century, but once having started, it proceeded with great rapidity.

The improvement in agricultural conditions thus helped the growth of the individualistic movement in the Russian countryside; but the process of reorganization of tenures, by putting a considerable strain on the peasants concerned, prevented them to a certain extent from immediately reaping the full benefits of the situation. Though the actual cost of enclosures, as far as the legal and technical procedure were concerned, was borne by the State, the radical reorganization of farming involved in the reform of tenure imposed a more or less heavy burden on the proprietor of the holding. This was especially so in the case of self-contained enclosed farms of the *hutor* type, which necessitated the transfer of the cottage and farm buildings from the village to the holding: a circumstance which, in numerous cases, caused the peasants to choose the less technically perfect form of *otrub*, since it permitted them to avoid this expense. The Russian peasant, beneficial as the transformation was to him and to farming, was unavoidably bound to feel a severe financial strain at the beginning. It will be remembered that, at the close of the first decade of the enclosure movement, at the end of 1916, it involved 10% of the peasant population of Russia. Besides, numerous villages, though they had not actually enclosed their land, had introduced important improvements in their methods of cultivation, while all over Russia peasants have been buying improved implements, seeds, live stock, etc., and thus investing their available money in their industry, with a view to future profits. The peasants have been generally spending more on the improvement of their land and its cultivation than ever before, especially in the case of enclosed holdings. Characteristic in this respect was the increase in capital improvements, as shown by the growth of borrowing for this purpose. The average annual amount of loans for such improvements, which was generally very small, did not exceed 304,000 roubles in 1904–8, while in 1909–13 it rose to 1,400,000 roubles, mostly on account of credits obtained by owners of newly enclosed holdings.[1] The re-equipment of peasant farmers in Russia with improved agricultural implements, which was proceeding apace, along with the general development of farming during the years immediately preceding the war, also involved them in considerable expenses, whose importance can be appreciated from the following statistical records, showing the numbers of steel ploughs, drills and harvesters in the possession of the peasants in various parts of European Russia in 1910 and 1917.[2]

	Steel Ploughs		Drills		Harvesters	
Regions	1910	1917	1910	1917	1910	1917
Northern . .	2,305	95,392	19	67	20	227
North-Eastern .	36,085	86,852	2,017	6,753	9,778	17,704
Petrograd * . .	43,109	173,127	221	671	42	700
Moscow . . .	188,510	840,360	974	2,839	1,136	8,986
Ukraine . . .	611,270	747,271	19,030	70,323	13,761	32,142
Central Agricultural †	191,123	453,709	1,948	10,331	4,035	17,100

* Exclusive of the province of Pskov.
† Exclusive of the province of Voronezh.

The figures above refer only to part of Russia, because the Census of 1917 did not cover the whole country. Besides the regions enumerated above and corresponding to the geographical divisions adopted throughout the present study, it included also a number of separate provinces, in which the general trend of development was similar to that revealed by the above table. The very rapid increase in the number of modern implements, which the peasants were substituting for the old traditional equipment their ancestors have used for hundreds of years, was characteristic of the agricultural evolution of twentieth-century Russia. The country was in the course of rapidly being modernized and was investing most of what it could spare into the business of its own economic transformation. Though the improvement in agricultural conditions had certainly helped the Russian peasantry to bear the unavoidable strain of this transformation more easily, the fact that the energies of the Russian village have been so largely engaged in the process of reorganization made the immediate

[1] A. Tumenev, *From Revolution to Revolution* (Rus.), p. 174.

[2] The figures for 1910 are based on a general census of agricultural implements in European and Asiatic Russia taken by the Central Statistical Committee in 1910 and published in 1913. For 1917—on the results of the Agricultural Census of that year, published in 1921. The present table was compiled from figures given according to provinces in A. Tumenev, op. cit., pp. 196–9.

advantages of the situation and the actual gains of the countryside less conspicuous than they might have been otherwise.

The rapidity of this transformation was to a large extent due to the policy of the Government, which, after 1906, besides enclosures and land-settlement in European and Asiatic Russia, involved a far-reaching programme of measures intended at assisting the technical progress of peasant farming and at improving the economic position of the peasantry. While, hitherto, the attention of the Government had been directed almost exclusively to the extension of the area of land in the possession of the peasants, now, partly on its own, partly in co-operation with the Zemstvos, it began energetically to develop its activities in this direction. The Zemstvos began to evolve an organization of agricultural assistance to the peasants since the early 'nineties, by maintaining agricultural and veterinary experts, agricultural stores, experimental farms, studs, etc. By 1904, out of 34 Provincial Zemstvos, 33 had established special consulting boards on agricultural matters; out of the 360 district Zemstvos, similar boards existed in 246. Provincial agricultural experts were maintained by 32 Provincial Zemstvos, while 254 district Zemstvos employed agricultural experts in their districts. The number of their agricultural stores, which, in 1891, was only 50, reached 317 in 1904, with an aggregate capital of about 7 million roubles.[1] Though the work of the Zemstvos developed with considerable rapidity, their efforts alone were not sufficient to meet the needs of the situation, and under Stolypin, from 1906 onwards, the Government entered the field with great zeal. Technical assistance to peasants was rendered either by the Ministry of Agriculture maintaining its own staff of agricultural experts in connection with the Land-Settlement Commissions and the colonization authorities, or by means of Budget grants to the Zemstvos for the extension of their organization of agricultural advisers. The rapid growth of the number of agricultural experts of the Ministry, on the one hand, and of the Zemstvos, on the other, may be seen from the following figures, unfortunately available only for the period from 1909 to 1912:[2]

	Agric. Experts of the M. of A.	Agric. Experts of the Zemstvos	Total
1909	721	1,820	2,541
1910	978	2,363	3,341
1911	1,581	3,604	5,185
1912	2,340	4,930	7,270

[1] Prof. A. I. Chuprov, *The Needs of Small Farming*, 3rd ed., pp. 123–4.
[2] *Agronomichesky Zhurnal*, No. 8, 1913, p. 171.

By 1914, the total reached 10,000.[1] Lectures on agricultural subjects were being organized, and courses of instruction held, in various parts of Russia, and both their numbers and their attendance increased rapidly. Thus, in 1907, courses were held in 21 provinces and attended by a total of 36,000 persons; in 1908, they extended to 35 provinces and involved an aggregate audience of 48,000; in 1909, in 52 provinces, courses were attended by nearly 234,000 persons.[2] The aggregate sales of agricultural machinery from Zemstvo stores increased as follows:[3]

1907	6,790,700 roubles
1908	6,954,000 "
1909	9,924,700 "
1910	11,767,700 "
1911	12,056,500 "

The State, hand-in-hand with the Zemstvos, extended its activities for the improvement of peasant farming continually. An idea of this development may be given by the analysis of the Estimates of the Department of Agriculture, as the section of the Ministry responsible for this branch of work, for the period from 1911 to 1914:[4]

Analysis of the Estimates of the Department of Agriculture

	1911	1912	1913	1914
	In thousands of roubles			
Agricultural education	3,159	3,884	4,890	5,667
Experimental farms	1,687	2,398	3,787	4,587
General agricultural assistance	2,644	3,405	4,731	5,318
Agricultural assistance in the districts of land-settlement	4,000	5,300	6,000	6,500
Agricultural assistance in the districts of colonization	510	1,200	1,550	2,232
Agricultural experts	1,547	2,272	3,154	6,548
Maintenance of the Department, etc.	2,818	3,420	5,087	6,075
Total	16,365	21,879	29,879	34,927

It may be seen from the above table that, in the course of the years immediately preceding the war, the Budget expenditure for the improvement of peasant farming in Russia increased with a rapidity even greater than that of the growth of Budget grants for colonization.

The most important feature of the situation, however, was probably the awakening of the peasantry to the necessity of raising the standards of their farming and their increased interest in anything that might help them in achieving their object. The attitude of the Russian countryside to progressive currents, which pene-

[1] A. Tumenev, op. cit., p. 174. [2] *Ibidem*, p. 175. [3] *Ibidem*, p. 180.
[4] *Agronomichesky Zhurnal*, Nos. 9–10, 1913, p. 139.

trated into it in one way or another, had changed considerably, and new ideas found the ground much more favourable to their reception than ever before. The passage of time, which had replaced the old generation by a new one, free from the mental legacy of the period of serfdom; the spread of education and the widening of the peasant's outlook by closer contact with the world outside the immediate vicinity of his village; the improvement in agricultural conditions, following the inception of the industrial revolution, which made technical progress worth while: all these factors had contributed to this great psychological change. The peasant realized the necessity, as well as the practicability, of better farming; he became more ready to learn, especially if he could observe the results of new methods in practice. He realized that, to achieve real improvement, he had to combine personal initiative with co-ordination of efforts, and so he was led to the twin developments of enclosures and co-operation.

So much has been written about the co-operative movement in Russia in various languages, that here I shall deal with it briefly, dwelling only on those of its aspects which concerned the economics of peasant farming most intimately.

From the economic view-point, the importance of the co-operative movement was that, in the first instance, it organized the peasant farmers into units adapted to the conditions and requirements of the modern competitive system, without restricting their freedom of action; besides, it taught the masses self-help and responsibility and contributed to their general and technical education.

The co-operative movement in the villages is, indeed, strictly matter-of-fact in its whole outlook. Whatever the views of the enthusiasts of the "co-operative commonwealth," in the eyes of the peasant it is but a means by which he can bring himself abreast of the progress of modern economic organization; a talisman giving the dwarf a chance in the competitive struggle of giants.[1] Accordingly, the co-operative movement cannot exist in a community not yet sufficiently commercialized; neither can genuine co-operation exist in a Socialist or Communist State, as the experience of Soviet

[1] It is difficult to agree with Professor Tugan-Baranovsky, who says: "In its capitalistic body the co-operative union conceals a soul hostile to Capitalism." . . . "The co-operative movement is one of the forms of self-defence of working classes against the onslaughts of Capital" (*Social Foundations of the Co-operative Movement*, p. 96). While true of the ideas which inspired certain pioneers of the co-operative movement, this statement hardly applies to peasant co-operation as such, which is rather a means of adaptation to Capitalism, than of class warfare.

Russia had amply proved. The co-operative movement is the product, as well as the necessary corollary, of Capitalism;[1] and in Russia its growth accompanied the industrial revolution and the awakening of the individualist spirit in the countryside. While the country still lived under conditions of predominantly natural economy and self-supplying maintenance farming, the attempts to implant co-operative organizations in the villages, made in the 'sixties and 'seventies by well-meaning intellectuals, met with distrust, ridicule and ultimate failure. When, however, by the close of the last century, conditions had changed and Russia had begun rapidly to develop along capitalistic lines, the ground became more favourable to co-operative organization, and the movement took roots among the peasantry.

Of the various forms of co-operation, co-operative credit institutions, which accounted for the largest membership, were probably the most important in their effects on the economic life of the Russian countryside. They supplied one of the most urgent needs of the peasantry at all times and in all countries, among whom usury had always been rampant; moreover, they enabled the peasants to find money for improvements and helped them in marketing their staple produce and in buying wholesale the necessary supplies.

The early attempts at organizing co-operative credit on a large scale dated from the beginning of the 'seventies, when the *Petersburg Society for the Encouragement of People's Credit*, with a branch in Moscow, was founded. Its Chairman was Prince A. I. Vassilchikov, a large landowner of progressive views and well-known writer on farming and the peasant problem. Under the aegis of the Society, a considerable number of co-operative credit associations, modelled on the lines of those inaugurated in Germany by Schulze-Delitzsch, was organized. Implanted by intellectuals and by the Zemstvos, they found no response among the peasantry, and most of them gradually withered away. The time was not yet ripe. Conditions were not yet available in the countryside, under which money could be profitably invested by the peasants in their husbandry, and productive borrowing, on which alone credit co-operation can thrive, could not develop. By the close of the century, however, the commercialization of the Russian village had advanced sufficiently to create a demand among the peasants for facilities of cheap

[1] A different view is expressed by Prof. Totomiantz (*Principles of the Co-operative Movement*, p. 8), who traces the beginnings of the co-operative movement to the Middle Ages and even to antiquity. His reasoning, however, would appear to be based on the confusion of co-operation in the strict sense, and of the simple combination of labour, which latter, indeed, had existed from times immemorial.

credit for purposes other than consumption; and now it was the State that came forward to meet this demand, by helping the organization of co-operative credit institutions among the rural population. The *Law of June 1st*, 1895, authorized, besides the existing type of Loans and Savings Associations (*ssudosberegatelnoie tovarishchestvo*), the establishment of a new type of co-operative credit institutions, namely the Credit Associations (*kreditnoie tovarishchestvo*), far better adapted to the requirements of the peasantry. The former type, modelled on the Schulze-Delitzsch pattern, in Russia, as in Germany, proved to be convenient rather for the better-to-do craftsmen and small traders in towns and boroughs, than for peasants, since it required the members to take shares in the capital of the association out of their savings. In Germany, an alternative type of association, answering the needs of the peasantry, had been evolved by Raiffeisen in the 'sixties, and the Law of June 1st, 1895, in the new type of Credit Association, had reproduced, on the whole, the outstanding features of this form of organization, with certain modifications dictated by the peculiarities of Russian village life. Thus, the Russian law dispensed with share capital entirely, substituting for it the joint guarantee of the members of the association, which served as security for the creditor, from whom they borrowed the amount necessary to set the scheme going. The Government, through the State Bank, assumed the rôle of that creditor, providing the capital on easy terms of repayment; besides, also through the Bank, it assisted and controlled the conduct of the business of the associations. In 1904, a special Department of Small Credit was established in the State Bank, which, with a staff of Inspectors of Small Credit, attached to the local branches of the Bank, was charged with assisting the organization of co-operative credit institutions and supervising their operations. At the same time, the Government issued model statutes of credit institutions of both types, adherence to which reduced the formalities involved in their establishment to a minimum. Assisted by the Government both financially and technically, credit co-operation in Russia began to expand very rapidly, the new type of Credit Association becoming by far the more popular form of credit institution in the countryside. The number of co-operative credit institutions in Russia, within her pre-war frontiers, which, in 1901, did not exceed 837, reached the figure of 14,502 at the beginning of 1915. Of this total, 10,687 belonged to the type of Credit Association. The total membership of all co-operative credit associations, in 1915, exceeded 9 millions of households, thus accounting for nearly one-third of the population of Russia. This rapid growth of credit co-operation in Russia, especially among the peasantry, was partly due to the increased demand of the peasants for cheap credit facilities; partly to the efforts of the pioneers of co-operation, who came from the ranks of the intellectuals or of the peasantry; but probably the movement owed most of all to the assistance of the Government, through the State Bank and its staff of Inspectors. It is, perhaps, going too far to say, with Tugan-Baranovsky,[1] that the whole organization of co-operative credit in Russia owed its existence and its success to the officials of the State Bank, but the importance of the part played by the State in its development, often deliberately minimized, cannot be denied.[2]

In the life of the Russian village credit co-operation played a most important part, not only by its extent, but by the great variety of services it rendered to the average peasant farmer. It was, indeed, as several statistical investigations have shown, the average peasant farmer who formed the real back-bone of the co-operative movement.[3] It was not merely a coincidence that the group of peasantry most active in the development of co-operation, was identical with that which was in the forefront of the enclosure movement, since the two movements, as I have pointed out, were the necessary corollaries of each other. To an ever-increasing extent, partly under the pressure of the State Bank, partly of its own accord, credit co-operation concentrated its attention on the financing of production.[4]

In its fight against the village usurer, according to numerous reports from various parts of Russia, credit co-operation was very successful, and in those localities in which it existed the former "benefactors" of the peasantry were effectively being driven out of business, in so far at least as the membership of the associations was concerned.[5]

[1] Op. cit., p. 372.

[2] To quote Dr. E. Fuckner, a German expert on Russian co-operation in his book *Die russische Genossenscheftsbewegung*, 1865–1921, p. 41: "Now that an unbiased judgment can be passed on the effects, either good or bad, of this supervision (viz. of credit associations by the State Bank, which lent them their original capitals—G.A.P.), one is bound to admit, in opposition to the negative attitude common in Russian co-operative circles, that this control had beneficial results. Care was exercised that loans should only be used for productive purposes, as distinguished from consumption. The Inspectors were good teachers in book-keeping, banking technique and the conduct of such operations as buying and selling on commission for members, advances against merchandise, etc." Dr. Fuckner had made a very thorough study of the movement in which he had worked in Russia for years.

[3] Tugan-Baranovsky, op. cit., pp. 379–82.

[4] S. Maslov, *The Co-operative Movement among the Peasants* (Rus.), p. 30.

[5] Tugan-Baranovsky, op. cit., pp. 385–8; Maslov, op. cit., pp. 21–2.

A very important branch of the work of co-operative credit institutions was their buying and selling on commission for their members, which enabled the latter to obtain their supplies wholesale, at lower prices, as well as to dispose of their own products more advantageously. These operations, to a greater or lesser extent, were being carried on by nearly every type of co-operative organizations, but credit co-operation, being so widespread and well-adapted to this kind of work, was by far the most important of such intermediaries. A very extensive field of activities confronted it in the sphere of co-operative selling of the staple produce of peasant farming, and here its work, though still only in its infancy, was developing continuously. As a rule, this branch of business developed gradually in connection with the grant of loans on the security of cereals or other agricultural products deposited in the association's warehouses. Advances on grain, on the eve of the war, formed part of the regular business of over 1,000 co-operative credit institutions, which, in 1913, owned between them 498 granaries and several large elevators belonging jointly to several associations. These figures are sufficient to show the still very limited extent to which the volume of such business was confined on the eve of the war. The scheme set on foot by the State Bank a few years before the war, involving the construction of a large network of elevators and granaries in the centres of grain trade and in the cereal-growing districts, and the handing over of such storage facilities in the rural areas to co-operative organizations for actual exploitation, had scarcely been started, when the war brought its development to an end. The business, which required considerable financial resources for both the original outlay and running expenses, was being slowly developed by the co-operative associations. The actual marketing of grain by credit institutions was usually effected in one of the three following ways. Sometimes, the associations accepted the grain as security for a loan, and left the peasant to sell it himself; here, the association did nothing, except the granting of a loan, and the transaction resolved itself into one of advance against merchandise. Its only advantage for the peasant was that he was enabled to keep his produce, instead of selling it immediately on a depressed autumn market. In other cases, the association bought the grain from their members at fixed prices and then disposed of it on its own account, any profit or loss on the transaction accruing to the institution. Finally, the most typical form and the one which alone could be properly described as selling on commission for members, was that in which the credit institution accepted the grain as security for a loan, and then sold it on the owner's account. Some institutions preferred one of these forms, some another. Thus, an investigation concerning the transactions of forty-seven credit associations in the South of Russia showed that twenty-three of their number were selling grain on commission, while the other twenty-four preferred to deal on their own account. In another group of institutions it was found that, of the total volume of grain handled, 44% were dealt with on the advance system, the actual selling being done by the owners; 37% were sold on commission and only 19% on the associations' own account.[1] In this field of activities, though co-operation had already been doing a certain amount of useful work for the peasant farmer, it had only just begun coming to grip with the problem whose adequate solution required a number of years of sustained efforts and vastly increased financial resources.

Practically from the beginning, the need has been felt among co-operative credit institutions for combination on a larger scale, involving whole districts and provinces. Accordingly, the first Union of Co-operative Credit Associations was founded in 1902 in the district of Berdiansk in Southern Russia. It was followed by several other Unions, namely in Melitopol, Kiev, Ekaterinburg and Blagodarnoie, but the movement developed slowly, owing to the restrictions imposed on the operations of the unions, which were not permitted to accept deposits, to borrow or to grant loans, as well as to the extremely reserved, not to say negative, attitude of the Government to such combinations, against which it seemed prejudiced on financial and political grounds. This attitude, however, was abandoned by the publication, in May, 1911, of model statutes of Unions of Co-operative Credit Institutions, by which the restrictions with regard to borrowing, lending and deposits were removed, and the formalities of organization simplified. Though, even now, the procedure remained long and rather complicated, this constituted a great step forward, and in 1912 and 1913 seven new Unions have been formed in the Kuban, Nizhny-Novgorod, Zlatoust, Ekaterinoslav, Terek, Plotzk and Lokhvitza.[2] On the whole, however, before the war the combination movement had not reached any marked development. Indeed, in 1914, the ten Unions then in operation represented between them only 558 credit institutions, with an aggregate capital of 563,842 roubles, a reserve of 11,853 roubles, deposits amounting to 2,730,515 roubles and

[1] Maslov, op. cit., p. 56.

[2] "Federations of Co-operative Credit Societies in Russia in 1913," *Monthly Bulletin of Ec. & Soc. Intel., Int. Inst. of Agr.*, No. 11, 1914, pp. 40–55.

advances reaching 3,418,968 roubles.[1] It is true that, during the war, the number of Unions and their membership increased enormously, until, after the March Revolution of 1917, in October of that year, the credit Unions numbered 140, while, besides these purely credit combinations, there existed 210 mixed unions, including, along with credit associations, consumers' societies and other co-operative organizations.[2] Yet, by that time the whole political, social and economic life of Russia had already reached so advanced a state of dissolution, that normal activities on the part of this enormous co-operative apparatus were impossible.

The distribution of co-operative credit institutions over the territory of European Russia reveals certain characteristic features of the movement. The figures given in the table below refer to the year 1912, the last for which complete data, including membership and the extent of operations, were available in sufficient detail. The statistics include, besides rural co-operative credit institutions, also those in towns and boroughs, but these latter belonged almost exclusively to the type of Loans and Savings Associations, following the Schulze-Delitzsch model, while the Credit Associations were practically confined to rural districts, and it is, therefore, their distribution that is of particular interest in this connection.

Table showing the Distribution of Co-operative Credit Institutions in European Russia in 1912

Regions	Number of Associations	Membership, Jan. 1, 1912	Loans granted in 1911, in 1,000 rbls.
Northern :			
Credit Associations . . .	83	23,633	679·8
Loans and Savings Associations	7	1,061	50·9
Total . . .	90	24,694	730·4
North-East :			
Credit Associations . . .	503	318,737	11,550·1
Loans and Savings Associations	47	18,496	1,369·7
Total . . .	550	337,233	12,919·8
Petrograd :			
Credit Associations . . .	99	44,138	1,216·0
Loans and Savings Associations	87	39,807	3,086·7
Total . . .	186	83,945	4,302·7
Moscow :			
Credit Associations . . .	794	420,575	15,792·9
Loans and Savings Associations	175	61,874	3,580·2
Total . . .	969	482,449	19,373·1

[1] *Recueil Statistique*, 1913–17, Livr. 2e, Moscou, 1922, p. 127.
[2] Fuckner, op. cit. p. 47.

Regions	Number of Associations	Membership, Jan. 1, 1912	Loans granted in 1911, in 1,000 rbls.
Western:			
Credit Associations . .	86	41,814	2,271·5
Loans and Savings Associations	210	101,272	14,403·1
Total . . .	296	143,086	16,674·6
South-West:			
Credit Associations . .	319	147,914	6,437·0
Loans and Savings Associations	244	187,500	21,315·0
Total . . .	563	335,414	27,752·0
Ukraine:			
Credit Associations . .	430	269,474	12,995·9
Loans and Savings Associations	126	67,793	9,199·0
Total . . .	556	337,267	22,194·9
Central Agricultural:			
Credit Associations . .	775	527,891	21,837·0
Loans and Savings Associations	97	55,492	3,746·9
Total . . .	872	583,383	25,583·9
M. Volga:			
Credit Associations . .	480	247,760	9,246·5
Loans and Savings Associations	55	25,114	2,279·3
Total . . .	535	272,874	11,525·8
Eastern:			
Credit Associations . .	393	273,356	9,320·7
Loans and Savings Associations	10	3,978	251·1
Total . . .	403	277,334	9,571·8
New Russia:			
Credit Associations . .	504	301,024	22,288·3
Loans and Savings Associations	165	123,394	17,236·9
Total . . .	669	424,418	39,525·2
South-East:			
Credit Associations . .	342	211,738	17,596·5
Loans and Savings Associations	137	90,253	13,017·3
Total . . .	479	301,991	30,613·8
Caspian:			
Credit Associations . .	100	39,614	3,151·9
Loans and Savings Associations	20	6,068	761·5
Total . . .	120	45,682	3,913·4
European Russia:			
Credit Associations . .	4,908	2,867,668	134,384·1
Loans and Savings Associations	1,380	782,102	90,307·6
Total . . .	6,288	3,649,770	224,691·7

A study of the preceding table tends to show, in the first instance, the far higher development of credit co-operation in the more purely agricultural parts of Russia, producing regular surpluses for the market, and more particularly in the cereal-growing belt, in which, owing to the strictly seasonal movements of cash, the peasants' need for credit facilities was greatest.

In the extreme North, the number of associations, their membership and turnover were very small, even in the more agricultural province of Vologda. The same applied to the Petrograd region, in which the peasants depended for their living rather on outside earnings than on their land, the latter being for them rather a source of produce for consumption, than of cash, which they obtained elsewhere. On the other hand, the North-Eastern region, though situated outside the black-earth belt, was predominantly agricultural and practically self-supporting with regard to foodstuffs, and its peasant population depended for money primarily on their land; accordingly, here credit co-operation was highly developed. The relatively highly-developed region of Moscow, in which many different sources of non-agricultural earnings combined with a more varied production of peasant farming, than was the case in the more purely cereal-growing districts, could not boast a large co-operative credit apparatus. Here, the peasant could more easily lay his hands on money, either by selling his labour, or such of its produce as could assure a more rapid turnover, and his need of credit facilities was less pressing than that of his cereal-growing brother-peasant in the black-earth, who depended on grain. His milk or butter, his vegetables, his poultry and eggs, as well as the products of the domestic industries, arts and crafts, highly developed in the Moscow region, sold more or less steadily, either for several months in a stretch, or even all the year round. Whatever his income, it had at least the not unimportant merit of being more or less evenly distributed over the whole year. In the Western region, as well as in the South-West and, to a lesser extent, in the Ukraine, there existed a relatively large number of Loans and Savings Associations: a phenomenon due to the presence in these localities of numerous small boroughs with an almost entirely Jewish population of craftsmen and small traders. Credit Associations in the Western region were very few, and their membership and turnover small, especially in the provinces of Minsk and Vitebsk, while the province of Mogilev, in this respect, bore a very close resemblance to the neighbouring Moscow region. Characteristically, the co-operative movement did not seem to take root among the White-Russian population of these provinces, though all the material conditions for its development would appear to have been present. How far the inertness of the peasantry in these, as well as in the neighbouring Lithuanian provinces of Vilno, Grodno and Kovno, may have been due to their having been for centuries under the heel of alien Polish landowners—a yoke which had left deep traces on the character of the people—can only be surmised. No one who had the opportunity to observe the White-Russian peasants of the provinces of Minsk and Vitebsk, could escape the impression of the effects of such social and religious domination of a country by an alien upper class and an alien Roman Catholic clergy, which had literally sapped the moral and physical energy, the manliness and natural dignity of the people. In the South-Western region, credit co-operation was somewhat better developed, though here, especially in the provinces of Kiev and Podolia, the small size of peasant holdings and the dependence of the peasantry on outside earnings, from which source they derived most of their money incomes, did not favour the growth of the movement. It was strongest in the province of Volynia, whose peasants were more dependent on their own agricultural production. In the Ukraine, credit co-operation was well developed, especially in the two Southern grain-exporting provinces of Kharkov and Poltava. The latter, in this respect, resembled to a certain extent the adjoining provinces of New Russia, with their extensive cereal farming and their highly developed credit co-operation, which financed and partly conducted the marketing of grain for its members. From New Russia, through the South-Eastern and the Central Agricultural region, to the Middle Volga and, beyond it, to the provinces of Samara and Ufa, that is throughout the whole cereal-growing black-earth belt, credit co-operation was an essential element in the economy of the peasant farmer, and was, accordingly, well-developed in membership and turnover. The greater importance of co-operative credit institutions for the peasantry of the Southern steppes, with their extensive production of cereals for the market, which these institutions helped in financing, may also be seen from the table below, showing the average amounts of advances made by Credit Associations to their members in 1911 in various parts of Russia:

Region	
Northern	28 roubles
North-Eastern	36 ,,
Petrograd	28 ,,
Moscow	38 ,,
Western	54 ,,
South-Western	44 ,,

Region	
Ukraine	48 roubles
Central Agricultural	41 ,,
Middle Volga	37 ,,
Eastern	34 ,,
New Russia	74 ,,
South-Eastern	83 ,,
Caspian	79 ,,

It may be seen that it was in the Southern steppes, with their commercialized extensive cereal farming, that the peasants would appear to have been making the greatest use of co-operative credit institutions; neither was this surprising, seeing that practically the whole money income of the peasant farmer in these parts of Russia was derived from the cereal crops. Moreover, it was precisely in these provinces that the marketing operations through the credit associations were most developed, which, unless the business was transacted on the associations' own account, naturally tended to swell the amount of advances.

Besides credit co-operation, there were other branches of the co-operative movement which, either direct or indirect, assisted the progress of peasant farming in Russia. It may, indeed, be said that even those co-operative organizations, whose connection with the agricultural industry seemed remote, were often used to serve its ends in one way or another. Thus, consumers' societies were often used, besides their original and direct purpose, as buying agencies, through which their members purchased the necessary implements and supplies wholesale. A co-operative association, whatever its nature, as a rule became the social centre of the village, and was invested with a variety of functions. Consumers' societies, co-operative dairies and other associations kept reading-rooms, organized lectures, etc., thus combining their economic function with educational work, which, when it was genuine and not used as a cloak for revolutionary propaganda, could and did play an important part in the awakening of the countryside.

On the eve of the war, the co-operative movement in Russia, besides credit institutions, included consumers' societies, agricultural associations of various kinds, co-operative dairies and a few associations for special purposes, mostly connected with domestic arts and crafts. The numbers of associations belonging to each of these several groups, as well as that of agricultural societies, which, though, strictly speaking, they were not co-operative associations, were often made to serve some of the purposes of the latter and generally played an important rôle in the progress of peasant farming, on January 1st, 1913, in various parts of European Russia, are shown in the following table:—

Number and Distribution of Co-operative Organizations and Agricultural Societies in European Russia on January 1st, 1913

Regions	Credit Assns.	L. & S. Assns.	Consum. Soc.	Agric. Assns. (gen.)	Agric. Assns. (spec.)	Co-op. Dairies	Agr. Soc.	Other Assns.
Northern .	88	7	526	2	—	169	45	35
North-Eastern	591	53	653	2	7	10	260	27
Petrograd	136	103	502	6	12	27	291	91
Moscow .	1,004	232	846	7	14	67	564	55
Western .	113	230	120	13	—	2	132	3
South-Western	395	267	1,655	35	5	9	80	9
Ukraine .	529	137	730	15	3	1	497	14
Central Agricultural	901	115	261	5	7	5	203	5
Middle Volga .	592	62	189	9	—	1	159	14
Eastern .	459	12	141	6	4	5	95	10
New Russia	526	172	486	12	3	1	193	18
South-Eastern	400	146	320	88	2	—	84	36
Caspian .	129	26	41	1	3	—	24	5
European Russia .	5,863	1,562	6,470	201	60	297	2,627	322

Incidentally, this table, if compared with the one showing the distribution and position of co-operative credit institutions in 1912, gives an idea of the rapidity with which the co-operative movement had been growing on the eve of the war. Indeed, in the course of a single year, from January 1st, 1912, to January 1st, 1913, the number of Credit Associations in European Russia had increased from 4,908 to 5,863, a rise of nearly 20%.

Of the total number of Consumers' Societies shown in the table, roughly 75 to 80%, or about 5,500 were rural.

The beginning of consumers' co-operation in Russia may be traced to 1865, when the first Consumers' Society was founded in the village Kinov, in the Urals. It was in the course of the same year that the first isolated attempt at the organization of co-operative credit had been made in the province of Kostroma by Luginin. The examples in both cases found little emulation, and throughout the 'seventies and 'eighties the efforts of pioneers met with meagre response. By 1880, there existed in Russia only 117 Consumers' Societies; by 1890, their number increased to 260. The movement began to extend in the 'nineties, with the result that, in 1900, the societies numbered 897. It was in the twentieth century, and more particularly since 1906, that the growth of consumers' co-operation became rapid, and that the movement began to acquire a considerable influence on the economic evolution of the Russian countryside. The peasants, who, hitherto, failing competition among rural traders, had been entirely at their mercy, and were usually exploited by them in the most ruthless manner, saw in co-operation a means of escape from this peculiar sort of usury. Wherever consumers' societies were established, the prices of necessaries fell considerably, the reduction being especially great in the most out-of-the-way corners of the country, in which the traders had hitherto enjoyed a more or less complete monopoly. Thus, in some parts of the province of Perm, the reduction in prices caused by the establishment in the villages of consumers' societies, was estimated at 30%. In the province of Novgorod, it was put at 20 to 40%, according to locality.[1] The saving in the peasant's household budget, due to such reductions, can easily be imagined and suffices to explain the progress made by the co-operative movement in this direction.

Agricultural Associations (*selskokhoziaistvennoie tovarishchestvo*) were a type of co-operative organization with a view to joint purchase of implements and supplies and to marketing the products of their members or of joint production, whose development in Russia was very recent. The first agricultural association was founded in Libau in 1890, and until the close of the last century, when this type of co-operation was imported into Russia proper, it had been confined to the Baltic and Western provinces. In Russia, before the war, it began to spread, especially in the black-earth belt, where it was found very convenient for the marketing of grain. Yet, at the outbreak of the war, this particular form of co-operative organization, in Russia proper, was still in its infancy. The total number of agricultural associations formed for general purposes, in 1913, did not exceed 201, of which 100 were in New Russia and the South-East, that is in the two principal wheat-growing regions, and fifty in the South-Western and the Ukrainian provinces. Among the special agricultural associations the most important part was played by those of flax-growers, which, in 1915, formed an All-Russian Union of Flax-Growers for the purpose of marketing their produce and the improvement of production. The Union, besides special flax-growers' associations, also included a large number of credit associations, through which the peasants marketed their flax.

A very important branch of the Russian co-operative movement, which had attracted considerable attention abroad in the course of the last two decades or so, were the dairy associations or *artels*. Though the first attempts to establish co-operative dairy associations in the provinces of Tver and Yaroslav dated from the 'sixties,

[1] Tugan-Baranovsky, op. cit., p. 390.

the movement had failed to take root in European Russia until the current century. In 1871, there existed twenty-five co-operative dairies in the provinces of Tver, Viatka, Yaroslav, Novgorod and Terek, established for the production of cheese and assisted financially by the local Zemstvos and the Imperial Free Economic Society, which encouraged the movement in every way. The experiment was a short-lived one, and the dairies collapsed, the co-operative cheese factories being sold to private capitalists, in whose hands the production of cheese in Russia was finally established on a sound basis. Here, as in other spheres, the Russian village was not yet ripe for co-operative organization, and the production of cheese was hardly a suitable branch to begin co-operative dairying with. Further development had to wait. And the impetus to the growth of dairy co-operation, when eventually it had come, came not from the old country, but from Siberia, where the movement began to expand with great vigour among the peasants of the provinces of Tobolsk and Tomsk since the early years of the twentieth century. Indeed, in these localities, far removed from consuming centres, butter-making provided the peasant farmer with a far more easily marketable commodity than cereals, and conditions, therefore, were especially favourable for its growth. Accordingly, co-operative dairies which, at the close of the last century, represented only a small fraction of the total number of dairies in Western Siberia, multiplied with great rapidity, once the movement was set on foot by its energetic pioneers, Sokulsky and Balakshin, in the later 'nineties. In 1900, the provinces of Tobolsk and Tomsk had 1,022 private dairies and only thirty-two dairy *artels*, the latter representing 3% of the total; in 1905, the numbers respectively were 1,942 and 347, and the percentage rose to 18%; in 1910, there were 3,109 privately-owned and 1,337 co-operative dairies, the latter having risen to 43% of the total.[1] In 1907, a Union of Siberian Dairy Artels was founded, which, at first, in 1908, included only sixty-five dairies, with a total turnover of some 2,880,000 roubles. Gradually expanding, in 1914 it increased its membership to 804 dairies and its turnover to 20,208,000 roubles.[2] In 1908, of the 2,841,000 poods of butter exported from Siberia, 131,000 poods or 4·6% had passed through the Union; in 1913, the last normal year of its operations, the Union accounted for 631,000 poods or 14·2% of

[1] Tugan-Baranovsky, op. cit., p. 397.

[2] These and the following figures referring to the activities of the Union are taken from Prof. G. Schwittau, *Russian Co-operative Organizations on the Foreign Market* (Rus.), pp. 63 ff.

the total exports of Siberian butter, which amounted to 4,443,000 poods. Besides butter, the Union also helped the marketing of other agricultural produce, whose aggregate value, in 1914, reached 4,274,523 roubles. It acted also as a purchasing agency on behalf of its constituent associations. In 1908, its purchases of machinery, implements and supplies for its members represented a value of 17,000 roubles, while it had also bought for them 127,000 roubles' worth of various necessaries. In 1913, the figures respectively were 421,000 roubles and 2,285,000 roubles: a not inconsiderable increase in five years.

Thus, the growth of dairy co-operation in Siberia on the eve of the war had proceeded with great rapidity. As in the case of credit co-operation, here also, though to a much smaller extent, the State, through the Ministry of Agriculture, lent a hand to the movement in its early stages. In 1902, Balakshin, the organizer of dairy *artels*, seeing that, to ensure their success, more capital and a certain amount of expert assistance were needed, submitted to the Ministry a memorandum on the position, the prospects and the needs of dairy co-operation in Siberia, and from that time to 1908, when the movement became strong enough to stand on its own feet, and the Union of Siberian Dairy Artels had been formed, the Government helped it both financially and technically.

In European Russia, the co-operative dairy movement, which had proved so utter a failure when tentatively started in the 'sixties, began to revive since the close of the last century, though, compared with Siberia, its progress has been relatively slow. Yet, the situation, so unfavourable to it before, had now changed considerably, and in many localities dairy farming appeared to be the only means of raising the insufficient yield of peasant husbandry. Moreover, in the modern type of co-operative dairy, cheese-making, which requires more specialized skill and a larger amount of capital than the simpler branches of dairy production, was practically abandoned and left to specially-equipped large factories, which had raised it to a very high standard of perfection, and the production was confined to butter, fresh milk and cream. In the development of co-operative dairying the first place in European Russia was held by the province of Vologda, where, since the construction of the Northern railways, first from Moscow, and later from Petrograd, dairy farming had been greatly encouraged by the extensive facilities of marketing the produce in the two capitals. The province of Vologda, in 1913, was responsible for 168 of the total number of co-operative dairies in European Russia proper. Next came the province of Moscow with only twenty-eight co-operative dairies and that of Novgorod with twenty, while in the black-earth belt, with its cereal-growing bias and its under-developed dairy branch, the movement was practically non-existent. In the greater part of European Russia, still living under the three-course system which, as I shall have the occasion to explain fully later, in dealing with the evolution of stock-farming, was extremely unfavourable to progress in the numbers and quality of cattle, the movement could not be expected to develop until the conditions of arable farming were modified. In localities where the peasants possessed few cows, the appearance of a dairy, co-operative or private, far from being a blessing, could be looked upon rather as a curse, since the peasants, always pressed for cash, were not unnaturally liable, when opportunities presented themselves for regularly disposing of their milk, to encroach unreasonably on the amount essential for the family, and especially for the children, with very injurious consequences for the health of the growing generation. Accordingly, the growth of dairy co-operation in European Russia raised problems of great national importance, and its encouragement, in this respect, required more circumspection than that of any other form of co-operation.

Much older than the co-operative movement in Russia was the development of Agricultural Societies (*selskokhoziaistvennoie obshchestvo*). In the agricultural evolution of the period under review, the part played by these societies was so important that, though they were not co-operative organizations in the technical sense, they may be considered together with these, as representing the combination of individual efforts for the achievement of certain common objects. The object of agricultural societies consisted in helping the technical and economic progress of farming in its various branches by research, by the exchange of information, by the organization of agricultural exhibitions, the maintenance of experimental establishments, the issue of publications, etc. Except the fees collected from their members and any sums they may have obtained as donations, etc., which formed their resources, these societies had no funds. Capital, in the proper sense of the word, they had none, and therefore if they engaged in purchases or other business transactions on behalf of their members, these operations had to be financed by arrangements *ad hoc*. Some of these societies were very large and important, and their influence extended over the whole country; but the great bulk were quite small, their activities being limited to their immediate neighbourhood in some obscure corner of the Empire. The largest institution of this class was the *Imperial Free Economic Society*, established in 1765, under

Catherine II, which, by the time of the revolution, could boast of an unbroken record of just over 150 years. Its influence on Russian farming, especially during the crucial periods of its development, such as the epoch of the Emancipation or the agrarian crisis of the closing years of the last century, by the shaping of public opinion, has been very considerable. Another large society of national importance was the *Imperial Agricultural Society of Moscow*, founded in 1819. Other agricultural societies, whose work and publications were widely known in Russia generally, were the *Agricultural Society of Kharkov*, the *Agricultural Society of Southern Russia*; besides, there existed a number of large societies of more or less local importance, mostly established since the 'sixties. By the close of the last century, the establishment of agricultural societies had been much facilitated by the publication, in 1898, of model statutes, which permitted greatly to reduce the formalities of their authorization. Until the close of the nineteenth century, the agricultural societies had not penetrated into the midst of the peasantry, and the 300 such societies which existed in 1898 were all large bodies, with more or less extensive areas of activities. The small rural agricultural societies, which consisted almost exclusively of peasants, with, perhaps, a few intellectuals, and whose objects were more immediately practical, and thus, to a certain extent, similar to those of genuine co-operative organizations, began to develop in the course of the current century. They were most numerous in the Ukraine, the region of Petrograd, the North-East, New Russia and in the Northern half of the region of Moscow. Their total number in European Russia, in 1913, reached 2,627. No information is, unfortunately, available concerning their membership.

To sum up the position and progress of peasant farming in Russia on the eve of the war, it may be said that, in its evolution, two principal developments stood out conspicuously. These two movements which distinguish this period as a definite epoch in the history of the Russian peasantry, were the enclosures, on the one hand, and the growth of rural co-operation, on the other. Different or even, to some, mutually antagonistic as these two movements may have appeared, in point of fact they were intimately connected with each other; nay, even represented the obverse and the reverse sides of the same phenomenon. That phenomenon was the rise of the individualist spirit, which had been roused by the economic growth of the country and the rapid commercialization of peasant farming in the early part of the twentieth century. The old communal organization, as a relic of a past stage of natural economy, was doomed to dissolution. In the modern competitive age the individual was coming into his own, and in the Russian village we saw him casting away the survivals of an era gone by, which only impeded the freedom of his movements. But, while he ensured free scope to his enterprise in production, in his contact with the outside world, in his dealings with the market, the peasant farmer had to seek strength in combination; and thus he was led to co-operative organization.

PART III

RUSSIA'S AGRICULTURAL PRODUCTION

CHAPTER VIII

GENERAL CHARACTERISTICS OF RUSSIA AS AN AGRICULTURAL COUNTRY; THE ORIGIN AND DISPOSAL OF THE SURPLUS

AMONG the great cereal-exporting countries of the world, Russia's position is to some extent peculiar. From most of her competitors on the world grain market she differs in not being a new country: a circumstance which exercises a considerable influence on the development of her agricultural production. Unlike her younger competitors across the seas, Russia had not evolved her agricultural industry in modern times, in response to the growing demand of the world, in course of rapid industrialization, for the products of the soil. The element of speculative investment, which had played so conspicuous a part in the agricultural development of the New World, where it has been responsible for the creation of large capitalistically-organized farming concerns, had played no rôle in the agricultural history of Russia. Russian farming has been developed gradually, mostly by small peasant producers, and has been growing spontaneously, as an essential function of the complex body of the nation, which it was called upon to feed and support. Accordingly, even down to the modern commercial age, the maintenance factor continued to play a prominent, if not actually a decisive part in its evolution: a feature characteristic of a country of small peasant farmers.

It is, indeed, to her age, on the one hand, and to her being essentially a peasant country, on the other, that Russia owes, in the first instance, the peculiarities of her agricultural evolution, as compared with other cereal-exporting countries inhabited by the White Race.

For an agricultural grain-exporting country, European Russia, on the eve of the Great War, had a relatively very dense population. The density of population per square mile in European Russia and the principal agricultural countries overseas, if compared, makes the position in this respect perfectly clear: [1]

European Russia, exclusive of Poland, 1897 .	53·5 per sq. mile
United States, 1910	30·9 ,, ,, ,,
Canada, 1901	1·4 ,, ,, ,,
Australia, 1901	1·3 ,, ,, ,,
New Zealand, 1901	7·4 ,, ,, ,,
South Africa, 1904	10·9 ,, ,, ,,
Argentine, 1895	3·4 ,, ,, ,,
Brazil, 1910	7·0 ,, ,, ,,

These figures refer to the whole territory, including the land unsuitable for agricultural exploitation in any form, of which the proportion in Russia, with her extensive tracts of Arctic barrens, swamps and wastes, is extremely large. The population per square kilometre of the area suitable for cultivation, according to a calculation made by the German statistician Dr. Ballod for the year 1912, worked out approximately at the following figures: [2]

	Temperate Zone	Tropical Zone
Russia, European and Asiatic . .	38·5	—
British Empire	23	44
France	48	10
China	100	—
United States	23	40
German Empire	120	8
Belgium and Holland	220	12
South America	7	3
Spain and Portugal	45	9
Turkey	18	—
Austria	72	—
Italy	120	2
Scandinavia	44	—
Balkans and Rumania	46	—
Switzerland	175	—
Japan	108	—

Though the above figures cannot pretend at being anything more than rough approximations, since only in a few cases had sufficiently exact data been available for their calculation, they can yet give some idea as to the relative position in this respect of a number of countries. With regard to Russia, here taken as a whole, with the sparsely-populated territories of Siberia, the Asiatic steppes and the Far East, they would appear to show quite plainly that, by the density of her population, of which the vast majority is engaged in agriculture and lives from the soil, she

[1] *Statistical Abstract for the several British Oversea Dominions and Protectorates*, 1903–17; and *Statistical Abstract for the Principal and other Foreign Countries*, 1901–12.

[2] Dr. Carl Ballod, *Grundriss der Statistik*, 1913, p. 74.

resembles rather the less highly industrialized, among the old countries, than any of the cereal-exporting and cattle-raising countries of America and Australasia, not excepting even the United States.

While she had thus to support a relatively dense population, Russia's agricultural output, per unit of cultivated area, was very small, owing to the low standards of her peasant farming, to which I have had the opportunity to refer in the preceding chapters. Neither did the large estates in the belt of extensive cereal farming differ much in their yield from the land cultivated by the peasants.

The yield of the principal cereal crops in Russia, as compared with other countries, in 1910, is shown in the table below: [1]

	Wheat	Rye	Barley	Oats
	In poods (36 lb.) per dessiatin (2·7 acres)			
Russia	48	51	57	53
Austria	86	89	89	75
Hungary	88	79	66	64
United Kingdom	142	—	124	120
Belgium	146	147	170	196
Bulgaria	71	62	78	53
Germany	130	113	123	123
Holland	152	114	172	140
Spain	67	57	82	55
Italy	58	—	—	—
Serbia	60	52	54	47
France	72	61	88	82
Switzerland	149	139	136	128
Sweden	152	107	116	106
India	48	—	—	—
Japan	95	94	142	—
Algeria	51	108	51	78
Tunisia	22	—	20	83
Argentine	43	—	—	69
Canada	72	77	88	84
United States	64	67	81	87
Australia	64	—	64	59
New Zealand	130	—	121	103

These figures are too eloquent to need commenting upon. They help us to visualize a vast country, the greatest cereal-producer in the world, trying to keep abreast of the demands of her growing population and of the increasing claims of her position as a Great Power, which subjected her economic and financial system to a severe strain, on the meagre proceeds of peasant husbandry, by turning under the plough every available bit of agricultural land in its European provinces and gradually extending the cultivated area beyond the Urals. Under the pressure of her growing financial

[1] *Recueil des données statistiques et économiques sur l'industrie agricole de la Russie et des pays étrangers*, 1913, p. 113.

burdens and of the claims of her balance of payments, Russia had, at all costs, to produce a net surplus of grain available for disposal on the world market, and this she did, year after year.

The problem of the origins of marketable surpluses of grain and of the classes of producers which, to a varying extent, contribute to the community's supply of the basic foodstuffs, is one of considerable theoretical, as well as practical, interest, and it had been approached by several economists and statisticians from different angles.[1] Here, this problem cannot be considered in all its details, involving many technical questions of little interest to the general reader, and I shall have to confine myself to giving a broad outline of the position.

In Russia, the marketable surplus of cereals was produced by the large farms, on the one hand, and by peasant family farmers, on the other. It will be remembered that, on the eve of the war, the former accounted approximately for about 10% of the crops in European Russia, and the latter for 90%. Obviously, it was the large farmer and the wealthier peasant, who possessed a relatively large holding of his own or leased additional land for cultivation, that produced primarily for the market. It is on these two classes of agricultural producers that the real marketable surplus of cereals actually depended. Yet, along with them, throughout the black-earth belt, much corn was being brought to market by peasants, whose holdings were too small to allow of the production of a genuine surplus, but who, being driven to it by their urgent need of cash, achieved this end by substituting less valuable cereals

[1] An interesting attempt at ascertaining the marketable surpluses of cereals and potatoes on holdings of various sizes in Prussia was made by the Prussian statistician A. Petersilie in his " Zahlen für das Gewicht der Landwirtschaft und ihrer Betriebagrossenklassen," *Zeitschrift des Preus. Statist. Landesauts*, 1914. In Russia, the first statistical investigation of this kind was made by L. Marres in his article on " The Production and Consumption of Grain by the Peasants " (*Influence of the Yield of Crops and Grain Prices*, 1897). His conclusion was that " all the needs of the market are met out of the production of large estates and of a small group of wealthier peasants," which latter he estimated at 15·9% of the total peasant population of European Russia. The most interesting recent statistical investigation, referring to the period 1909–13, is that of Prof. N. Kondratiev, in *The Grain Market and its Control during the War and the Revolution* (Rus.), Moscow, 1922. He calculates the " marketing ratios " of the principal cereals on large and peasant farms according to provinces, as well as the aggregate surplus of wheat, rye, oats and barley in European and Asiatic Russia. The latter he estimates at about 1,180,532,000 poods p.a. in 1909–13, or 579,031,000 poods of wheat, 215,675,500,000 poods of rye and 385,825,500,000 poods of oats and barley. The peasants' share in the marketable surplus he puts at 79·9% for wheat, 83·9% for rye and 73·1% for oats and barley, though he points out that these percentages are unavoidably somewhat raised by the methods he had to use in his calculations.

for wheat, or potatoes for cereals, in their own diet. By this simple expedient the aggregate marketable surplus was increased far beyond what it should have been, considering the sizes of peasant holdings from which it was derived and the theoretical "norms of consumption" of peasant households established by statisticians. On the eve of the war, indeed, nearly every average peasant of the cereal-growing belt had some grain to put on the market; and though only too often the surplus was obtained at the expense of the producer's own diet, it must be admitted that the growing interest of the peasant farmer in the market undoubtedly had the effect of stimulating agricultural progress among the peasantry.

A. The Russian Agricultural Market

The way in which the available surplus of agricultural produce was disposed of on the market, either home or foreign, played a most important part in the development of Russian farming: a part, to which I have had the opportunity to refer in dealing with the agricultural geography of Russia.

The internal market for agricultural products, besides the cities and industrial centres throughout the country, comprised also, at least in respect of the principal cereals, the rural population of the under-producing provinces which, to a greater or lesser extent, depended for its supplies of grain on imports from the cereal-growing provinces of European and Asiatic Russia. The capacity of the home market, which thus included a large number of agricultural population, was necessarily subject to large variations according to the yield of crops in the under-producing belt, leaving a larger or smaller margin to be met by imports. Some idea of the extent of the Russian internal market, as compared with her exports abroad, in the case of the four principal cereals, may be obtained by deducting the average net exports of wheat, rye, oats and barley in 1909–13 from the average net marketable surplus as estimated by Professor Kondratiev for the same period. The table showing the approximate distribution of the net surplus of the four principal cereals as between the foreign and the home market is given below:

Distribution of the Net Surplus of the four principal Cereals as between the Foreign and the Home Market in 1909–13

	Wheat	Rye	Oats and Barley	Total
	In millions of poods (1 pood = 36 lbs.)			
Total net surplus	579·0	215·7	385·8	1,180·5
Exports abroad	260·4	41·8	289·3	591·5
Sales at home	318·6	173·9	96·5	589·0
Share of the home market	55·0%	80·6%	25·0%	49·9%

Thus, on the eve of the war, roughly half the net surplus of the four principal cereals was disposed of on the home market. Especially great was the demand of the latter for wheat, whose consumption was gradually increasing both in towns and among the rural population of the under-producing provinces. Rye, as the staple bread-stuff of the great bulk of the Russian peasantry, recorded the highest percentage of home sales. Not only, as I have said, had it to be brought in large quantities into the under-producing provinces, but within the wheat belt itself it formed a most important article of trade, many peasants preferring to sow wheat, as the more profitable cereal, and to depend on the market for their own supplies of rye.

It is only natural for the capacity of the home market for agricultural produce to increase with the industrialization of the country, and unless the latter is in the process of rapid colonization and agricultural development, this brings about the growing encroachment of internal demand on exports, and the consequent reduction in the volume of the latter. Such has been the case with most countries of Western Europe which, down to the industrial revolution, had been still mainly agricultural and often possessed exportable surpluses of grain, but have since become heavy importers of foodstuffs for their teeming industrial populations. Within the last fifty years, a similar evolution has been taking place in the United States, which, though still in the front rank of agricultural exporters, has been developing industrially with such a rapidity that the relative weight of her agricultural exports, as compared with internal consumption, has been continually declining. The same process was bound to accompany the industrial development of Russia since the close of the last century; and statistical evidence of this evolution was not lacking. Thus, the statistics of traffic on the Russian railways, which distinguish between freights consigned to destinations within the country, on the one hand, and those consigned to termini on the frontiers or to exporting harbours, on the other, point to a continuous increase in the transport of cereals to the former, at the expense of the latter. The trend of the evolution may be seen from the figures on the opposite page, which refer to the period 1895 to 1912.[1]

Though the period was very short, it may be seen that, while there has been a marked increase of traffic in both directions, the

[1] Kondratiev, op. cit., p. 18. The same subject is discussed in detail, and figures given for the period from 1876 to 1902, by P. Liashchenko, op. cit., pp. 203 ff., as well as by B. Brutzkus, *Economics of Agriculture*, pp. 133–4; especially table on p. 134.

Evolution of the Grain Traffic on the Russian Railways, 1895–1912

	For Internal Destinations		For Frontiers and Ports	
	Millions of Poods	Per cent	Millions of Poods	Per cent
1895–1900	250·73	43·4	327·44	56·6
1900–1905	327·90	44·1	416·80	55·9
1906–1910	434·50	48·4	463·00	51·6
1911–1912 [1]	452·08	46·8	513·46	53·2

demand of the home market would appear to have been growing faster than exports: a development most important for the progress of Russian farming, which depended primarily on the increase in the purchasing capacity of the internal market and the varied nature of its demand.

From what has been said in the preceding pages, it may be perceived, indeed, that by reason of her historical evolution and her modern agrarian organization, Russia can hardly be said to belong to the type of agricultural countries which can thrive and develop culturally and economically on the export of the products of farming. The essentially exporting agricultural country, unless situated in the tropical or sub-tropical zone and able to produce large quantities of valuable cereals, such, for instance, as rice, at low cost, even in spite of intensive systems of cultivation, must necessarily be one affording facilities of extensive farming and cheap production. Accordingly, if its staple exports are cereals, it must be also a country of relatively large holdings and of production on a large scale, which would permit the producer, in spite of the low yield per unit of area, due to extensive systems of cultivation, and of low spot prices enabling the exporter to compete on the world market, to obtain a sufficiently large margin for his own maintenance in decent conditions and for defraying the growing expenses of a community advancing politically and culturally. These are the outstanding features of the younger cereal-exporting countries of thc New World, in which the sizes of individual holdings, even in the case of the family farmers, are large enough to permit their cultivators to live in conditions of reasonable comfort, or even to prosper, on the yield of extensive systems of farming. The homesteads of the United States and the 160-acre holdings given to settlers in Canada, have their parallel in the relatively large holdings of 15 dessiatins (about 40 acres) per male soul,

[1] It should be noted that the years 1911 and 1912 were marked by extensive failures of crops in a large part of European Russia and in Siberia, which fact may account for the break in the trend of evolution during 1911–12, as years quite abnormal in this respect.

allocated to the peasants in the Russian colonial regions of Siberia and other Asiatic dominions, but cannot certainly be compared with those of the peasant farmers of European Russia, which were under 30 acres per family on the average. In all the exporting countries of the New World, moreover, a considerable part of the land is exploited on a large scale and on capitalistic lines; and it is from such large, capitalistically-organized, agricultural concerns, that the great bulk of the staple agricultural exports is derived. A country must, indeed, have enough room and to spare, to enable such farming concerns to arise, as the Coleman-Fulton Pasture Co. in the United States, with its 80,000 acres and a working population of 4,000, or its numerous smaller brethren, serving to feed the huge meat-packing and exporting industry of America.[1] Europe, even in its less densely populated Eastern regions, has no room for this form of agricultural mass-production. Historically, to a greater or lesser extent, the European continent had developed agriculturally, as well as economically, on the foundation of peasant husbandry; and, with scores of millions of population living on small holdings owned by them, their future was bound to depend entirely on the raising of the standards of farming and of its yield per acre. Russia, with her relatively dense peasant population, dependent on holdings of a very moderate size, could only develop and grow economically under conditions which would stimulate the intensification of peasant farming; and since the latter involves increased costs of production, spot prices of agricultural produce had to rise, before it became possible. Production for export, in competition with young countries of extensive farming, did not contribute to intensification, especially in a country of small peasant farmers which, owing to the scattering of its production, was at a disadvantage as compared with its competitors on the world market. Even with a highly-organized co-operative system of marketing, which enables the peasant to do away with some of the drawbacks of production on a small scale, the collection and sorting of products in small lots, brought by innumerable producers, is still bound to bear heavily on the spot prices under any conditions. Besides, the small peasant farmer always thrives best on varied production, which enables him to put every bit of his holding and every scrap of its produce to the most profitable uses. This necessarily involves production in small quantities for a near market, the articles produced being mostly of perishable nature. It is true that butter, eggs and poultry may also be produced for export abroad, and so, to a certain extent, they actually were; but unless some new

[1] Prof. J. Boyle, *Agricultural Economics*, pp. 60 ff.

wave of prosperity were to sweep the importing countries of the world, raising the standards of life of their peoples to an unprecedented level and vastly increasing their effective demand for such products, such exports would be too small to provide any real solution of the problem. What was really needed, was the growth within Russia herself of large and prosperous consuming centres, brought into being by the development of industry and trade: a process which set in for good since the close of the last century, and became especially pronounced since the Japanese War and the revolution of 1905. A glimpse of the resulting increase in the capacity of the home market, though indirect, has been afforded, in the case of the principal cereals, by the statistics of grain freights on the Russian railways. More direct evidence, and one involving, besides, a greater variety of products, may be obtained, however, from the study of the total turnover of freights on the Russian railways, on the one hand, and of price movements on the home market, on the other; more particularly, of the latter.

The increase in the transport by rail of some of the principal agricultural products, exclusive of cereals, in 1906–10, as compared with 1901–5, is shown in the table below:[1]

	1901–5 Yearly	1906–10 Averages	Increase per cent
Hay, millions of poods	26·1	35·1	34·5
Hops, 1,000 of poods	337	450	33·5
Fresh vegetables, millions of poods	90	161	78·9
Fruits and berries, millions of poods	14·7	23·3	58·5
Tobacco, poods	9,279	10,503	13·0
Cattle, 1,000 of heads	1,186·4	1,275·8	7·5
Sheep, 1,000 of heads	283·1	412·9	45·8
Pigs, 1,000 of heads	742·7	809·0	8·9
Meat, various, 1,000 of poods	6,621	8,398	26·9
Pork and lard, 1,000 of poods	2,391	3,002	25·5
Poultry, killed, 1,000 of poods	840	921	9·6
Butter, 1,000 of poods	5,350	6,742	26·0
Milk and cream, 1,000 of poods	3,449	4,501	30·5
Fat and tallow, 1,000 of poods	2,749	3,178	15·5
Hides and skins, millions of poods	14·6	18·5	26·7
Eggs, millions of poods	11·7	13·2	12·8

The figures given above have the obvious defect of including, besides the products intended for the home market, also those handled by the Russian railways on their way across the frontiers. Moreover, it must be borne in mind, that of the total supplies sent to market, a certain percentage, most important in the case of such products as fruit, vegetables, poultry, eggs or dairy produce, escaped

[1] Compiled from statistics of traffic on the Russian railways given in the *Recueil des données statistiques et économiques sur l'industrie agricole*, 1913, pp. 593–9.

registration in the statistical records of the railways through being transported to their destination by road or by one of Russia's many internal waterways. The table, accordingly, should be considered bearing in mind these reservations, but with regard to certain of its items, which by the nature of things could not enter largely into foreign trade, its meaning is unequivocal. Thus, the 30·5% increase in the transport of milk and cream is significant; and no less interesting are the figures for fresh vegetables and fruit showing advances of 78·9 and 58·9% respectively. Though butter was under suspicion, since its exports had increased by over 40% during the same period, the statistics of its consumption, available for the city of Moscow for the last three years before the war, point to so rapid an increase as to suggest that the home market for butter was growing very fast. Indeed, the *per capita* consumption of butter in Moscow, which was 8·1 Rus. pounds in 1911, rose to 9·3 pounds in 1912 and reached 10·6 pounds in 1913.[1]

By far the best indication, however, of the real trend of development of the Russian internal market for agricultural products is probably given by the movement of wholesale prices in the large Russian consuming centres. The table below shows the price movement for three consecutive sections of the period under review, in the form of relatives for each of the commodities included: [2]

Price Movements on the Russian Agricultural Market

	1901–5	1906–10	1911–13
Wheat flour, Petrograd	100	127	140
Rye flour, Moscow	100	138	133
Beef, Moscow	100	126	146
Mutton, Moscow	100	134	153
Pork, Moscow	100	129	134
Fat, beef, Moscow	100	113	121
Butter, salt, Moscow	100	116	120
Eggs, Moscow	100	114	122
Malt, Moscow	100	133	147
Linseed oil, Moscow	100	102	114
Hempseed oil, Petrograd	100	105	115
Hides, ox and cow, Odessa	100	128	149
Wool, merino, Moscow	100	127	130

In the case of some of the articles mentioned in the above table, the price movements in Russia can be compared with those of the world market for the periods 1901–5 and 1906–10, no figures for 1911–13 being, unfortunately, available to the writer. The relatives for the articles for which comparison is possible, are as follows: [3]

[1] B. Brutzkus, *Economics of Agriculture*, p. 62.
[2] Compiled from quotations in the *Recueil des données statistiques et économiques sur l'industrie agricole*, published in 1917 (last issue of the series).
[3] Calculated from the same source as the preceding table.

	1901–5	1906–10
Wheat flour, London	100	111
Rye flour, Danzig	100	119
Beef, London	100	100
Mutton, London	100	100
Pork, London	100	105
Fat, beef, London	100	111
Butter, salt, London	100	112
Eggs, London	100	108
Malt, London	100	111
Hides, ox and cow, Hamburg	100	125

The comparison of the two tables shows that, in the case of all these products, the prices on the home market have been rising much more rapidly than those of the world market, with the result that production for the former was encouraged in preference for exports. In the vicinity of large towns and industrial centres and along the railways which connected them with the countryside, this had exercised a powerful influence on the development of farming.

While economic conditions, since the close of the last century, have been shaping themselves, on the whole, very favourably to the progress of Russian farming, there still remained much to be done in the way of marketing organization to enable the farmer to reap the full benefit of the improvement. The Russian farmer, and especially the peasant producer, suffered much from the difficulty of marketing his produce, the lack of any adequate marketing organization and the consequent exploitation of the agriculturist by numerous middlemen, who inserted themselves between him and the consumer, or rather the wholesale merchant or large exporter. The cereal trade of Russia, indeed, though, at the top, it was conducted by large dealers in the great Corn Exchanges of the capitals, the Volga towns and the exporting harbours, at the bottom, where small local dealers collected grain from the peasants in lots of a few poods or sacks apiece, presented a picture of utter anarchy and petty exploitation.[1] In olden days, before the abolition of serfdom and the construction of railways, the system of marketing was generally simpler and involved fewer stages, the trade being concentrated in a few hands and in a few trading centres, mostly situated on the Volga or on other large waterways of the producing belt. The principal sellers were large landowners, and the transaction, therefore, even in primary buying, involved relatively large quantities of more or less uniform grain. The advent of the railways, following on the Emancipation, which resulted in the appearance on the grain market of hosts of small peasant producers,

[1] *Vide* J. Rubinow, *Russia's Wheat Trade*, U.S.A. Dptm. of Agr., 1908, p. 12.

whom the pressing need for cash was forcibly driving into commercialization, changed the situation completely.[1] Instead of the few large centres of gravitation of local grain surpluses, there appeared a large number of railway stations, to which the grain could be brought; and the scattered production, coupled with the producers' urgent need for cash, made the grain trade a happy hunting ground for shoals of greedy and unscrupulous middlemen. The lack of organization in the grain trade affected not only the prices obtained by the farmers on the spot, but those on the ultimate market as well, owing to the difficulty of proper sorting and the resulting impurity and uncertain standards of quality of the grain.[2]

The first attempts to improve the organization of the grain trade dated from the 'eighties. In 1884, the South-Western Railways, then still controlled by a private company, inaugurated the granting of credits against consignments of grain for export. These credit operations were originally financed by commercial banks, but later the State Bank also took a hand in these transactions. These credits, however, benefited primarily the large grain dealers, and their ultimate effect on the position of the actual producers was practically nil. In 1888, the railways were generally authorized to grant credits on grain on behalf of the State Bank, for which they acted as agents; they were also allowed to organize the marketing of grain handed to them and to build their own granaries or elevators for this purpose. From 1893 on, the State Bank had been extending its financial participation in the grain trade, with a view to gaining control over the market. In combination with the railways, also to a large extent State-owned, the Bank began to develop a system of credits and a network of railway elevators;—a scheme which had originated with a Commission, appointed at the close of the 'eighties to consider the measures by which the further decline of grain prices could be stopped. The necessity of erecting more elevators was emphasized again by the Commercial and Industrial Congress which took place in Nizhny-Novgorod in 1896, on the occasion of an All-Russian Exhibition held there. Yet, little progress had been made in the matter, mostly owing to the conflicts between the interests of the various groups concerned. Thus, the Government wanted, as far as possible, to ensure for itself, mostly through the State Bank, a measure of effective control over the grain trade, which would enable it to combat the exploitation of the actual producer by middlemen, as well as to prevent the depreciation of Russian grain on the world market, resulting

[1] Liashchenko, *Essays*, pp. 140 ff.
[2] *Vide* J. Rubinow, op. cit., pp. 15–20.

from its adulteration and lack of grading. The producers, represented by the Zemstvos, were generally in favour of public control over the trade, but they preferred it to be exercised not by the State, but by themselves within their respective provinces. In other words, they wanted the elevators to be controlled by the Zemstvos. The railways, to which had belonged the initiative of taking an active part in the marketing of the Russian grain surplus, and which, by the close of the last century, with the assistance of the State Bank and of commercial banks, had already evolved a considerable system of granaries, agencies and credits, preferred any further progress to follow the same lines. The grain dealers, though they also recognized the desirability of better organization and equipment of their trade, were strongly opposed to Government interference, in any shape or form, in the course of a business which they looked upon as concerning themselves alone. Accordingly, when, in 1899, the Government called together a conference of the interests concerned in the grain trade, namely the landowners, the grain merchants and the railways, the decision reached was a rather ineffective compromise between their conflicting points of view. Indeed, though the principle of State control of elevators was agreed to, no one was made ultimately responsible for actually taking any definite measures; and, while recognizing the necessity of establishing standard grades of cereals, the Conference stopped short of making grading compulsory.[1] A few years later, the local Committees on the Needs of the Agricultural Industry insisted on the necessity of a thorough re-organization of the grain trade, with a view to protecting the producers. In the meantime, the actual construction of elevators proceeded slowly, most of them being built either by the railways or by the State, in which latter case they have generally been handed over to the railways for exploitation. Here, as in the matter of Siberian colonization, the Government has been working hand-in-hand with the railways, mostly owned and controlled by the State. In 1900, the total number of elevators at the disposal of the railways was 62, with a total capacity of about 19 millions of poods and 198 granaries for about 25 millions of poods; facilities almost negligible, considering the volume of grain trade in Russia.[2] The real progress in the organization of the Russian grain trade had yet to come, and signs of it

[1] Rubinow, op. cit., p. 21.

[2] *Ibidem*, p. 22. Rubinow compares Russia's position to that of the United States, where, in 1906, Chicago alone possessed 89 elevators with a capacity of 63,545,000 bushels. Of this total, the 19 " regular " elevators were practically equal in capacity to the 62 elevators and 198 granaries in the possession of the Russian railways in 1900 (p. 23).

became apparent during the decade immediately preceding the war. The disadvantages of the existing conditions have been brought home to the grain dealers and exporters by the effects they exercised on the course of prices of Russian grain on foreign markets, and a marked tendency had appeared towards eliminating unnecessary intermediaries, leading to the large exporting firms establishing, through their own agents, direct contact with the producing areas. The railways, acting either on their own account, or as agents for the State Bank, which extended its financing operations, became more active through their commercial agencies, thus helping the concentration of the surpluses in their hands, as well as the handling and ultimate marketing of the cereals. The joint-stock commercial banks also began increasingly to interest themselves in the financing of the grain trade, and thus became more directly concerned in its organization. Locally, among the scattered peasant cereal-growers, the co-operative movement and the gradual development of co-operative marketing, though still in its infancy, also pointed to the existence of a tendency towards the substitution of organized dealing for the former anarchy. During the same decade, and especially since 1910, the Government also stepped in with a scheme of a vast network of elevators, to be built under the general direction of the State Bank in the large centres of grain trade, as well as of smaller elevators and granaries in the cereal-growing localities. The large elevators, on completion, were to be handed over for actual exploitation to the railways, while the smaller ones in the countryside were to be placed under the control of local co-operative organizations. The scheme was only just started, when the outbreak of the war brought its development to a close. Yet, on the eve of the war, though still far from sufficient to meet the requirements of the grain trade, the available storage facilities were much greater than they had been at the close of the last century. The railways, in 1913, had 75 mechanically-equipped elevators with an aggregate capacity of 26·75 millions of poods and 4,808 granaries for 188·5 millions of poods. In the various ports of shipment, accommodation was available for a total of 103·6 millions of poods. Besides, there were the first 9 mechanical elevators of the State Bank, erected in accordance with the new scheme, for 5·7 millions of poods. With the elevators of the Zemstvos, which, in 1911, numbered over 40, with a total capacity of 20 millions of poods, and the granaries of co-operative institutions, numbering 432 and taking a total of about 6·4 millions of poods, on the eve of the war the aggregate storage accommodation for grain in course of marketing, at its various stages, reached approximately 350 to 400 millions of poods.[1] The progress thus accomplished in the early part of the twentieth century, though it left still much to be done, was important in that it facilitated the solution of two of the principal problems in the organization of the grain trade, namely, on the one hand, the concentration, sorting and cleaning of the cereals sent to market and, on the other, the financing of both the producer and the dealer.

The credits granted by the State Bank and by the joint-stock commercial banks against grain delivered to the elevators and granaries, increased with considerable rapidity. Thus, the total amount of advances made by the State Bank on the security of agricultural products, mainly consisting of cereals, during the period 1908 to 1913, were as follows:

	1908	1909	1910	1911	1912	1913
			In millions of roubles			
Advances direct by the State Bank	49·2	71·1	104·3	122·7	111·9	143·6
Advances by the Bank through Zemstvos, etc.	11·8	15·2	39·7	72·5	77·7	100·7
Advances by the Bank through the Railways	12·9	10·2	7·3	9·8	8·4	9·6
Total	73·9	96·5	151·3	205·0	198·0	253·9

The above table, besides the general increase in the amount of credits, points also to the direction in which the evolution was proceeding. It may be seen that the increase was confined to the amount of advances made direct by the State Bank, which enabled it to obtain more effective control over the course of marketing, and to loans through intermediaries, such as Zemstvos, commercial banks and co-operative institutions, while the volume of credits granted through the railways was rather diminishing: a development probably due to the fact that, in this direction, much more had already been achieved before. Moreover, it was mostly through intermediaries other than the railways that credits granted by the State Bank were made available, in the first instance, to producers, as distinguished from dealers. In the amounts of credits granted by the State Bank to dealers, against forwarding documents or on discount of bills, the increase was much less marked, the totals under both these heads being:

1909	83·2 million roubles
1910	99·9 " "
1911	95·7 " "
1912	117·2 " "
1913	100·6 " "

[1] Kondratiev, *The Grain Market and its Control during the War and the Revolution*, pp. 29–30.

Up to the last, however, the greater part of the advances made direct by the State Bank in one form or another went into the hands of the grain trade, as distinguished from the producers. The distribution of the total advances in the course of the years immediately preceding the war was as follows:

	1910	1911	1912	1913
		Percentages		
Grain dealers	56	56	66	52
Producers	44	44	34	48

By this means the State Bank endeavoured to obtain more complete control over the conditions of handling and marketing the grain, while by means of advances granted through intermediaries it developed a system of credits to the actual producer and, in the first instance, to the peasant farmer, to whom direct credits were not available.

Besides the State Bank, credits were granted on grain and other agricultural produce by the commercial banks and the railways. The advances of the banks, which, at the close of the last century, did not exceed the total of about 200 million roubles a year, reached nearly 1,000 million roubles on the eve of the war. The amounts invested in financing the grain trade by the railways, apart from the sums put at their disposal by the State Bank for that purpose, in 1908–10 averaged about 50 to 60 million roubles a year. It may be seen that, while the State Bank played a very important part in the provision of financial resources for the grain trade, on the eve of the war by far the greater part of the total funds necessary for the marketing of the Russian grain surplus was provided by the private credit apparatus. Here, as elsewhere, with the progress of Russia's capitalistic development, capital could be seen extending its sphere of activities and taking over from the State, one by one, those essential economic functions, which, hitherto, failing the necessary means, private enterprise had been unable to assume. More and more, in the course of the years immediately preceding the war, the grain trade, at the top, in the large and important centres of commerce, was becoming dependent on private sources of credit, on the money market at large, while the resources of the State Bank were increasingly being diverted to those stages of the grain traffic which were least accessible to private banks, while they were most in need of money and organization, namely to the stages involving the transfer of his grain to the marketing agencies by the producer and its handling on the way from the producer to the trade centres.

Though some progress in the organization of marketing had been made in the early part of the current century, much more still remained to be done to secure to the actual producer a fair share of the price his grain fetched on the ultimate market. On the eve of the war, the problems of grading and sorting the grain were not yet solved, since only part of the cereals passed through elevators and granaries supplied with the necessary equipment. During the grain-marketing seasons, congestion at some points of the railway lines, by which the grain was transported, was still a frequent occurrence, and yearly resulted in heavy losses. Even apart from the defects of the marketing organization, moreover, the Russian producer had to contend with certain inherent difficulties of the country's geographical situation and of its agrarian régime. The remoteness of the producer from the consumer, either at home or abroad, and the necessity of long hauls over rough country roads, tended to keep spot prices down. Water transport was only available for part of the marketing season. Besides, natural waterways, convenient for shipping, are far less developed in the wheat-growing belt of Russia, than outside it, in the under-producing regions. In the case of grain shipped abroad, the proportion of the cost of transport attributable to railway freights to the harbour was very high, in spite even of the low rates, while loading expenses in the ports were far in excess of those incurred by Russia's competitors overseas, mostly owing to lack of adequate mechanical equipment.[1] The effects of the extreme scattering of production over millions of peasant cereal-growers, which even a highly developed co-operative organization of marketing would hardly be able to eliminate completely, are so obvious that to them it is superfluous to refer. In the early part of the twentieth century, Russia was only beginning to tackle the problems involved in the marketing of her agricultural surplus, and gradually from various quarters efforts were being made to improve its organization. The accumulation of national capital was making itself increasingly felt, through the commercial banks, in this important branch of Russia's economic life; but she still stood in need of large sums of fresh capital for investment in the equipment of her harbours, the construction of elevators, auxiliary railways and roads; and, so long as for this outlay she had to depend primarily on the State, with its many financial commitments, progress in this respect was bound unavoidably to be slow.

[1] " It is officially estimated that the port expenses alone in various Russian ports amount to from 2·5 to 6 copecks per pood of wheat " (Rubinow, op. cit., p. 24). He also quotes striking figures concerning the time employed in loading grain on board in Russian, as compared with the American, ports.

B. *Russia's Agricultural Exports*

Though, as I have endeavoured to show, the internal market for agricultural produce in Russia has been expanding with considerable rapidity in the early part of the twentieth century, the home demand was still far from sufficient to absorb the whole of the annual surplus of the production of farming. For about one-half of his surplus of the principal cereals and a large part of his other produce, the Russian farmer was dependent on foreign custom. Moreover, though, as pointed out above, Russia does not belong to the type of country best adapted, by its natural and economic conditions, to the part of agricultural exporter, that rôle was forced on her by her relatively backward state of industrial and capitalistic development and her consequent dependence upon foreign capital and imported manufactured goods. The producer, not yet able to dispose of his output at home, had to seek customers abroad, while the Government had to encourage agricultural exports by all possible means, in order to provide for the needs of the country, as well as its own, in foreign products and foreign exchange. Before the war, Russian Government loans, railway debentures guaranteed by the State and municipal securities held abroad were estimated at about 4,200 million roubles and involved an annual disbursement in foreign currencies of approximately 180 million roubles. Besides, large sums of foreign capital were invested in various Russian industrial concerns, mostly in connection with mining, metallurgy, oil and textiles, and this also involved the necessity of yearly remittances abroad. The Russian balance of payments, therefore, was heavily adverse, and its deficit could only be met out of a large surplus of exports over imports. In the course of the last quarter of a century before the Great War, along with the general growth of the volume of her foreign trade, the margin which, on the balance of trade, remained in favour of Russia, tended gradually to increase, as may be seen from the table below, showing the average exports and imports for 1891–1913:[1]

	Exports	Imports	Balance
	Millions of Roubles		
1891–1900	659·8	535·4	+ 124·4
1901–1910	1,073·1	887·4	+ 185·7
1911–1913	1,543·4	1,235·8	+ 307·6

From an analysis of the distribution of exports and imports according to the nature of the articles of which they consist, one

[1] M. Sobolev, "The Foreign Trade of Russia," in A. Raffalovich's *Russia; its Trade and Commerce*, p. 301.

can see at a glance the origins of the balance which served to meet Russia's liabilities on the foreign markets. Taking an average year, such as, for practical purposes, was the year 1910, it may be seen that the composition of the balance was as follows : [1]

	Exports	Imports	Balance
	Thousands of Roubles		
Animals, living . .	30,120	10,791	+ 19,329
Articles of food . .	927,517	191,462	+ 736,055
Raw and semi-manufactured materials . . .	436,934	554,386	− 117,452
Manufactured articles .	54,514	327,807	− 273,293
Total . . .	1,449,085	1,084,446	+ 364,639

The meaning of the table is perfectly clear. If anything, it rather tends to minimize the part played by agricultural products, since some of them, including flax and hemp, which yielded a large balance in favour of Russia, are hidden from sight in the group of raw and semi-manufactured materials. The overwhelming importance in the Russian trade balance of the agricultural exports is obvious. They were necessary and, indeed, indispensable, for Russia's economic stability and progress. Even a rapid development of the home market for agricultural produce, beneficial as it would be for the progress of Russian farming, would not enable Russia, for generations to come, to dispense with the production of a large surplus of foodstuffs for export abroad, since, even then, she would still have large liabilities to meet on account of loans and investments, and any excessive encroachment of internal consumption on the exportable margin would be bound, therefore, to have injurious effects on the financial and monetary situation. It was, and still is, most important, accordingly, that agricultural production should keep pace with the simultaneous growth of the demands of the internal and foreign markets: a requirement necessitating, in the first instance, an improvement in the methods of farming and an increase in the yield of all its branches. With regard to the standards of cultivation and to the yield of the crops there was ample room for improvement. Indeed, had Russia raised the yield of her crops to the level not even of Germany, Holland or Belgium, but to the much lower one of Austria—a proposition involving nothing intrinsically impossible—she would be in a position to meet, without strain, the demands of a home market roughly twice as large as the present, without in any way encroaching on

[1] *Statistical Abstract for the Principal and other Foreign Countries in each year from* 1901 *to* 1912, p. 99.

her exportable margin. The actual development of farming in Russia, in the case of large estates as well as in that of the peasants, pointed to the existence in the Russian countryside, on the eve of the war, of strong progressive currents, which brought the raising of the yield of the Russian agricultural industry to hitherto unprecedented figures well within the range of immediate possibility. Indeed, this had actually taken place in large numbers of individual cases, and, along with the reform of peasant tenures, which was proceeding apace, the improvement in this respect, helped by co-operative organization, agricultural education and better equipment, was bound to spread and to raise the averages for the whole country conspicuously.

Dealing with Russia's agricultural exports, it is necessary, in the first instance, to point to their steady increase in volume and value before the war. The table below shows the increase in the value of Russia's foreign trade in agricultural products in the early part of the current century:

	Exports	Imports	Balance
	Average Value per annum in Thousands of Roubles		
1901–1905	721,659·5	255,270·6	+ 466,388·9
1906–1910	916,831·4	387·280·2	+ 529,551·2
1911–1913	1,175,897·1	448,299·2	+ 727,597·9

Closely allied to farming, though not included in the present study, is forestry, on which a considerable part of the incomes of both landed proprietors and peasants in the wooded provinces of Russia depended. The foreign trade in timber and other products of forestry was also increasing apace. The net exports, which averaged 55·1 million roubles in 1901–5, increased to 105·8 millions in 1906–10 and reached 140·4 million roubles in 1911–13.

Among the products of arable farming exported abroad, the most important by far were cereals, of which Russia was the world's greatest exporter. The evolution of the export trade in the separate cereals followed different courses. Thus, rye, which, until the second half of the last century, had been the most important among the cereals exported from Russia, has been declining in importance ever since, its consumption in continental Europe, to which it used to be exported in large quantities, being on the decrease. The decline continued into the current century, with the result that, of the principal cereals, rye held the last place throughout the period from 1901 to 1913. Oats were exported in somewhat larger quantities, but their exports tended also, on the whole, to decline, though their fluctuations during this period, as well as in the past, were rather erratic and seemed to depend more on the current changes in prices than on any inherent tendencies in their production or consumption. Used as feeding grain, oats could more easily be made available for export by means of substitution, whenever prices made this expedient, than the bread-making cereals, and this may have at least partly accounted for the seemingly capricious variations in the quantities exported. Wheat and barley were the two staple articles of Russian grain exports, and the movements in them, during the period dealt with, tended characteristically to compensate each other. In the case of wheat, the growing demand of the home market and the rapid rise in prices in Russia tended to reduce the exportable surplus, and the net exports of wheat, which amounted to 229·6 millions of poods *per annum* in 1901–5, declined to 222·2 million poods in 1906–10 and to 193·5 millions of poods in 1911–13. At the same time, the exports of barley increased rapidly, the net exports being 122·8 millions of poods in 1901–5, 179·9 millions of poods in 1906–10 and 222·4 millions of poods in 1911–13. The evolution of the export trade in these two cereals was significant, not to say ominous. What it pointed to was the failure of Russian farming in keeping abreast of the combined call on its production of wheat of the home and the foreign market, and the consequent substitution in exports for the more valuable wheat of the less valuable barley: a development which, if continued, was bound ultimately to exercise an adverse effect on the balance of trade. This deterioration in the main articles of export, though it may have been to some extent accentuated at the time by the rapid rise in barley prices on the world market, which stimulated the cultivation of this particular cereal, certainly had deeper causes in the increasing diversion of the available surplus of wheat for consumption in Russia. Other cereals, such as maize, sarazin, millet, etc., as well as wheat and rye flour, played a less important part in the export trade.

Russia's net exports *per annum* of the principal and other cereals during the period 1901 to 1913 are shown in the table below:

Russia's Grain Exports, 1901–1913

	1901–5	1906–10	1911–13
	Thousands of Poods (36 lbs.)		
Wheat	229,590·6	222,238·0	193,523·8
Rye	75,706·4	36,716·6	33,202·4
Barley	122,839·2	179,937·2	222,434·8
Oats	76,507·2	55,646·2	56,388·1
Sarazin	1,983·2	1,505·0	3,958·7
Millet	930·2	533·0	773·9
Maize	35,023·4	35,519·0	54,102·7

The volume and value of the net exports of all grains and flour during the same period are shown below:

	Volume in Millions of Poods	Value in Millions of Roubles
1901–1905	562·9	435·9
1906–1910	551·5	503·5
1911–1913	595·4	561·7

It may be seen that, on the whole, the increase in the net exports of grain and flour during the years immediately preceding the war was not very marked. Indeed, it lagged far behind that of the total turnover of Russia's foreign trade, and of her exports in particular, the difference having been made good by the growth of trade in other products of the agricultural industry, as well as by the expansion of the exports of timber and other products of forestry, to which I have pointed above.

Of the other products of arable farming, flax and hemp played an important part in the foreign trade of Russia. The net exports of flax in various forms increased from 13·6 millions of poods in 1901–5 to 16·2 millions of poods in 1906–10 and 17·9 millions of poods in 1911–13. The value of the flax exported in 1911–13 averaged 93·4 million roubles *per annum*. The exports of hemp were practically stationary and amounted to 3·0—3·5 millions of poods *per annum* to the value of about 19 million roubles.

A very important item in the export trade of Russia was sugar, whose net exports in the early part of the current century were as follows:

	Millions of Poods	Value in Millions of Roubles
1901–1905	9·6	20·2
1906–1910	11·1	25·7
1911–1913	19·8	49·9

With regard to the exports of live animals and animal products, the developments of the period preceding the war displayed certain characteristic features.

As a horse-breeding country, Russia has always been a large exporter of horses. These exports have shown a considerable increase in the course of the last decade preceding the war, during which Russia had largely contributed to the remounting of the armies of both the Entente and the Triple Alliance. The net exports of horses from Russia are shown below:

	Numbers Exported in Thousands of Heads	Value in 1,000 Roubles
1901–1905	44·6	4,594·8
1906–1910	81·0	8,421·4
1911–1913	82·1	10,596·1

The export of live pigs, mostly to Germany, was a not unimportant branch of Russia's foreign trade, as may be seen from the table below:

	Numbers exported in Thousands of Heads	Net Value in 1,000 Roubles
1901–1905	55·6	2,843·8
1906–1910	16·2	3,067·6
1911–1913	64·2	7,230·3

On the balance, the foreign trade in live animals yielded Russia, as surpluses of exports over imports, the following amounts:

1901–1905	4,378,400	roubles a year
1906–1910	6,010,000	" "
1911–1913	7,811,200	" "

It may be seen that, on the whole, this branch contributed to the surplus on the balance of trade on a very modest scale. Indeed, in the early part of the current century, Russia had already been leaving behind that stage in her agricultural evolution in which the export of live animals could play anything but an entirely subordinate part in her foreign trade. With the transition to more intensive systems of farming, the export of live animals was bound to diminish in relative importance, giving place to that of animal produce.

Another article of Russia's exports, namely the concentrated feeding stuffs, such as oilcake, etc., though, during the period dealt with, it had been increasing in volume and value, could also hardly be expected to expand, or perhaps even to continue on the same scale, with the imminent spread of intensive stock farming. The export of concentrated feeding stuffs amounted to 60·7 millions of poods in 1901–5, to 70·9 millions of poods in 1906–10 and to 91·0 millions of poods, to the value of about 70 million roubles a year, in 1911–13. Under intensive conditions of stock farming and with the improvement in the quality of animals, which generally accompanied peasant enclosures, an increased proportion of these feeding stuffs was naturally bound to be diverted to farm uses and converted into valuable animal products for disposal on the market.

Among the animal products exported, the most important were butter and eggs. The net exports of butter are shown below:

	Quantity Exported in 1,000 Poods	Value in 1,000 Roubles
1901–1905	2,298·2	29,344·2
1906–1910	3,346·0	47,267·2
1911–1913	4,551·1	69,568·6

On the balance, the net surplus of exports of dairy products over imports averaged 28·8 million roubles a year in 1901–5, 46·9 millions in 1906–10 and 69·4 millions in 1911–13. In the case of butter, it will be noted that nearly the whole surplus was due to Siberian production (4,443,000 poods exported from Siberia in 1913), while the output of European Russia was disposed of almost completely on the home market. With the gradual intensification of stock farming in the European provinces of Russia, the improvement in the type of cattle and the growing diversion of the farmers' attention from meat to dairy production, characteristic of the progress of Russian farming in the course of the period dealt with, this branch of exports, in which hitherto, European Russia had played almost no part, could be expected to develop enormously.

Eggs and live fowl, the products of poultry farming, which was relatively well-developed in the cereal-growing provinces situated within more or less easy reach of the consuming centres or the frontiers, formed another important group of items in the export trade of Russia. Their exports were as follows:

	Eggs		Live Fowl	
	Number in 1,000	Value in 1,000 Rbls.	1,000 Heads	Value in 1,000 Rbls.
1901–1905 .	6,961·4	47,944·6	7,961·0	6,696·8
1906–1910 .	7,446·8	57,471·8	8,181·4	7,439·4
1911–1913 .	9,370·1	86,375·3	9,291·8	8,326·6

On the very eve of the war, the export of bacon from Russia, mostly to England, began to assume very considerable proportions. Thus, another important branch of trade in animal produce showed signs of development, the actual figures of the export trade in bacon pointing to a rapid increase: [1]

	Bacon Exported 1,000 Poods	Value in 1,000 Roubles
1902–1905	119·8	800
1906–1910	204·9	1,600
1911	419·0	3,800
1912	524·2	3,900
1913	718·0	5,100

It may be seen that, on the eve of the war, the export of animal products from Russia tended to increase in importance. This development of Russia's agricultural exports pointed in the right direction, because it enabled the farmer better to utilize the produce of his holding and, by introducing greater variety in the production and increasing the live stock, contributed to the progress of farming.

[1] J. Pelferov, "Agriculture," in Raffalovich's *Russia; its Trade and Commerce*, p. 38.

In this respect, a comparison of the development of the export of dairy products and eggs with that of cereals and flour at the beginning of the twentieth century is interesting. It is, accordingly, given below:

Evolution of Net Exports of

	Cereals and Flour		Dairy Produce and Eggs	
	Millions of Poods	Per cent	Millions of Poods	Per cent
1901–1905 . .	564	100	9·3	100
1906–1910 . .	551	98	10·9	117
1911–1913 . .	595	106	13·7	146

The period was too short to enable any definite conclusions to be drawn from the comparison of the above figures, but considering the general trend of the evolution of Russian farming in the early part of the current century, which involved the increasing diversion of wheat to the home, as distinguished from the foreign, market, as well as the tendency towards higher standards of farming, a greater variety of agricultural production and a fuller utilization of the available resources, the rapid increase in the exports of animal products, as compared with cereals, would appear to possess a certain significance. While, indeed, in the case of cereals, Russia, though she continued to increase the bulk of her exports, did so with intermissions and has been gradually substituting less valuable produce for wheat, in the exports of animal products and of such products of intensive arable farming as sugar, seeds, potatoes, tobacco [1] and flax, she was making considerable progress. This was the line of development traced for Russia by the pressing needs of her agricultural industry, and both the growing demand of her home market and the evolution of her foreign trade seemed to point to her taking that direction.

A few brief remarks should be made here concerning the distribution of Russian agricultural exports between the principal foreign markets.

The changes in the destination of Russian exports of grain, since the close of the last century, may be seen from the table on page 272, showing the percentages of the total exports to the credit of each of the principal importing countries. [2]

[1] The exports of seeds, potatoes and tobacco from Russia increased as follows, in thousands of poods:

	1901–5	1906–10	1911–13
Seeds, various	9,728·2	12,590·4	18,241·1
Potatoes	3,257·4	4,973·6	12,381·9
Tobacco	330·0	520·2	714·9

[2] M. Sobolev, "The Foreign Trade of Russia," in Raffalovich's *Russia; its Trade and Commerce*, p. 318.

	1893–97	1898–1902	1903–7	1908–12
Germany . . .	15·9	20·9	22·2	31·9
United Kingdom . .	26·4	20·3	21·7	15·3
Holland . . .	12·8	17·0	18·3	20·1
Italy	6·9	9·0	8·7	7·1
France . . .	9·4	8·4	6·7	6·3
Belgium . . .	3·4	2·2	2·8	2·8

In the United Kingdom, Russian grain was being ousted from the market by cereals imported from the New World and from the British Oversea Dominions, while German imports increased partly owing to increased consumption, partly as a result of the practice of free imports on custom permits (*Einfuhrscheine*), with a view to encouraging the German milling industry, which converted the grain into flour and then re-exported. With regard to the different cereals, the position varied greatly. Thus, in the case of wheat, Russia's largest customer was the United Kingdom, which was responsible, in 1908–12, for 19·5% of the total quantity exported; next came Italy with 18·8%; Holland with 12·6% and France with 12·2%. In the case of all these countries, however, between 1901 and 1913, there had been a certain diminution in their respective shares. On the contrary, Germany, which, with her system of agricultural protection, in 1903–7 took only 6% of the wheat exported from Russia, in 1908–12 raised her share to 8·5%. In Western Europe, since the close of the last century, the competition of oversea wheat was making itself increasingly felt, with the result that Russia found her usual outlets being restricted: a phenomenon due partly to the relative cost of transport by rail and by sea, but mainly accounted for by the better and more efficient organization of the handling and marketing of cereals in the countries of the New World. The growing exports of barley, used partly for breweries, but mostly for feeding live stock, went chiefly to Germany, which took nearly one-half of the total quantity exported. The share of the United Kingdom in the total, which, in the 'eighties, amounted to one-half, had been gradually declining since, until in 1908–12, it did not exceed 10·8%. In Germany, with the development, since the last quarter of the nineteenth century, of intensive cattle-breeding and dairy farming, to which her farmers had turned, along with agricultural protection, during the period of agricultural depression, and which has been growing ever since, there was a large market for feeding stuffs in general, and for barley in particular. Oats were exported, in the first instance, to the United Kingdom, though with them, as with wheat, Russia was faced with the keen competition of the New World, and lately of India as well, which tended gradually to reduce British imports of

Russian oats. The latter, indeed, had declined from 29·2 millions of poods in 1903–7 to 19 millions of poods *per annum* in 1908–12, the percentages of the total being respectively 43 and 29·2%. At the same time, exports to the Continent, and more particularly to Holland and Germany, tended to increase.

Flax, of which Russia is the world's largest producer, was exported, in the first instance, to the United Kingdom, which took approximately 30% of the total exports, thus meeting with Russian supplies about 75% of its total requirements on imported fibre.

Oilcakes, exported in large quantities, went mostly to the continental countries with intensive stock farming, and especially to Germany and Denmark.

Butter, up to 50% of the total exports, went to the United Kingdom. Next came Germany and Denmark. From the latter some of it was re-exported abroad, mainly to the British market.

Of the Russian exports of eggs, before the war about one-third went to the United Kingdom and about a quarter each to Austria and Germany.

Russian hides were mostly exported to Germany, which took 40% of the total exports, and to the United Kingdom, which accounted for about 20%.

Horses went mostly to Germany, whose imports from Russia, in 1907–11, averaged about 60,000 heads *per annum*, or approximately two-thirds of the total number exported. Nearly all the rest was exported to the United Kingdom, the share of the other countries being negligible. In this case, however, one has to do with a branch of Russia's foreign trade belonging to the past and having no possible future.

Tobacco was mostly exported in leaf, for manufacture abroad, the principal importers of Russian tobacco being Finland, France, Germany and Austria-Hungary. In Finland and Germany it was used mostly for the manufacture of cigars, while France and Austria-Hungary, in which there was a monopoly of the tobacco trade, bought large quantities in bulk. Accordingly, in 1911, of the 636,000 poods of leaf tobacco exported, Finland took 27%, France 23%, Germany 20% and Austria-Hungary 7%.[1] Besides, on the very eve of the war, an extension was becoming noticeable of the market for Russian tobacco (mostly of the varieties grown in the Caucasus) in Asia Minor and in Egypt, to which it was exported in increasing quantities.

[1] N. Malakhovsky & A. Isenberg, " The Cultivation of Cotton, Flax, Hemp, Sugar-Beet, Tobacco, Vine, Hops, Fruit," in Raffalovich's *Russia ; its Trade and Commerce*, pp. 78–9.

The principal market for Russian sugar was the United Kingdom, which sometimes took nearly the whole quantity exported from Russia.

To conclude this discussion of Russia's foreign trade in agricultural products, it is necessary to point out that, for the bulk of her produce, Russia certainly had, and would have for generations to come, an extensive market. Yet, to build wholly on the basis of the experience gained in the early part of the current century, and to rely on the progress in agricultural conditions which was observable at the time as a situation which was going to last, would certainly have been rash, even had peace not been broken. The wave of prosperity, as so many others before, was bound, sooner or later, to come to an end. It may even be said that on the very eve of the war, since 1913, certain signs began to appear pointing to a slowing-down in the rising trend of the agricultural markets. In the face of an imminent change for the worse in agricultural conditions, the problems of marketing and of the direction to be given to the further evolution of the exporting branches of the agricultural industry assumed particular importance. On the eve of the war, in both these directions Russia had been feeling her way towards improvement. It has been seen that, on the one hand, Russia was getting to tackle the problem of marketing, while, on the other, it would appear that she had been developing the export of the more valuable products of intensive farming. Slight as have been these indications of progress, they reflected the general trend of Russia's agricultural evolution during the decade immediately preceding the war. Faced with the necessity of maintaining and gradually increasing the scale of her agricultural exports, and yet ill-adapted by her agricultural organization to the position of an exporter of products of extensive cereal farming, Russia, on the eve of the war, could be seen in the process of starting the re-adjustment of the elements of her export trade in a way which would best accord with the agricultural conditions of a country whose prosperity depended on the development of intensive peasant husbandry.

CHAPTER IX

THE EVOLUTION OF ARABLE FARMING ON THE EVE OF THE WAR

THE outstanding feature of Russian farming on the eve of the war was its strong cereal-growing bias which, combined as it was in most parts of the country with the traditional three-course system, necessitating the fallowing every season of approximately one-third of the total arable area, tended to reduce both the extent and the yield of the crops. In those parts of Russia, in which, as in the Southern and Eastern steppes, the soil was so fertile that, for a number of years in succession, it could bear crops without regular fallowing, the latter was mostly dispensed with altogether and cereal crops sown on the same field until it showed signs of exhaustion: a practice which, though it reduced the fallow in these localities to so small a proportion, sometimes, as 4 or 5% of the total area, also had the effect, after a few years of application, of bringing the yield of Russia's richest black-earth down to the level of that of the poor, but more efficiently cultivated, sands of the regions of Petrograd or Moscow. This position, largely responsible for the aggravation of the "land-hunger" in Russia, as well as for the general poverty of the country, had to be remedied, and a more varied cropping, permitting the fuller and more scientific exploitation of the available agricultural area had to be substituted for the existing methods of farming, which, in only too many cases, were nothing better than a mere spoliation of land, tersely described by Roscher as *Raubwirtschaft*. It will be remembered that, until the twentieth century, there existed serious obstacles to evolution in this direction, among which the two most important were the open fields, on the one hand, and the slow growth of the home market for agricultural produce, on the other. During the decade immediately preceding the war, the enclosure movement, once set on foot, could be seen rapidly doing away with the difficulties involved in the old systems of tenure, while the industrial revolution and the growth of capitalism in Russia, which distinguished the beginning of the twentieth century, brought about a continuous expansion of the home market for the products of

-farming, and thus encouraged agricultural progress. The State and the Zemstvos have vastly extended their technical and financial assistance to the development of peasant farming, which had also been helped by the growth of the co-operative movement. Accordingly, in the early part of the current century there has been noticeable in Russia a distinct tendency, among both peasants and large farming landowners, towards the more complete and, at the same time, more scientific exploitation of the available agricultural area, involving the substitution of more varied rotation for the prevailing three-course system and a reduction in the proportion of cereals and of fallow. Even during the short period of fifteen years, to which refers the table given below, a certain progress in this direction may be observed, though it should be borne in mind that its extent is definitely minimized by the necessity of using averages for the years 1901–5 and 1911–15, instead of the actual figures for 1901 and 1915 which were not available to the writer. The table gives the areas of the principal crops, except grass, in thousands of dessiatins, as well as their respective proportions in per cent., written in each case in the form of a fraction, in which the area stands for numerator and the proportion for denominator.

Distribution of the Principal Crops, except Grass, in 1901–1905 and in 1911–1915

Regions		Cereals	Peas and Beans	Potatoes	Flax	Hemp	Sugar Beet	Total
		Thousands of dessiatins and percentages						
Northern	1901–5	769·9	11·3	24·7	31·7	2·2	—	839·8
		91·7	1·3	2·9	2·2	0·3	—	100
	1911–15	772·2	11·0	28·0	28·4	2·2	—	841·8
		91·7	1·3	3·3	3·3	0·4	—	100
North-East	1901–5	4,793·9	71·8	64·6	160·8	26·3	—	5,117·4
		93·7	1·4	1·3	3·1	0·5	—	100
	1911–15	5,249·1	58·8	62·0	162·1	23·8	—	5,555·8
		94·5	1·1	1·1	2·9	0·4	—	100
Petrograd	1901–5	1,390·0	23·0	90·6	146·9	2·1	—	1,652·6
		84·1	1·4	5·5	8·8	0·2	—	100
	1911–15	1,336·8	19·3	104·7	134·5	0·8	—	1,596·1
		83·7	1·2	6·6	8·4	0·1	—	100
Moscow	1901–5	6,536·7	121·9	510·5	346·1	87·2	2·0	7,604·4
		85·9	1·6	6·7	4·5	1·3	—	100
	1911–15	6,246·0	101·9	588·2	400·3	63·0	3·7	7,403·1
		84·4	1·4	7·9	5·4	0·9	—	100

Regions		Cereals	Peas and Beans	Potatoes	Flax	Hemp	Sugar Beet	Total
		Thousands of dessiatins and percentages						
Western	1901–5	2,254·3	53·1	273·0	75·0	25·4	—	2,680·8
		84·1	1·9	10·8	2·5	1·3	—	100
	1911–15	2,227·6	42·1	334·9	80·2	19·1	—	2,703·9
		82·4	1·6	12·4	2·9	0·7	—	100
South-West	1901–5	3,945·3	157·1	157·2	10·8	26·3	266·8	4,563·5
		86·4	3·4	3·4	0·2	0·7	5·9	100
	1911–15	4,046·2	216·5	183·5	9·5	17·0	359·0	4,832·6
		83·7	4·5	3·8	0·2	0·4	7·4	100
Ukraine	1901–5	5,240·8	53·3	228·4	57·8	100·8	83·8	5,764·9
		90·9	0·9	3·9	1·0	1·7	1·6	100
	1911–15	5,391·2	49·8	255·2	34·1	85·0	135·5	5,950·8
		90·6	0·8	4·3	0·6	1·4	2·3	100
Central Agr.	1901–5	8,630·1	185·2	419·8	54·7	254·7	71·2	9,615·7
		89·7	1·9	4·3	0·8	2·6	0·7	100
	1911–15	8,648·6	182·7	477·0	45·8	209·3	121·2	9,684·6
		89·3	1·9	4·9	0·4	2·2	1·3	100
Middle Volga	1901–5	6,498·9	230·0	131·0	62·9	62·6	—	6,984·5
		93·0	3·3	1·9	0·9	0·9	—	100
	1911–15	6,690·9	241·6	165·7	59·3	45·2	—	7,202·7
		92·9	3·3	2·3	0·8	0·7	—	100
Eastern	1901–5	5,569·1	84·7	62·6	29·5	23·9	—	5,769·8
		96·5	1·5	1·1	0·6	0·3	—	100
	1911–15	6,606·2	91·1	72·2	16·0	22·3	—	6,807·8
		97·0	1·3	1·1	0·3	0·3	—	100
New Russia	1901–5	8,934·2	13·2	120·7	73·9	16·6	8·7	9,167·3
		97·4	0·2	1·3	0·8	0·2	0·1	100
	1911–15	9,476·5	10·6	133·1	40·9	11·8	12·7	9,685·6
		97·8	0·1	1·4	0·3	0·2	0·2	100
South-East	1901–5	8,202·5	19·0	121·2	134·8	27·4	—	8,504·9
		96·4	0·3	1·4	1·5	0·4	—	100
	1911–15	10,852·9	10·5	119·3	37·8	11·8	3·5	11,035·8
		98·4	0·1	1·1	0·3	0·1	—	100

Regions		Cereals	Peas and Beans	Potatoes	Flax	Hemp	Sugar Beet	Total
		Thousands of dessiatins and percentages						
Caspian	1901–5	991·9	2·8	19·7	6·4	2·3	—	1,023·1
		96·9	0·3	1·9	0·7	0·2	—	100
	1911–15	1,536·7	3·0	25·7	17·1	1·7	—	1,584·2
		97·0	0·2	1·6	1·1	0·1	—	100
Eur. Russia	1901–5	63,757·6	1,026·4	2,224·0	1,190·4	657·8	432·5	69,288·7
		92·2	1·5	3·2	1·7	0·9	0·5	100
	1911–15	69,080·9	1,038·9	2,549·5	1,066·0	513·9	635·6	74,884·8
		92·2	1·3	3·4	1·4	0·8	0·9	100

The above table, however, does not contain any information concerning one of the most important elements of the rotation, namely grass crops, whose introduction and extension is usually the first and the most certain sign of the abandonment of the three-course system and of the transition to scientific arable farming. Failing statistics of grass crops for the periods dealt with in the preceding table, I have not been able to include in it this particular item; but an idea of the development in this respect may be obtained from a comparison of the areas sown with grass in 1901 and 1912, for which years statistical records are available. The proportion of grass per 100 dessiatins of the other crops had to be calculated approximately, by referring the areas of grass in 1901 and 1912 to the total areas under crops in 1901–5 and 1911–15 respectively: a substitution which could not affect the percentages to any significant extent. The results of this computation are given in the table on opposite page, which must be studied in conjunction with the preceding one, since only by the inclusion of grass crops can a true idea be obtained of the real trend of evolution of Russian arable farming on the eve of the war.

In judging of the evolution of Russian arable farming at the beginning of the twentieth century, as revealed by the two tables given, it is necessary to bear in mind that the figures for the whole country fail to give any idea of the real development, because the effects of changes in the separate regions are neutralized in the totals by the existence, in various parts of the country, of movements tending to compensate each other. Indeed, a closer study of the table points to the country being roughly divided into two main groups of localities, of which one shows a decline in the proportion of cereals, accompanied by an increase in grass and root crops, while in the other, along with a more or less rapid extension of the cultivated area, the proportion of cereals continues

Regions	Areas under Grass Crops 1901	1912	1901	1912
	Thousands of dessiatins		Percentages	
Northern	1·6	10·7	0·2	1·2
North-Eastern	3·3	35·9	0·1	0·7
Petrograd	54·2	87·7	3·2	5·2
Moscow	159·2	239·3	2·0	3·2
Western	68·7	137·6	2·5	4·9
South-Western	91·1	105·7	1·9	2·2
Ukraine	32·1	105·8	0·6	1·7
Central Agricultural	59·4	96·6	0·6	1·0
Middle Volga	24·0	12·0	0·4	0·2
Eastern	9·1	12·4	0·2	0·2
New Russia	14·7	65·6	0·1	0·7
South-Eastern	1·1	6·9	—	0·1
Caspian	—	0·9	—	—
European Russia	518·5	917·1	0·7	1·2

to increase, occasionally even at the expense of such crops as potatoes or grass. The first of these two groups comprises the central and Western provinces of Russia, more densely populated and more immediately subject to the economic influence of the growing home market and of foreign demand for the products of intensive farming. It includes the regions of Petrograd, Moscow, the Western, South-Western, Ukrainian, Central Agricultural and Middle Volga, forming, in a sense, the historical, cultural and economic nucleus of the country. In the first three of these regions, situated outside the black-earth, besides the reduction in the proportion of cereals, there has been a characteristic reduction in the actual area under the crops enumerated in the first table, compensated for by the simultaneous extension of grass: a circumstance which would appear to be an unmistakeable sign of the existence in the localities concerned of a tendency towards a general transition to improved rotation, due to the growing commercialization of the agricultural industry. It was only natural that these regions, ill-adapted to the cultivation of cereals, but compelled to concentrate on their production so long as they had to exist under a system of isolated natural economy, should be the first to take advantage of the growth of the home market for agricultural products in order to transfer their attention from the cultivation of cereals to dairy farming, stock-raising, potato and flax-growing, as well as other branches of production which, under local conditions, were more profitable to the farmer. Accordingly, on the eve of the war, the Petrograd, Moscow and Western regions possessed the highest proportion of grass crops, and the cultivation of grass, especially in the neighbourhood of the capitals, had to a considerable extent penetrated even into peasant farming, mostly under a four-course rotation. In the Moscow region, indeed, there existed localities where, since the close of the last century, the cultivation of grass had been adopted by the great majority of the peasants, notwithstanding the communal system of tenure, its advantages being so obvious that they succeeded in converting to it the bulk of the voters at the village assemblies, on which the change in the system of cropping depended. Similar cases were recorded in certain parts of the Petrograd region, while separate instances had occurred in many other districts as well.[1] Since 1907, with the progress of peasant enclosures, the extension of grass crops and the spread of improved rotation generally among the peasantry was strongly accentuated, improved systems of arable farming having been mostly introduced by peasant farmers following the enclosure of their holdings. Though, owing to the shortness of the time allowed for the development of enclosures, this tendency in peasant farming could not yet have exercised any noticeable effect on the statistics of crops, it was evident to every observer of the Russian countryside and found ample corroboration in the results of the statistical investigations of enclosures carried out in a number of districts in 1913. This investigation pointed out also to the extension of grass-sowing among the owners of enclosed holdings being particularly evident outside the black-earth, and especially in the Western and central provinces. In the eleven administrative districts for which official statistics are available, the position in this respect was as shown in the table on page 281.[2]

Seeing that the enclosure movement only began in 1907, and was, accordingly, scarcely six years old by the time of the inquiry of 1913, the rapidity of the technical transformation of peasant farming resulting from the reorganization of tenures, especially in the provinces outside the black-earth, was indeed striking and clearly showed how really vast were the prospects of agricultural progress which had been opened before Russia by the agrarian reforms of Stolypin.

[1] Prof. A. I. Chuprov, *Small Farming and its Principal Needs*, pp. 117–8, quotes a number of such cases in the provinces of Moscow, Kaluga, Tver and Pskov. In the province of Moscow, in 1906, grass was regularly cultivated by 176 villages in the district of Volokolamsk, representing 59% of the total *nadiel*; in that of Zvenigorod by 197 villages with 45% of the peasant land; in the province of Kaluga, in 1906, grass was grown by 285 villages. In the province of Tver, in 1899, of the total of 10,036 villages 847 cultivated grass. In the district of Pskov, in 1901, about 20% of the peasants cultivated grass as part of the regular rotation of crops.

[2] Figures quoted from A. Tumenev, *From Revolution to Revolution*, p. 176, to whom the writer is indebted for interesting statistical information in connection with the progress of peasant farming on the eve of the war, otherwise unavailable.

Districts and Provinces	Percentage of Farms with Improved Rotation				Percentage of Farms with Grass Crops	
	Hutor Type		Otrub Type			
	Before Enclos.	After Enclos.	Before Enclos.	After Enclos.	Hutor Type	Otrub Type
Ostrovsky, Pskov	21·1	86·5	27·3	85·6	23·5	27·5
Sychovsky, Smolensk	62·6	91·0	60·7	89·7	32·8	33·7
Mologsky, Yaroslav	4·3	73·6	—	67·6	24·0	24·4
Rzhevsky, Tver	28·5	83·2	13·3	80·9	26·0	26·0
Bogoroditzky, Tula	—	42·9	—	13·5	—	—
Orlovsky, Orel	—	44·0	0·8	18·2	10·1	10·3
Bogodukhovsky, Kharkov	—	16·0	1·2	8·4	2·0	0·8
Kremenchugsky, Poltava	—	—	0·5	1·2	2·1	1·2
Berdiansky, Taurida	5·0	11·0	0·4	2·0	2·5	1·1
Nikolaievsky, Samara	—	5·5	—	1·1	—	0·3
Krasnoufimsky, Perm	—	71·8	—	37·3	10·6	5·0

In the regions of the same group, belonging to the black-earth belt, namely the South-Western, the Ukrainian, the Central Agricultural and the Middle Volga, more distant from the consuming centres and better adapted to the cultivation of cereals, the reduction in the proportion of cereal crops was mostly insignificant, except in the South-West, while the actual area sown with cereals had increased everywhere. Whatever diminution there had been in the proportion of cereals, was compensated for by an increase in the cultivation, in the first instance, of sugar-beet, and to some extent of grass and potatoes. With the exception of the spread of sugar-beet growing, the progress of arable farming in these regions was far less marked than in those of Petrograd, Moscow and the West. With regard to the extension of grass crops, apart from the progress here being much slower, there was a circumstance which may have contributed to the diminution of the share of this particular crop, as recorded in the table. Among the peasantry of these regions, grass-growing was still an exception, and, as the figures just quoted tend to suggest, the progress of enclosures in the black-earth zone had not been accompanied by the extension of grass crops to anything like the degree reached outside it. Grass crops, in these parts, were mostly, if not exclusively, confined to large estates, and the wholesale liquidation of landed property following the revolution of 1905 and the agrarian disturbances, which have been especially violent and widespread in the Northern part of the black-earth belt, may have partly accounted for the insignificant increase in grass crops in the Central Agricultural region and their actual diminution on the Middle Volga: phenomena for which it would be very difficult to find another explanation.

Outside the central group of regions, on the left bank of the Volga, as well as along the shores of the Black Sea and the Caspian, in the North-Eastern, Eastern, South-Eastern, Caspian and New-

Russian regions, arable farming, on the whole, in the twentieth century still continued to develop its cereal-growing branch, even at the expense, sometimes marked, of the other crops. Throughout this vast outlying belt, more sparsely populated than the central and Western parts of the country, there had been a more or less considerable increase in the total area under crops, with a rise in the proportion of cereals and a corresponding decline in that of the other crops. The farmer in these localities was still in a position to live, more or less, by the spoliation of the large reserves of natural fertility of the black-earth. In the North-Eastern region, outside the black-earth, what the cultivator lacked in the quality of the soil, he could take in the area, the holdings there being very large, as compared with the rest of the country. The South and South-East, still mainly engaged in the production of staple cereals for export, and not yet awakened to the possibilities of higher farming by the vicinity of large consuming centres, on the eve of the war were still looking for prosperity rather to the extension of cultivation than to the raising of the technical standards of farming.

After these brief observations concerning the general trend of evolution of Russian arable farming at the beginning of the twentieth century, I shall now proceed to the closer examination of the development of its principal branches.

In order of importance, cereals must come first for consideration. These include the following crops: first, the so-called "principal cereals," namely rye, wheat, oats and barley, and secondly the "minor cereals"—maize, millet, sarazin or buckwheat and spelt, cultivated in relatively small quantities. It will be remembered that, in European Russia, in 1911–15, the proportion of these crops averaged 92·2% of the total area sown with anything except grass, the variations according to regions extending over a range from some 81 to 98%. It was also to the export of cereals that, before the Great War, Russia owed her position on the world market, while in her home trade they formed the principal item.

Rye, from times immemorial the staple bread-making cereal of the Russian people, and until the second half of the nineteenth century the principal among the cereals exported abroad, was responsible for the largest proportion of the cultivated area even on the eve of the war. Yet, the share of rye in the grain exports from Russia was in the course of steady decline, while in the diet of the population wheat had been gradually gaining ground at its expense. Accordingly, between 1883–7 and 1911–13, during roughly a quarter of a century, while the total area of the four principal cereal crops in European Russia (fifty provinces) had increased from 59,982,000 to 72,504,000 dessiatins, or 20·9%, the increase in the cultivation of rye was only 0·7%. Oats came next with 8·4%, wheat with 76·5% and barley with 89·3%.[1] It may be seen that, with the gradual commercialization of farming, the more marketable crops have been consistently gaining on those which served primarily the purpose of immediate consumption or possessed a more restricted marketability. Moreover, the territorial expansion of Russian farming, throughout modern times, has been proceeding in the direction of regions best suited to the cultivation of wheat, such as the Eastern and South-Eastern steppes, Western Siberia and Asiatic Russia generally. The increase in the cultivation of barley was due, in the first instance, to the growing demand of the world market and the consequent encouragement of rapidly rising prices.

An idea of the development of the cultivation of the four principal cereals in the various parts of European Russia may be obtained from the table below, showing the movements of the areas of each of the crops concerned between 1901–5 and 1911–15.

Changes in the Area and Distribution of the Principal Cereals from 1901–5 to 1911–15

		Rye		Wheat		Oats	Barley
		Winter	Spring	Winter	Spring		
Regions		1,000 dessiatins and per cent. of the area under crops					
Northern	1901–5	320·4	0·5	—	22·6	295·0	131·4
		38·1	—	—	2·7	35·1	15·6
	1911–15	348·7	0·5	—	16·1	298·4	108·5
		41·4	—	—	1·9	35·4	12·9
	Increase or decrease	9%	—	—	−29%	0·3%	−2·9%
North-East	1901–5	1,969·5	46·8	1·5	499·0	1,814·0	351·1
		38·9	0·9	—	9·7	35·4	6·9
	1911–15	2,074·3	41·2	0·6	670·5	1,980·6	372·8
		37·8	0·7	—	12·2	35·6	6·7
	Increase or decrease	5%	−12%	—	34%	0·2%	−0·2%
Petrograd	1901–5	722·2	5·5	5·8	3·8	521·8	124·0
		43·7	0·3	0·3	0·2	31·5	7·5
	1911–15	734·5	4·5	4·5	1·7	469·7	125·4
		46·0	0·2	0·2	—	29·4	7·8
	Increase or decrease	2%	−18%	−25%	−55%	−10%	1%

[1] B. Brutzkus, *Economics of Agriculture*, p. 130, where he quotes statistics given by P. Liashchenko in his book on *Cereal Farming and the Russo-German Trade*, published by the Russian Ministry of Finance in connection with the impending revision of the Russo-German Trade Treaty.

Regions		Rye		Wheat		Oats	Barley
		Winter	Spring	Winter	Spring		
		1,000 dessiatins and per cent. of the area under crops					
Moscow	1901–5	3,644·7	8·8	16·3	46·3	2,277·4	237·4
		48·1	0·1	0·2	0·6	30·0	3·1
	1911–15	3,527·2	14·8	15·6	41·2	2,126·5	231·5
		48·2	0·2	0·2	0·6	29·0	3·1
	Increase or decrease	−3%	68%	−6%	−11%	−7%	−3%
Western	1901–5	1,226·9	25·3	20·8	27·3	557·4	213·1
		45·8	0·9	0·7	0·9	20·8	7·9
	1911–15	1,252·0	24·0	19·0	20·6	526·7	222·5
		46·6	0·9	0·7	0·7	19·6	8·3
	Increase or decrease	2%	−5%	−11%	−25%	−6%	4%
South-West	1901–5	1,039·5	12·9	951·9	121·3	822·8	407·1
		22·7	0·3	20·8	2·6	18·0	8·9
	1911–15	1,070·7	9·3	950·5	107·9	830·4	492·5
		22·1	0·2	19·6	2·2	17·1	10·2
	Increase or decrease	3%	−28%	—	−11%	1%	21%
Ukraine	1901–5	1,723·5	4·5	191·0	1,174·0	713·9	798·1
		29·9	—	3·3	20·4	12·4	13·8
	1911–15	1,655·2	8·0	235·2	1,315·0	769·0	865·5
		23·9	0·1	3·6	20·3	11·9	13·4
	Increase or decrease	−4%	−23%	23%	12%	7%	8%
Central Agr.	1901–5	4,085·2	3·5	221·4	508·4	2,278·8	290·1
		42·5	—	2·3	5·3	23·9	3·0
	1911–15	4,119·2	4·0	161·0	644·1	2,287·7	258·7
		41·3	—	1·6	6·5	22·9	2·6
	Increase or decrease	1%	14%	−27%	26%	—	−11%
Middle Volga	1901–5	3,122·3	5·3	10·7	1,035·0	1,548·2	92·0
		44·7	0·1	0·2	14·8	22·2	1·3
	1911–15	3,129·9	6·0	4·7	1,345·6	1,500·6	92·5
		42·5	0·1	0·1	18·3	20·5	1·2
	Increase or decrease	—	13%	−56%	30%	−3%	1%
Eastern	1901–5	1,701·1	8·4	1·8	2,307·8	749·3	101·8
		29·4	0·1	—	40·0	13·0	1·4
	1911–15	1,880·5	5·6	12·9	3,125·1	821·8	148·4
		27·6	0·1	0·2	45·9	12·1	2·2
	Increase or decrease	11%	−33%	617%	35%	11%	45%

Regions		Rye Winter	Rye Spring	Wheat Winter	Wheat Spring	Oats	Barley
		1,000 dessiatins and per cent. of the area under crops					
New Russia	1901–5	1,141·3	16·0	922·1	3,705·1	400·3	2,369·6
	1901–5 (%)	12·4	0·2	10·0	40·4	4·5	25·8
	1911–15	822·4	6·9	1,281·9	3,307·6	432·4	3,218·6
	1911–15 (%)	8·5	0·1	13·2	34·1	4·5	33·2
	Increase or decrease	−28%	−57%	39%	−11%	8%	36%
South-East	1901–5	1,028·9	31·5	1,872·9	2,762·1	529·8	1,577·7
	1901–5 (%)	12·0	0·4	21·7	32·1	6·1	18·3
	1911–15	943·3	22·9	2,294·9	3,897·6	612·1	2,675·8
	1911–15 (%)	8·3	0·2	20·3	34·4	5·4	23·6
	Increase or decrease	−8%	−28%	23%	41%	15%	61%
Caspian	1901–5	168·9	12·2	155·2	273·1	32·9	52·8
	1901–5 (%)	16·5	1·2	15·2	26·7	3·2	5·2
	1911–15	226·7	20·7	264·7	353·5	83·3	161·4
	1911–15 (%)	14·1	1·3	16·5	22·0	5·2	10·0
	Increase or decrease	40%	69%	70%	29%	153%	206%
European Russia	1901–5	21,894·1	181·2	4,371·4	12,485·8	12,541·6	6,742·2
	1901–5 (%)	31·5	1·1	6·3	18·0	18·0	9·7
	1911–15	21,792·6	168·4	5,245·2	14,846·5	12,748·2	8,974·1
	1911–15 (%)	28·6	0·2	6·9	19·5	16·8	11·8
	Increase or decrease	−1%	−7%	20%	19%	2%	33%

It may be seen that, in the case of rye, in the early part of the twentieth century there has been an actual diminution, though indeed small, not only in the proportion relatively to other crops, but in the area as well. Throughout the regions in which the extension of arable was most rapid, namely the Southern and the South-Eastern steppes, there has been a falling-off in the cultivation of rye, which found only partial compensation in its extension in the Eastern and Caspian regions. The area under oats had remained, on the whole, stationary, a certain increase in the regions of expanding cultivation having been counteracted by a decline in most localities outside the black-earth, where, hitherto, oats had played the part of a market crop *par excellence*: a rôle in which they have been largely superseded, in the course of the last decade or two before the war, by flax, dairy produce or other branches of production.

The extension of cultivation proceeded generally in the direction of the South and the East, into localities by nature best fitted for the growing of wheat, which was also the most valuable cereal on the home and foreign markets. Accordingly, the area sown with wheat of either the winter or the spring variety had increased considerably, the rise amounting for both kinds of wheat to roughly 20%. Winter wheat was almost entirely confined to the Western and South-Western provinces and to the shores of the Black Sea and the Caspian, where, indeed, it had been actually gaining ground on the spring crop. The centre of gravity of the wheat-growing belt had continued to shift in the South-Easterly direction, the extension of the area having been most marked in the Eastern, South-Eastern and Caspian regions. While in New Russia, where, hitherto, it had been expanding, it remained stationary, and the only noticeable change was a reduction in the area of spring wheat, compensated for by an increase in that of the winter crop. The greatest increase of all was recorded in barley, whose cultivation expanded with great rapidity throughout the South and South-East, from New Russia to the shores of the Caspian. In New Russia, its area extended at the expense of spring wheat, while in the rest of the wheat-growing belt, though barley did not actually encroach on this particular crop, it had certainly the effect of preventing its increase: apparently the result of the more rapid appreciation of barley on the world market, and one which, as I have pointed out, in dealing with Russia's agricultural exports, could hardly be described as desirable.

The total yearly production of the four principal cereals in European Russia, in 1911–15, was distributed among the thirteen agricultural regions as follows:

Regions	Rye	Wheat	Oats	Barley
	Thousands of Poods (36 lb.)			
Northern	16,907·2	721·3	13,639·7	5,681·5
North-Eastern	103,904·5	34,460·3	90,962·8	18,736·0
Petrograd	39,589·0	342·3	21,563·6	5,519·5
Moscow	182,187·3	3,070·1	111,702·6	11,376·6
Western	63,931·7	2,158·2	25,835·8	10,343·9
South-Western	82,150·8	85,585·7	63,520·8	36,438·9
Ukraine	106,586·8	88,161·0	52,206·0	50,017·8
Central Agricultural	240,657·4	42,231·7	128,782·8	12,861·4
Middle Volga	155,175·6	45,348·1	57,436·2	4,023·4
Eastern	88,788·7	116,470·0	33,059·8	5,824·7
New Russia	43,335·7	202,288·8	25,307·0	170,166·4
South-Eastern	33,122·5	280,708·2	31,677·3	148,952·9
Caspian	6,273·9	23,863·1	4,788·1	8,968·9
European Russia	1,162,611·1	925,408·8	660,482·5	488,911·9

The "minor cereals" cultivated in European Russia comprised spelt (*polba*), sarazin or buckwheat (*grechikha*), millet (*proso*) and maize. Their evolution during the period dealt with may be seen from the table below, showing the areas under each of these crops in 1901–5 and in 1911–15 according to regions, excepting the Northern in which none of these cereals was cultivated.

Evolution of the Minor Cereal Crops in European Russia

Regions	Spelt (*Polba*)	Sarazin (*Grechikha*)	Millet (*Proso*)	Maize
	Thousands of dessiatins			
North-Eastern:				
1901–5	29·2	91·3	1·5	—
1911–15	17·0	90·9	1·2	—
Petrograd:				
1901–5	—	6·9	—	—
1911–15	—	3·5	—	—
Moscow:				
1901–5	0·2	187·0	118·9	—
1911–15	0·7	115·5	87·6	—
Western:				
1901–5	—	163·6	19·7	0·2
1911–15	—	125·9	16·8	0·1
South-Western:				
1901–5	0·8	202·4	219·8	166·8
1911–15	—	207·0	212·3	165·6
Ukraine:				
1901–5	0·9	445·3	184·4	5·2
1911–15	—	366·0	167·7	8·6
Central Agricultural:				
1901–5	5·7	381·4	855·0	0·6
1911–15	1·0	307·6	824·8	0·5
Middle Volga:				
1901–5	153·7	146·5	384·4	0·4
1911–15	97·0	133·6	371·5	0·5
Eastern:				
1901–5	192·2	188·8	316·7	1·2
1911–15	86·0	251·6	273·2	1·1
New Russia:				
1901–5	1·2	15·7	132·7	230·2
1911–15	0·9	13·9	78·8	313·1
South-Eastern:				
1901–5	14·0	13·2	317·6	55·5
1911–15	2·5	6·1	295·4	103·3
Caspian:				
1901–5	1·0	4·4	134·4	157·0
1911–15	0·2	2·6	209·1	214·5
European Russia:				
1901–5	398·9	1,846·5	2,685·5	617·1
1911–15	205·3	1,625·0	2,538·4	807·3
Increase or Decrease	−48%	−12%	−5%	31%

It may be seen that, with the exception of maize, there has been a falling-off in the cultivation of every one of the minor cereals. The reasons for this decline were the same in every case, namely that the products for which the market was relatively narrow and of which the prices advanced more slowly than those of the great staple crops, were the first to go out of cultivation and to be superseded by other cereals, flax, potatoes, sugar-beet or grass, as the conditions dictated. Any extension there has been in the cultivation of spelt, sarazin and millet was confined either to localities of extensive farming in the East and South-East, or to regions in which natural conditions were especially favourable to their growing and where they were an important constituent of the people's traditional diet. Maize alone has been extending considerably, the increase being mostly due to the demands of the foreign market. Between 1901 and 1915, the exports of maize from Russia had increased considerably, and their development encouraged the extension of maize-growing in New Russia and the South-Eastern and Caspian regions, on whose agricultural output the foreign market has always exercised a very strong influence. The increase in the total area of maize crops reached approximately 31%, and its gradual extension, especially in the Eastern steppes, subject to frequent droughts and, as a rule, rather superficially cultivated, was generally welcomed by agriculturists, partly owing to the capacity of maize to withstand a shortage of water better than most other cereals, partly because it necessitated more thorough tillage.

The total yearly output of the minor cereals in European Russia, in 1911–15, averaged 121·4 millions of poods of millet, 58·6 millions of poods of sarazin, 65·2 millions of poods of maize, and 7·1 million poods of spelt.

The area under peas and beans had slightly increased, but the share of these crops in the total had diminished from 1·5 to 1·3%. Here, as in the case of the minor cereals, the extension of the area received little encouragement in the movement of prices, since the rise in the prices of this group tended to lag behind that of those products of arable farming which were in more active demand. The extension of the area of peas and beans was limited to a few localities, mostly along the Volga and in the South-West. Everywhere else these crops had declined not only in proportion, but in their actual area as well. Their output, which averaged 46·6 millions of poods a year in 1911–15, was apparently more than sufficient to meet the demand, and the conditions of the market would appear to have favoured rather the reduction of their cultivation, than any increase.

Next to the cereals, it is necessary to consider two of Russia's traditional fibrous staples, namely flax and hemp.

Flax has been cultivated from times immemorial in the North-Western parts of the Russian plain, in the dominions of the ancient republics of Novgorod and Pskov, which, in the Middle Ages, had been engaged in active trade in flax fibre with the Hanseatic League. For centuries, until the appearance and growth of cotton manufactures in Russia in the early part of the nineteenth century, flax had no competitor in the country and had a large market abroad. The development of the cotton industry was a severe blow to the Russian flax-grower, from which he could not recover until the export of flax fibre had been eventually revived as a result of the American Civil War, which interfered with Europe's supplies of raw cotton. The improvement was not of long duration, and in the 'seventies the European flax-growers, who had availed themselves of the opportunity given them to extend their cultivation, found themselves in a worse position than ever, and had to reduce their output of flax, limiting it almost entirely to the finest grades of fibre. In the supply of the coarser flax fibre, Russia, owing to the advantage she possessed in the cheapness of production, due to the cultivation of flax being almost entirely in the hands of the peasant family farmers, became practically a monopolist on the world market, with the result that Russia's flax exports have since been continually increasing. The principal flax-growing localities, in which flax was produced for fibre, were in the Moscow, Petrograd and North-Eastern regions and some adjoining provinces, to which the cultivation had spread from these principal centres. In the region of Moscow, the principal flax-growing district was situated in the North-Western part, at the junction of the three provinces of Tver, Moscow and Smolensk. In other parts of the Moscow region flax-growing was also widespread. A characteristic feature of the evolution of flax-growing here, as distinguished from most other localities, was its development in the course of the last few decades before the war on the basis of relatively high farming, as an element of a multiple rotation involving grass crops and potatoes. This feature pointed to the relatively advanced agricultural state of the region and made the extension of flax, which, under the three-course system, is generally injurious to the soil, very welcome, ensuring its lasting success. In the region of Petrograd, the province of Pskov was particularly noted for flax; and from there the cultivation of flax extended to the neighbouring districts of the province of Vitebsk, thus overflowing into the Western agricultural region. Here, the development of flax-growing had been proceeding more or less along the same lines as in the Moscow region, often being combined with that of grass and potatoes. In some other parts of the Petrograd region, and notably in the remoter districts of the province of Novgorod, however, flax-growing was still largely practised either, as in olden days, on fresh forest clearings (*liadi*), which at least had the merit of not interfering with the regular arable, or on part of the spring field in the three-course system. A third, very large, district of extensive flax-growing had its principal centre in the North-Eastern region; and here, on the eve of the war, the cultivation of flax had been expanding with great rapidity. The principal flax-growing locality in this area was the province of Viatka; and to the same group belonged the adjoining parts of the province of Vologda and certain districts of the provinces of Kostroma, Nizhny-Novgorod, Yaroslav and Vladimir. These three principal flax-growing areas, which may be referred to as the Central, the North-Western and the North-Eastern, accounted between them for about 80 to 90% of the total output of flax fibre in Russia, which, in 1911–15, averaged 25·3 millions of poods a year. It is important to note that, while the total area of flax in Russia, in the early part of the twentieth century, had decreased, in two at least of the three principal centres, namely in the Central and North-Eastern flax-growing districts, in 1911–15 it was more than 10% larger than in 1901–5: an increase which more than compensated for a set-back in the North-Western area. The development, in the twentieth century, of co-operative marketing of flax, by enabling the actual producer to take better advantage of the conditions of the market, has been helping the development of this branch of production very considerably. In the black-earth belt, in which the cultivation of flax had never been so widespread, and flax has been cultivated mainly for seeds, as an important oleaginous plant, there had been a general diminution of the area. While, in the Northern half of the country, flax was, and actually still is, the principal commercial crop of the peasant, in the black-earth belt the peasant farmer had other money-producing crops to choose from. Moreover, the climate and soil of the Northern and Western provinces are far better suited to the production of good flax fibre than the richer soil and the dry climate of the cereal-growing belt. The cultivation of flax for seeds, as a commercial proposition, was far less profitable, besides, than the production of fibre. Accordingly, the decline in the area of flax, very marked throughout the black-earth, was a perfectly natural development. It has been especially great in New Russia and the South-Eastern region, in which the sharp fall was due to the fact that, in localities with an abundance of virgin land, flax-growing on fresh soil generally yields exceedingly heavy crops, though it tends to exhaust the land rapidly. In the South and the South-East, where virgin land had only begun to give out in the twentieth century, until lately the area of flax had been very considerable, but on the eve of the war it showed signs of rapid reduction, along with the exhaustion of the reserves of land.

Hemp, another fibrous and oleaginous plant, also cultivated in Russia from times immemorial, had declined in all parts of the country. The demand for hemp, at home as well as abroad, has been steadily diminishing in the course of the last few decades, and although Russia's exports of hemp had not declined, they remained stationary, and the crop had been loosing ground in competition with more profitable plants. Accordingly, the area sown with hemp had declined from 657,800 dessiatins in 1901–5 to 513,900 dessiatins in 1911–15, and the decline was likely to continue. The cultivation of hemp was mostly concentrated in the Northern part of the black-earth belt, and the Ukraine, the Central Agricultural and the Middle Volga regions, in 1911–15, have been responsible between them for over 350,000 dessiatins, or rather more than 70% of the total area. The output of hemp fibre in European Russia, in 1911–15, averaged 21·7 millions of poods a year.

Passing now to the root crops, it is necessary to begin with potatoes, which, in order of importance, came next to cereals. In European Russia, potatoes accounted for 2,224,000 dessiatins in 1901–5 and had increased to 2,599,500 dessiatins in 1911–15, their share in the total area under crops being respectively 3·2 and 3·4%. These figures, however, are somewhat misleading, because of the compensating effects of opposite movements in different localities. Thus, in the spacious North-Eastern region, both the area and the proportion of potatoes had actually diminished. The same was the case in the South-Eastern region, while in the Caspian, though the area of potatoes had increased, the proportion had diminished. In the Eastern region, an extension of the area had left the proportion stationary. In the central and Western parts of Russia, however, where, with the greater density of population, potatoes were needed both as a crop yielding a maximum of food from a given area, and on account of their influence on the soil and its cultivation, the position was quite different. Throughout the whole country West of the Volga and the Don, as well as in the Northern and Petrograd regions, in the early part of the current century there had been a marked increase in the area and the proportion of potatoes. Combined with grass crops, the extension of the cultivation of potatoes involved in these localities a noticeable change for the better in the rotation, especially pronounced in the vicinity of large consuming centres. The proportion of potatoes, indeed, was highest in the Western, the Moscow and the Petrograd regions: the three regions in which grass crops were also most developed. In the Western, South-Western, Ukrainian and Central agricultural regions, the existence of numerous spirit distilleries also encouraged the cultivation of potatoes. On the whole, on the eve of the war, the distillation of spirit consumed roughly some 13% of the total potato crop of European Russia proper (forty-five provinces), potatoes having been increasingly substituted for cereals in this branch of industry. The total output of potatoes in European Russia, in 1911–15, averaged 1,213·5 millions of poods a year.

The other root crop, whose importance could hardly be overestimated, was sugar-beet, of which the area, in 1901–5, was 432,000 dessiatins, or 0·5% of the total crops; in 1911–15 it reached 635,600 dessiatins, and its share in the total rose to 0·9%. Before the war, as a producer of sugar, Russia ranked second, Germany coming first, Austria-Hungary third and France fourth, and had the yield of beet in Russia been as large as that of Germany or Austria-Hungary, she would have been far in front of all the other producers. The yield of Russian beet crops, however, averaged only 1,081 poods per dessiatin, while in Germany it reached 1,863 poods, in Austria-Hungary 1,620 poods and in France 1,666 poods.[1] A few words should be said of the history of the sugar industry in Russia. It originated at the beginning of the nineteenth century in the provinces of Tula and Orel, where it had been started by a few landowners, but local conditions did not favour the extension of beet-growing in this part of the country. Accordingly, the cultivation of sugar-beet only began to take root in the 'thirties, when one of its pioneers, Count Bobrinsky, after having unsuccessfully tried beet-growing in the province of Tula, decided for a warmer climate and, in 1836, transferred his activities to his other estate, Smiela, in the province of Kiev. From this time on, the cultivation of sugar beet began rapidly to spread over the South-Western and the Ukrainian provinces. The development of the industry, concentrated entirely in the hands of large landowners, though affected by the crisis following the Emancipation of peasants, was, probably, one of the first branches of large farming to recover from its effects. It continued to expand, until, in the 'eighties, a

[1] Figures for the five years, 1908 to 1912, quoted according to B. Brutzkus, *Economics of Agriculture*, p. 215.

point had been reached at which the producers saw themselves faced with a continuous fall in the prices of sugar. The attempts at organization among the producers of sugar, with a view to controlling the output and the prices, having proved abortive, the great importance of the industry for the country compelled the Government, at first reluctant to do so, to intervene, and in 1895 it made it compulsory for all owners of sugar factories to join a syndicate for the control of production and marketing. Moreover, it imposed on the syndicate the so-called system of "sugar norms," which consisted in the Ministry of Finance fixing for every year ahead the quantity of sugar to be put on the home market, which was then distributed by the syndicate among its members according to the capacity of their respective factories. Of the output above the norm every factory had to keep a certain quantity in reserve, with a view to issuing it on the market if the prices of sugar should rise above the normal. The balance of the sugar produced could either be exported abroad or sold in Russia, provided in the latter case that the producer paid on the amount disposed of in this manner a double excise duty: a condition which effectively prevented the producers from taking this course. The result was that the only outlet available for the sugar industry, which permitted it to expand, was export abroad, and the sugar "norms" actually amounted to a premium on exports. In the world sugar market, the position, by the close of the last century, was already one of disorganization, and the policy thus adopted by Russia made the existing confusion worse confounded. Accordingly, when, in 1902, alarmed by the crisis in the West-Indian cane-sugar industry, Great Britain initiated the Brussels Sugar Convention, the proposal met with general response, and the Convention adopted a course involving the practical prohibition of export bounties on sugar in any shape or form. Thus, import duties on sugar were limited to Fr. 5·50 per 100 kg. over and above the excise levied at home, while all sugar imported from countries in which bounties were in operation was made to pay customs duties equivalent to the premium. Russia refused to sign the Convention and, seeing her Western markets closed to her sugar, was compelled to have recourse to measures intended at preventing a further increase in production. Yet, to a large extent, her position was relieved by the extension of her exports of sugar to the Eastern markets, unaffected by the Convention, as well as to Finland, in which Russian sugar had been accorded preferential treatment. Besides, her home consumption of sugar has been increasing continually. As a result, the Russian sugar industry had not been much affected by the restrictions imposed in virtue of the Brussels Convention on the exports of sugar. In 1907, on the instance of Great Britain, where the effects of the Convention were to produce an excessive rise in the prices of sugar, the restrictions with regard to the import of sugar from countries having export bounties were repealed, and in 1907 Russia had notified her adhesion to the Brussels Convention, without abandoning her system of "norms," on condition that, in the course of the next five years she would not export to Western Europe, in the aggregate, more than one million tons of sugar: a quantity actually exceeding Russia's export of sugar before 1903. Moreover, by special agreements, in 1911, 1912 and 1913, the signatories of the Convention authorized Russia to increase her exports of sugar by certain specified quantities in excess of the annual quota. The result was that, on the eve of the war, within the limits of her actual production of sugar, Russia ceased to feel the effects of any restrictions and was in a position even to proceed with the extension of sugar-beet-growing with confidence in an expanding market at home and abroad.

Sugar-beet cultivation, before the war, was mainly concentrated in the South-Western provinces of Kiev and Podolia and in the Ukraine, which jointly accounted, in 1911–15, for 494,500 dessiatins or about 79% of the total area sown with beet. Next came the Central Agricultural region, in which the cultivation of sugar-beet had been spreading rapidly, reaching as far East as the provinces of Voronezh and Tambov. On a smaller scale, it showed signs of extension in New Russia, namely in the province of Kherson, where, however, labour conditions were not favourable to its expansion. The province of Tula, in the South of the Moscow region, was the northernmost point reached by this crop. On the eve of the war, sugar-beet made also its first appearance on the Kuban, in the South-Eastern region. Until relatively recent times, the cultivation of sugar-beet had been confined to large estates, but since the beginning of the current century it began to spread among the peasants as well, and some progress had been achieved in this direction in all the beet-growing districts, the crop being generally well suited to the conditions of small family farming. In 1916, in the Ukraine, about 21% of the area sown with beet belonged to peasants; in the South-West, about 20%, and in the Central Agricultural region about 13%. A curious feature of sugar-beet-growing, peculiar to the South-Western region, was that in a number of cases, large growers leased the land of the neighbouring villages with a view to extending their plantations, thus reversing the usual position.

Fodder-roots, in the early part of the twentieth century, were still very rare. The only information available concerning these crops is that contained in the Agricultural Census of 1916, according to which the total area under fodder roots was 25,900 dessiatins, or under 0·1% of the total. Of these, 16,000 dessiatins were accounted for by large farms, while the rest was scattered among peasant farms. The two regions in which the cultivation of roots had achieved any importance at all were the Ukraine and the Central Agricultural, the former having about 9,000 dessiatins, and the latter 8,500 dessiatins.

A not unimportant part in the black-earth zone was played by the cultivation of various kinds of melons, such as water melon, melon, vegetable marrow, etc. These crops were especially widespread in the South and South-East of the black-earth belt. The total area of these vegetable crops in European Russia, in 1916, according to the Agricultural Census which contained the first statistical information on the subject, exceeded 400,000 dessiatins, of which approximately 338,000 dessiatins, or roughly 85%, were in New Russia, the South-Eastern and the Caspian regions. The Ukraine, the Central Agricultural, the Middle Volga, the Eastern and the Moscow regions, in the order named, accounted for practically the whole of the rest. The staple crops were water melons and melons, which, especially in the South-East, were of remarkably high quality and fine flavour.

Among the important minor crops of the black-earth zone, tobacco, hops and sunflower should be mentioned.

The tobacco produced in Russia belonged to two distinct classes, namely, on the one hand, the high-grade varieties, grown in the South of European Russia and in Transcaucasia from Turkish and American seeds, and, on the other hand, the coarse native tobaccos, such as *makhorka*, *bakun*, etc., used for cheap smoking mixtures and grown in the Northern part of the black-earth belt, more particularly in the Ukraine and the Central Agricultural region. As a producer of tobacco, before the war Russia ranked third, the United States being first and the West Indies second. Both the fine and the coarse varieties of Russian tobacco were exported abroad, the latter mostly to Egypt. The growing of coarse tobacco was essentially a branch of peasant family farming and formed one of the important sources of cash in the peasant household budgets of the producing districts. Fine tobaccos were cultivated in the Crimea and in the South-East, especially on the Kuban. In the returns of the area sown with tobacco no distinction was, unfortunately, made between the varieties grown ; but, generally speaking, it may be assumed, without risking a serious error, that all crops outside the Crimea and the Kuban belonged to the coarse variety, and that those of the Crimea and the Kuban, as well as of Transcaucasia, which latter, however, is outside the scope of this study, represent the fine grades of tobacco. Of the total area of tobacco crops, which, in 1906–10, in European Russia averaged 47,943 dessiatins, 15 to 16,000 dessiatins, on this assumption, would appear to belong to the higher grades, being situated in the Crimea and on the Kuban. In the production of coarse tobacco, the Ukraine came first with 21,603 dessiatins; then came the Central Agricultural region with 5,348 dessiatins, of which 3,726 were accounted for by the province of Tambov, while in the Eastern region the province of Samara had 1,688 dessiatins of tobacco. The rest was scattered over the South-Western provinces, New Russia and other localities, with a score or so of dessiatins to every province. The plantations were generally very small, the average size in European Russia, exclusive of the Northern Caucasus, being just over 0·1 dessiatin. The largest plantations were on the Kuban (2·3 dessiatins on the average) and in the Crimea, where, however, the average did not quite reach 0·5 dessiatins. Indeed, in the South, in the case of the higher grades of tobacco, the cultivation was usually carried on on a somewhat larger scale, with the investment of relatively large amounts of capital, while in other parts of Russia, being essentially a peasant crop, it was handicapped by a shortage of capital, which prevented its extension and improvement. On the whole, this branch of production, owing to its character of a peasant crop, was dependent to a very large extent on the circumstances of the individual producer, and accordingly subject to great variations from one year to the next, with the result that it was difficult to form a definite judgment as to the trend of its evolution. Yet, on the whole, conditions have been shaping themselves very favourably to the extension of tobacco-growing in Russia, owing to the increase in the consumption and export of both varieties of tobacco, as well as to the extension of co-operative credit and marketing which enabled the cultivator more easily to surmount his financial and commercial difficulties.

The total output of tobacco in European Russia (forty-five provinces) in 1906–10, averaged 4,929,470 poods a year, of which total just under one-third belonged to the higher grades.

Hops were grown on a considerable scale in a few well-defined localities. Of these, the most important was the South-Western region, with the centre of hop-growing in the province of Volynia, which in 1911, had 1,536 dessiatins of this crop. Next in the extent of cultivation came the North-Eastern hop-growing district, in which the two provinces of Kazan and Kostroma, in 1911, accounted for 1,042 dessiatins. The so-called Guslitzky hop-growing district, comprising the provinces of Moscow, Vladimir, Riazan, and Tula, in 1911, had 266 dessiatins. The White-Russian provinces of Mogilev, Minsk and Smolensk came last, with about 84 dessiatins of hops between them in 1911. As in the case of tobacco, the hops grown belonged to two distinct varieties, namely, fine hops from imported seeds, cultivated mainly in the South-West, and inferior native hops, of which the production was mainly concentrated in the North-Eastern district. Of the total output of hops, which fluctuated enormously according to season, approximately four-fifths belonged to the finer variety. The exports of hops, accordingly, were also subject to wide variations. In the course of the early years of the twentieth century, therefore, Russia, on the balance, had been an importer of hops in 1901–5 and 1906–10, and only in 1911–13 she had a net export amounting to 42,600 poods of hops a year.

Sunflower was cultivated, as an oleaginous plant, in increasing quantities, mostly in the South-Eastern quarter of European Russia. The total area of this crop, according to the Agricultural Census of 1916, amounted to just over one million dessiatins, and the output of sunflower seed reached about 40 to 45 millions of poods. The extension of the cultivation of sunflower was generally looked upon with favour, not only because of the commercial value of the produce, but owing to the influence of this crop on the soil also. Indeed, as an intensive crop, requiring careful tillage, hoeing and repeated weeding, it raised the standards of cultivation conspicuously: a matter of particular importance in the South-Eastern part of Russia, where they were otherwise notoriously low.

To conclude the chapter on Russian arable farming, a few observations should be added concerning the general trend of its evolution on the eve of the war. The period dealt with was one of rapid progress in Russia's economic development, and the statistical records of the growth of trade, industries, banking, in fact of every branch of economic activities, bore witness to a striking expansion, which made the Russia of 1914 far different not only from that of the last century, but even of 1904. In these branches, where man was absolute master, human enterprise had done wonders in the course of a single decade. Wherever either capital or technical assistance were needed, they were easily obtained from abroad, and the development had proceeded unchecked. Modern economic organization made the transfer of capital in industries, trade and transport, easy and smooth. In agriculture the conditions were different, and it is hardly surprising, therefore, that the statistical records of its evolution, though they certainly show unmistakable signs of improvement, fail to impress the mind with the magnitude of the achievements. Rather, indeed, one had the impression that, on the whole, the changes which had taken place between 1901 and 1915 were relatively insignificant: an impression largely due to the fact that, following ancient tradition, Russia, in the early part of the twentieth century, while she raised the standards of farming in the older and more densely populated districts, continued to develop the extensive forms of cereal cultivation in the still available spaces of her Eastern and South-Eastern outskirts. In the more highly developed parts of European Russia, the progress of farming technique was unmistakable, and if its statistical records fail to strike the eye in the same way as those of industrial expansion, this failure is due to the essential characteristics of the agricultural industry. Farming, especially in a peasant country, is not a mere trade of the people, but rather the whole life of the farmer; and this life, woven of customs, current interests, ignorance of the world and deep-rooted distrust of sudden changes, does not easily lend itself to dramatic transformations. Nor is the wherewithal to finance such transformations as easily obtainable to the peasant farmer as it is to the trader or manufacturer in the cities. Even when progressive currents get under way, as they actually did in Russia with the extension of the enclosure movement, which signified a complete revolution in both tenures and systems of farming, they are bound to take time, and their effects cannot be immediately brought to light in the statistics of crops. The enclosure movement, rapid as it had been; the growth of rural co-operation, great as had been its influence on peasant farming on the eve of the war; the re-equipment of peasant farming with improved instruments of production and the adoption by the peasant farmers of improved methods of cultivation, obvious as they have been to any competent observer of the Russian country-side, could not, within less than ten years, produce any striking changes in the statistical records of Russian farming as a whole. There had been a great process of development going on and full of promise for the future, but much of it, in the agricultural statistics of Russia, was bound to remain unrecorded, as does the latent heat of certain critical physical processes, though, on their being completed, the matter emerges in a hardly recognizable form.

CHAPTER X

RUSSIA'S LIVE STOCK AND THE EVOLUTION OF STOCK FARMING

THE evolution of stock farming is always closely connected with that of arable, and there would appear to exist a certain general scheme of development to which, in all countries of the temperate zone, it had consistently conformed in the past. In the infancy of countries and nations, stock-raising occupies a position of pre-eminence among the economic activities of the human race, still living under semi-nomadic conditions. This stage of economic evolution may be seen not only in the familiar stories of the Bible and in other historical records of early civilizations, but also in modern times, in the countries of recent colonization, at the beginning of their agricultural growth. With the increase in population, the plough begins to encroach on the pastures, and the herds either diminish or are driven farther afield, into regions still undeveloped, to which, however, they are eventually followed by the victorious ploughman. When the victory of arable is complete, and most of the available land is turned to agricultural uses, stock farming recedes to the background, there to play a modest auxiliary part besides the more productive arable branch. In the history of European farming, this transition was roughly synchronous with the triumph of the three-course system of cropping, which dated from the eighth or ninth century of our era, and the evolution of stock farming, since then, for about a thousand years, had depended primarily on the conditions created by this system, accompanied, as a rule, by open-field tenure. Everywhere, the combination of three-course cropping and open fields, with the increase of population, only occasionally checked by such calamities as the Black Death of the fourteenth century or the Thirty Years War of the seventeenth, had eventually resulted in a reduction of natural meadows and pastures and the ever-increasing dependence of the live stock on stubble and meagre fallows, and its numbers and quality had suffered accordingly. Everywhere, therefore, the closing period of the predominance of three-course cropping and open-field tenure had been marked by a heavy decline in stock farming, and it was only since the spread of improved rotation, with grass and root crops used as fodder, and the economic expansion which increased the demand for animal products, that a turning point had been reached, and live stock began again to assume greater importance. This important change originated in Holland about the sixteenth century, whence, in the course of the seventeenth, it penetrated to England. In Germany, imported from the same source, the transition began in the second half of the eighteenth century, and though it had been delayed in its progress by the slowness with which the liquidation of medieval survivals in agrarian relations had proceeded, it had changed the face of German farming in the course of a few decades. Once again, in most countries of Europe, live stock, and primarily horned cattle, came into its own in the general scheme of the agricultural industry, to the inestimable benefit of the arable branch itself.

Now, turning to Russia in the early part of the twentieth century, one sees her just beginning to emerge, in her agriculturally more advanced districts, from the stage of three-course husbandry, with all the consequences of that system for the live-stock branch of farming. In the bulk of the country, which was passing through the last stages of a decaying three-course system, the facilities for the feeding of cattle were reduced to an absolute minimum. In certain localities, more highly developed economically and possessing better marketing facilities for animal produce, as well as for various products of arable farming, the worst had already been left behind, and along with improvement in the rotation of crops, one could already discern a more or less marked advance in the numbers and quality of cattle. At the other end of the scale, in the Eastern and South-Eastern districts of extensive cereal farming, with facilities still available for wild grazing, there still existed large herds of steppe cattle, which, however, were being continually shifted farther East by the unrelenting extension of arable.

In proceeding to the statistical study of stock farming in Russia, the student must be fully aware of the difficulties and pitfalls which await him on his quest. The evolution involved the quantitative as well as the qualitative aspects of the live-stock branch of the agricultural industry, while the statistical data available on the subject refer only to numbers: a circumstance which, in this particular field of work, greatly reduces the value of figures. While, in the statistics of crops, there is only the possibility of an error in figures to be guarded against, and this, even in indifferent records, is generally made relatively insignificant by the compensatory influence of large numbers, in the statistics of live stock, besides errors, there is the ever present and unavoidable defect of the heterogeneity of statistical units. Accordingly, while, on the whole, such statistics may serve, if cautiously used, for the purpose of verifying and illustrating the reasoning, they can hardly be relied upon as, in themselves, providing the necessary basis for definite conclusions. In a word, the statistical evidence in this case is mostly of an auxiliary nature. In Russia, moreover, the statistics of live stock, with the exception of those of horses, could not be described as satisfactory. With regard to horses, there existed the data of the so-called Military Horse Censuses, carried out from time to time with a view to providing information concerning the resources available for the Army in case of mobilization. For the rest, the only available data were those derived from the current statistical records of the Central Veterinary Department, whose accuracy was open to doubt. These data, though, indeed, they could be relied upon to give a more or less correct idea of the trend of development in the numbers of live stock, have on many occasions been found, following intensive local investigations by the Zemstvos, to differ considerably from the real position. Far more reliable were the statistics of live stock collected by the various Zemstvos in the course of their statistical investigations; but, referring as they did to separate provinces and districts, as well as to different times, and mostly confined to peasant farming, they were of little value in studies involving the whole country. These limitations of the statistical material to be used in the present chapter it is essential to bear in mind in drawing conclusions from the tables given below.

After these preliminary remarks, I shall now proceed to the study of the actual evolution of Russian stock farming in its various branches.

As a horse-breeding country, Russia stood easily foremost in the world, the total number of horses in the Russian Empire (exclusive of Finland), in 1914, reaching 35 millions. Next to Russia came the United States with about 24 millions of horses. The majority consisted of farm horses (roughly 60%); saddle-horses accounted for some 30–35%, and racing breeds for the rest. Among the farm variety, by far the most numerous were the small, hardy peasant horses, which, though they did not belong to any clearly defined type, had some common features engendered by the conditions in which their characteristics had been evolved. Though not strong, the breed is very sturdy and possesses great endurance, while its needs are extremely modest, which makes of it probably the best type of animal for the conditions in which it has to live and to work.

Some of the more highly developed and typical breeds of peasant horses are, in the North and North-West, the Finnish horse (*shvedka* or *finka*) : a small, strongly-built animal, greyish-brown in colour, with a lighter, nearly white, mane and tail. Generations of crossing had infused a considerable proportion of this blood into most of the local varieties of peasant horses in the Northern and North-Western provinces of European Russia, with, on the whole, highly beneficial results. Farther East, in the North-Eastern region and its neighbourhood, such crossings with Finnish horses have resulted in producing two distinct types of excellent peasant horses, the *viatka* and the *obvinka*. Ill-advised attempts at improving these breeds by the infusion of other blood, however, in the course of the last few decades, have greatly contributed to the degeneracy and gradual disappearance of both these types. In the East and South-East, various breeds of Tartar, Bashkir and Kyrghiz horses have helped in evolving certain local types of peasant horses, in whose gait the characteristic side-trot (*inokhod*) of the horses of Eastern nomads has sometimes left distinct traces, making them pleasant to ride. In the central provinces, probably as a result of crossing with Polish and Lithuanian breeds (*zhmudka*), the peasant horses, though small, are yet somewhat higher than in the North and East, and some of the types (e.g. the *kozlovka*) are rather good in stature and appearance, and were in great demand for hackney work in the cities and as carriage and mail horses throughout the country. The noblest product of Russian peasant horse-breeding, however, was the *bitiug*, bred by the peasants of the district adjoining the river Bitiug in the province of Voronezh from the original cross, as tradition has it, with heavy Danish and Dutch sires imported to the locality by order of Peter the Great two hundred years ago. The recent history of the breed, highly valued for its qualities and exterior, has been disappointing, a rapid degeneration having set in at the close of the last century, partly due to the reduction in facilities for horse-breeding in the locality, which accompanied the agrarian crisis, partly, as some authorities hold it, to the wholesale export to South America of the best stallions and mares.[1] Both causes must have had their share in the deterioration of the *bitiug*, and on the eve of the war a perfect specimen of the breed was not easy to find. Saddle-horses have mostly been bred in the Eastern

[1] The present writer has been given this explanation by a great authority on Russian horse-breeding, N. N. Levshin, a large landowner and noted horse-breeder of the province of Tula, who told him the story of the wholesale export of the *bitiug* to South America about the close of the last century. As far as I remember, he had published a monograph on this breed of horses and its history.

and South-Eastern steppes, on the rivers Don, Manych and Sal, as well as in the Northern Caucasus and the Lower Volga: the regions from which, besides the excellent Don Cossack horses, much of the material for the yearly remounts of the regular cavalry was obtained. The breeding of saddle-horses and trotters, the former mostly from English and Arab blood stock, was also carried on in various parts of Russia, partly by private breeders, partly in Government studs, such as the Khrenovsky in the province of Voronezh, the Bielovodsky in the province of Kharkov and several others, maintained by the State for the purpose of ensuring the maintenance of the standards of breeding in the country: a matter of such importance in Russia, that its control was entrusted to a special Department of State Horse-Breeding, independent of any of the Ministries.

The tendencies of the development of Russian horse-breeding, and the general evolution of this branch of live stock according to localities during the decade immediately preceding the war may be seen from the table below, showing the number of horses in various regions in 1904 and 1914, as well as their relative density at the two dates:

Numbers of Horses in European Russia in 1904 *and* 1914

Regions	Numbers of Heads 1904	Numbers of Heads 1914	Change per cent.	Per 100 des. of Crops 1904	Per 100 des. of Crops 1914
Northern	399,620	435,033	8·8	48	52
North-Eastern	1,670,662	1,763,454	5·5	33	32
Petrograd	664,872	649,406	− 2·3	40	41
Moscow	2,782,159	2,701,500	− 6·5	37	37
Western	1,211,587	1,157,369	− 4·5	45	43
South-Western	1,795,820	1,854,039	3·2	39	38
Ukraine	1,862,718	1,653,876	−11·3	32	26
Central Agricultural	3,160,015	2,890,484	− 8·6	33	29
Middle Volga	1,694,364	1,663,990	− 1·8	24	23
Eastern	2,019,379	2,003,200	− 0·9	33	29
New Russia	1,837,025	1,983,353	7·9	20	20
South-Eastern	1,781,242	2,259,407	26·8	21	20
Caspian	369,472	411,927	11·5	36	26
European Russia	21,248,935	21,427,438	0·8	31	28

During the ten years from 1904 to 1914, the total number of horses in European Russia had increased 0·8%, while the extension of the area under crops reached 8%. The result was the diminution in the number of horses per 100 dessiatins under crops from 31 to 28. Considered according to regions, the changes in the numbers and density of horses characteristically reveal the influence of local economic conditions.

Starting from the Northern region, one can see there an increase of 8·8%, while the increase in the area under crops did not exceed some 0·5%. The horse here, as a rule, is not so much used in cultivation, as for haulage, mostly of timber, which accounts for the very large number of horses relatively to the cultivated area. The increase should be attributed, in the first instance, to the development of the timber trade in the Northern provinces of Russia on the eve of the war.

In the North-Eastern region the increase amounted to 5·5%, but it had not kept pace with the extension of cultivation, with the result that the density of horses relatively to the area under crops had diminished from 33 to 32 per 100 dessiatins. In this region, the most agricultural in the North, with large peasant holdings and extensive cultivation, there was still no question of arable encroaching on grassland and tending thus to reduce the numbers of live stock, but the remoteness of the locality, which prevented, as yet, the rapid development of forestry, did not encourage the keeping of many more horses than was strictly necessary for purely agricultural purposes.

In the Petrograd region there has been a 2·3% reduction in the number of horses, though their density had increased from 40 to 41 per 100 dessiatins of crops, owing to a reduction in the cultivated area. Of the three provinces of this region, that of Novgorod alone had actually increased the number of horses: a phenomenon due to the very rapid development of forestry in this particular province, of which it had always been one of the mainstays.

More important reductions, amounting respectively to 6·5 and 4·5%, had taken place in the Moscow and the Western regions. In the former, the reduction was compensated for by a parallel decline in the area under crops, with the result that the density had remained unchanged at 37 per 100 dessiatins. Here, the decline was probably due to a variety of causes, among which the reduction of natural meadows in some localities, the improvement of railway communications in the vicinity of consuming centres and the more widespread use of rail transport over short distances, as well as the increased advantage, under modern conditions, derived from the keeping of dairy cows, as compared with extra horses, must have all been instrumental in bringing about the decline. In the Western region the reduction, probably due to similar causes, was accompanied by an extension of cultivation, and the number of horses per 100 dessiatins under crops had accordingly diminished from 45 to 43.

In the South-Western region, an extension of the area under

crops, accompanied by a growing substitution of horses for oxen in ploughing and heavy haulage, was probably the principal among the reasons responsible for the increase of 3·2% in the number of horses, whose density per 100 dessiatins of crops had yet fallen from 39 to 38.

In the Ukraine, the Central Agricultural and the Middle Volga regions one enters that belt of the Northern black-earth, in which the agrarian crisis, at the beginning of the twentieth century, was most severe, and agrarian overpopulation most general. It was, accordingly, here, that the pressure of population brought about the continuous extension of arable at the expense of grasslands, and since the horse in these regions, as distinguished from the North, was almost exclusively used in husbandry, the diminution in the number of horses has been especially great. In the Ukraine, it reached 11·3%; in the Central Agricultural region 8·6%, while on the Middle Volga, the least affected of the three by "land-hunger," it was 1·8%. In the former two regions, the density per 100 dessiatins of crops fell respectively from 32 to 26 and from 33 to 29, while in the middle Volga region the decline was only from 24 to 23.

In the Eastern region, along with a large extension of cultivation, there has been a slight decrease in the number of horses, amounting to 0·9% and due to the reduction in extensive horse-breeding in the province of Ufa. Owing to the rapid extension of arable, the relative decrease, per 100 dessiatins of crops, was rather marked, the density having diminished from 33 to 29.

In New Russia there has been an increase of 7·9%, accounted for entirely by the extension of cultivation and the increasing substitution of horses for oxen in farm work. The density had remained unchanged at 20 per 100 dessiatins of crops.

The largest increase had taken place in the traditionally horse-breeding South-Eastern region, the home of the Don Cossack charger, as well as of several other breeds of saddle horses. The number of horses in this region increased 26·8%; yet, the extension of cultivation was so rapid, that the density of horses had actually diminished from 21 to 20 per 100 dessiatins of crops. In this region, however, which accounted for nearly two-fifths of all the draught oxen in European Russia, and where nearly all the farm work and haulage was done by oxen, this proportion only serves to show the rapidity with which the extension of arable had been reducing the facilities for horse-raising. How far the latter had been affected by the extension of cultivation, may be seen from the change in the proportion of young horses, under 4 years, which, in 1904, reached 36·4%, and in 1914 had fallen to 28·9% of the number of horses above this age: a heavy diminution, indeed, for a single decade.

In the Caspian region extensive horse-breeding was also developed on a large scale. Here, as in the South-Eastern region, though the number of horses had increased, there has been a heavy decline in their density, which had fallen from 36 to 26 per 100 dessiatins of crops. The proportion of young horses had also declined from 51·8% in 1904 to 42·2% in 1914.

The days of extensive horse-breeding in the South-Eastern parts of European Russia, where it still continued to play an important part at the beginning of the twentieth century, would appear to have been numbered, and the industry was visibly shrinking. From the South-East, since the close of the last century, its centre of gravity has been shifting towards the East, into the Asiatic steppes, but even there the rapid colonization and ploughing-up of the land, formerly the roaming ground of nomad Kyrghiz tribes, while it increased the numbers of horses, tended to reduce their breeding considerably. Indeed, the proportion of young horses to those over 4 years of age in the Asiatic Steppes (provinces of Turgay, Akmolinsk, Semipalatinsk and Semirechie), which reached 83·1% in 1904, declined to 77·1% in 1914. Yet, the total number of horses in these provinces, as well as in Siberia (provinces of Tobolsk, Tomsk, Yenissey and Irkutsk), had greatly increased, with the result that the young stock, though its proportion had diminished, had actually increased in numbers. The figures below show the numbers of horses in these two regions of Asiatic Russia in 1904 and 1914:

	1904	1914	Increase per cent.
Asiatic Steppes	1,331,859	2,956,976	122%
Siberia	2,921,704	3,961,007	35%

This increase in Asiatic Russia was the natural effect of the agricultural development of the country under extensive conditions of farming, and it certainly tended to make good for the decline in horse-breeding and in the actual numbers of horses in the European part of the Russian Empire. But, while it increased the numbers of farm horses, it reduced, at the same time, by the extension of arable, the facilities for the breeding of large herds of horses under nomadic or semi-nomadic conditions, and the horse-breeding industry, in its extensive form, had to diminish accordingly. The large exports of horses abroad, as well as the enormous demands for the remount of the Army, the cities, etc., on the eve of the war, have greatly assisted the maintenance, if not the increase, of this branch of stock farming, which had not yet been seriously affected by the mechanization of road transport. Under changed conditions, with the development of motor transport and the unavoidably heavy fall in foreign demand, though Russia herself, and her farming in particular, would still remain dependent, on the whole, on horse traction for many years to come, in the long run horse-breeding, except for the limited purpose of replacing worn-out stock, which would mostly be done out of local sources, and of providing a yearly draft for the remount of the Army and a few sporting mounts or trotters for racing, may certainly be looked upon as doomed to extinction.

It was in the case of horned cattle that the extent of the crisis brought about by the prevalence of a system of farming which had long outlived its usefulness, as well as the tendencies towards improvement on the eve of the war, have been most pronounced.

From what has been said of the evolution of Russian farming in the past it will be remembered that, until the later part of the eighteenth century, Russia had been a country almost entirely cultivated by peasants. Large farming, in Russia proper, was practically non-existent. Only in a few monasteries and in still fewer large estates did the owners of the land cultivate it on a large scale on their own account. Accordingly, the cattle of native Russian varieties had been evolved by peasants, under primitive conditions of peasant husbandry, and it was this type of animal that, even in the twentieth century, formed the great bulk of the stock of the Russian farms. Superimposed on it, as a small aristocracy of cattle, were the improved breeds scattered over the country, mostly in the estates of farming landowners. The presence of this stock, though occasionally useful in improving the local types of cattle by the infusion of better blood, could not, on the whole, achieve much in this direction, partly owing to the purely technical difficulties of proper selection of breeding material, partly because of the great difference in the conditions of maintenance, which made finer and more delicate stock mostly unsuitable for the average peasant farmer. In some cases, indeed, such ill-advised attempts at improvement, far from exercising any beneficial effect, resulted rather in spoiling a well-established local breed, which, though certainly poor in physique and in milk, as compared with good foreign animals, was yet much better fitted to the conditions it had to live in. The foreign breeds most common in Russia were Tyrolian, Dutch and English, while more recently Danish stock has been gaining in popularity. The Government, since the close of the last century, by the establishment of studs, the importation of pure stock from abroad and the grant of credits to breeders for the purchase of pedigree animals, had begun to assist the improvement of cattle and had been extending its work in this direction systematically. By the joint efforts of the Government, the Zemstvos and of those landowners who, besides the goodwill, also possessed the necessary knowledge in the matter of breeding, in the early part of the twentieth century some beginning was made in the real improvement of peasant cattle by crossing with scientifically selected breeds, though the work had scarcely started as yet, and its results largely depended on the progress of the enclosure movement.

It would, however, be wrong to assume that the native peasant breeds of cattle were all necessarily of poor quality. Several types of Russian cattle, considering the conditions in which they lived, could bear comparison with most foreign breeds, barring those artificial products of scientific in-breeding which, though certainly far superior in weight and milk output, could not possibly exist without constant care, attention and fixed rules of diet. The best known types of Russian cattle are the *kholmogorsky*, originating from the Kholmogorsky district of the province of Archangel and fairly common in the North of Russia, and the *yaroslavsky*, from the province of Yaroslav, the two being probably the best milkers in Russia. The *yaroslavsky* cattle is very popular in the central and North-Western provinces. The local breeds of Vladimir and Kargopol (province of Olonetz), as well as Zyrian native cattle, are also among the best for their milk output. As meat-stock, the best are the fine Kyrghiz and Kalmuk cattle of the Eastern steppes, bred under conditions of extensive wild grazing. A type apart is the *Ukrainian grey cattle*, used primarily as working oxen, but also distinguished by the high quality of its meat. This type, unfortunately, had been on the decline for the last few decades, its deterioration having partly been due to the development of the agrarian crisis, partly to haphazard crossing with other breeds. It must be said that, on the whole, given proper conditions of upkeep and due attention to scientific selection and breeding, these, as well as many other little-known local types of Russian cattle, could certainly produce excellent stock for milk and meat. At the time dealt with in the present study, however, the general position of this branch of farming in Russia could only be described as highly unsatisfactory. But signs of progress began to appear, on the one hand, in the more economically advanced districts of the country and, on the other, in the marked improvement of cattle which, as a rule, followed on the enclosure of peasant holdings.

The table below gives fairly clear indication concerning the trend of development of this branch of stock farming in Russia during the decade immediately preceding the war.

Numbers of Cattle in European Russia in 1904 and 1914

Regions	Numbers of Heads 1904	Numbers of Heads 1914	Change per cent.	Per 100 des. of Crops 1904	Per 100 des. of Crops 1914
Northern	885,539	976,890	10·3	105	116
North-East	2,139,330	2,473,375	15·6	42	45
Petrograd	1,181,927	1,348,897	14·1	72	85
Moscow	3,772,788	4,125,784	9·3	50	56
Western	2,209,551	2,400,599	8·6	82	89
South-West	2,214,461	2,701,838	21·9	49	56
Ukraine	2,920,369	2,267,507	—22·4	50	35
Central Agricultural	2,835,053	3,059,690	7·9	29	31
Middle Volga	1,860,090	2,082,067	11·9	27	29
Eastern	1,047,184	2,053,534	96·1	18	30
New Russia	1,932,706	1,785,424	— 7 7	21	18
South-East	4,059,643	4,107,987	1·1	48	36
Caspian	1,362,955	1,343,346	— 1·6	133	84
European Russia	28,421,596	30,726,938	8·1	41	40

It may be seen that, taking European Russia as a whole, from 1904 to 1914, though there had been an increase in the number of cattle, amounting to 8·1%, the density had diminished from 41 to 40 per 100 dessiatins of crops. A closer study of the table reveals, however, that the diminution was strictly local, and that, as a matter of fact, an analysis of the development in separate regions would appear to suggest the existence of three main types of evolution.

In the first instance, there is the relative decline in the numbers of cattle in the Southern and South-Eastern steppes, the region of extensive cereal farming. Then, in the more densely populated and more highly economically developed localities of the centre and West of Russia, one can perceive more or less definite signs of progress in the numbers of cattle and its density relatively to the area under crops. Finally, there is noticeable an increase in the numbers of cattle on the Northern and Eastern fringes of European Russia, in localities fitted for stock farming by their natural conditions and still unaffected to any significant extent by the development of agrarian overpopulation.

The first type of evolution could be observed in the South-Eastern and Caspian regions, as well as in New Russia. Here, the evolution presented a clear-cut issue between arable and stock farming, and the former has been getting the better of the cattle, partly owing to the geographical situation of the zone, far distant from Russian consuming centres and lying close by the Black Sea, partly owing to the natural conditions and the properties of the soil which enabled these districts to produce valuable cereals for export abroad. In New Russia, on the eve of the war, the development had reached a point beyond which the reduction in cattle could not proceed without serious injury to arable farming itself. In this region, though the diminution in the numbers of cattle had not been actually due to the three-course system, which had never been prevalent in New Russia and the Southern steppes generally, the decline was the result of what essentially amounted to the same, namely the survival of extensive systems of cultivation beyond the limits of their expediency, accounted for largely by economic conditions which inclined the farmer towards one-sided concentration on the production of grain. Throughout this vast belt, from the Caspian shores in the East to the Dniester in the West, one could see displayed the consecutive stages of the evolution of farming from the wild grazing of its primitive semi-pastoral forms to extensive cereal cultivation: a development unavoidably accompanied by a gradual shrinkage of cattle-breeding.

The opposite tendency, that is the increase in the numbers of cattle and in its density relatively to the arable area, was in evidence in two different groups of localities, representing two entirely distinct types of evolution. On the one hand, there were the more economically advanced parts of European Russia, in which this progress involved the intensification of this branch of stock farming; on the other, the outlying districts in which it involved the mere extension of herds. At the head of the former group, most representative of genuine progress in Russian stock farming, stood the Petrograd, Moscow and Western regions, in which the cattle had considerably increased in numbers, absolutely and in relation to the cultivated area, as well as improved in quality, under the influence of the growing demands for dairy and other animal produce of the two capitals and of the industrial districts, and the South-Western provinces, with their high standards of farming and the vicinity of Kiev and the Western frontier of Russia. Very slight signs of similar improvement could be observed in the two regions most typically representative of the three-course system in Russia and most severely affected by the agrarian crisis, namely the Central Agricultural and the Middle Volga, in which the density of cattle, very low relatively to the cultivated area, had increased slightly. The Ukraine should be included in the same group, in spite of the extremely heavy decline in the numbers of cattle, accounted for wholly by the province of Chernigov, in which, between 1904 and 1914, it had diminished by over one-half, owing to some purely local causes, such as failure of crops or cattle disease, which I am not, unfortunately, in a position to specify. In all these localities, to a varying extent, dairy farming was gaining ground on meat-cattle-raising and, especially in the vicinity of large consuming centres, the live-stock branch of farming was assuming intensive forms. Foremost in this respect were the two neighbouring provinces of Tver and Smolensk, in the Moscow region, situated within easy reach of both capitals. To the North of Moscow, relatively intensive forms of dairy farming prevailed in the provinces of Yaroslav and Vladimir, while in its more extensive forms the industry extended to the province of Kostroma and, along the Northern railway, over the adjoining districts of the province of Vologda; to the North-East, it reached as far as the province of Viatka. South of Moscow, it was far less developed, farming in these provinces being mainly arable; yet, on the eve of the war, especially along the railways, it has been extending gradually here also.

The third group comprises the Northern, North-Eastern and Eastern regions, in which cattle-breeding and, in the North and North-East, dairy-farming also, had been increasing under extensive conditions. Here, the population was relatively sparse, and the abundance of natural meadows and pastures favoured the growth of this branch of farming. The increase was most conspicuous in the Eastern region, in which the number of cattle had nearly doubled, and in spite of a large increase in arable, its density relatively to the area under crops had increased from 18 to 30 per 100 dessiatins. Here, indeed, the conditions were still present, which, given the stimulus of rising meat prices, were bound to result in the extension of cattle-raising; and the rise recorded in the early part of the current century had not failed in producing that effect.

While, in the central and Western provinces of European Russia, dairy farming had been gradually gaining ground on the raising of meat cattle, the latter was being shifted eastwards, beyond the Volga and, farther still, to Siberia and the Asiatic Steppes, from where cattle and meat were being brought to European Russia in increasing quantities. In the four provinces of Siberia proper, namely those of Tobolsk, Tomsk, Yenissey and Irkutsk, the number of cattle had increased from 3,525,872 in 1904 to 4,954,454 in 1914: an increase of about 40%. Owing, however, to the rapid colonization of Siberia and to the extension of arable, the density of cattle had diminished from 95 to about 85 per 100 dessiatins of crops, while the growing exports of butter in the twentieth century encouraged the extension of dairy farming at the expense of meat-production. The latter branch, on the eve of the war, had been expanding most rapidly in the Asiatic Steppes. Thus, in the provinces of Turgay, Akmolinsk, and Semipalatinsk, the number of cattle had increased from 706,460 in 1904 to 2,140,827 in 1914, or nearly 230%. Here, under conditions of extensive grazing, the raising of meat cattle could still develop unhampered by those restrictions of space which have been gradually driving it from Europe to Asia, until, at the beginning of the twentieth century, its last foothold on European soil was East of the Volga. Elsewhere it has been dwindling away rapidly, under the pressure of arable in the South and in competition with dairy farming in other parts of European Russia.

In Russian sheep-breeding it is necessary to distinguish the common sheep of various native breeds, on the one hand, and the fine-fleeced varieties loosely described as merinos, on the other. The former are essentially a product of peasant stock-farming, while the latter have been introduced to Russia by large landowners, and to the last have been bred exclusively on large estates.

The breeding of merinos, which began to attract the attention of the Russian Government, as a possible contributor to the export trade, since the days of Peter the Great, could not develop to any extent before the annexation by Russia of the Southern steppes, which had provided the necessary facilities for its growth. At the beginning of the nineteenth century, along with its other attempts at expediting the colonization of New Russia, the Government granted vast tracts of land in the steppes to several foreign sheep-breeders, on condition that they should start breeding merinos in Russia. The attempt proved a great success, and this induced a number of large landowners in the Southern steppes, who were on the look-out for anything that might increase the yield of their newly-acquired estates, to take up merino-breeding, which some among them, by paying careful attention to scientific selection and crossing, have raised to a high pitch of perfection. Accordingly, throughout the first half of the nineteenth century, in spite of the increasing competition of fine wool from overseas, mostly from Australia, on the world market, and the consequent fall in wool prices, merino-breeding in Russia had been growing continually. From some 800,000 merinos in 1808, the number increased to 8 millions in 1856 and reached 12·3 millions in 1881. Yet, since the 'seventies, though it had continued to expand, under intensive conditions of farming, in the Baltic Provinces and in Poland, in Russia proper fine sheep-breeding began to shrink with great rapidity. The colonization of New Russia, accompanied by the extension of steam transport by land and by sea, which provided new outlets for the cereal production of the steppes, opened new prospects before the South-Russian farmer, urging him to give up merino-breeding in exchange for the extensive cultivation of wheat. By 1896, following this evolution, the number of merinos fell to 5·3 millions; in 1906 it was 4·6 millions and in 1910 it did not exceed 2·2 millions. Like other extensive branches of the agricultural industry in Russia, fine sheep-breeding, besides diminishing in extent, had also been shifting eastwards, with the result that, on the eve of the war, in 1914, it was almost entirely confined to the South-Eastern and Caspian regions. In New Russia, which, in the past, had been the centre of merino-breeding, only the province of Taurida still possessed any considerable numbers of fine-fleeced sheep. This shifting of merino-breeding has been characteristically reflected in the transfer of the centre of fine-wool trade in Russia, which, in the first half of the nineteenth century, had been in Romny and Poltava; thence, for the next few decades, it had migrated to Kharkov, only to move again, in the twentieth century, to Rostov-on-Don.[1]

Russian merino-breeding had, indeed, been affected, besides local causes, peculiar to Russia, also by certain general developments whose consequences it shared with all the rest of the sheep-breeding world. The latter consisted in the impossibility, under existing conditions and prices, to make pure-bred fine-fleeced sheep pay: a difficulty from which, in other countries, a way out had been found by evolving heavier types of sheep, in which the somewhat coarser fleece was combined with excellent flesh, and thus making the animals serve a double purpose. In Russia, this solution was made unpracticable by the extremely small consumption of mutton, which it would take generations of progress to increase sufficiently. The attempts made in this direction, which have been responsible for the vogue which, in Russia, in common with Germany and some other continental countries, had been enjoyed for a time by the heavier types of sheep, and more particularly the Rambouillet, have failed in stemming the decline, though they have greatly changed the actual composition of the dwindling flocks still referred to by the misleading generic term of merinos.

It may be said that, on the eve of the war, Russian sheep-breeding was confined to the more or less coarse-woolled small native varieties, bred mostly by peasants. Indeed, in 1916, in European Russia the peasants had 56·3 millions of sheep, while large estates accounted only for 4·5 millions, of which about 2 millions belonged to the so-

[1] B. Brutzkus, *Economics of Agriculture*, p. 176.

called merinos. The number of sheep to every 100 dessiatins under crops averaged 60 in large estates and 99 on peasant land. According to regions, it varied, in 1916, in the case of large farms, from a maximum of 1,280 in the Caspian region to a minimum of 3 in the South-Western, pointing to the part played by fine-fleeced types, bred in the South-East, in the case of large farming. Among the peasants, the minimum was 12 in the Northern region, and the maximum, also in the Caspian region, did not exceed 290.

Some of the native breeds of sheep deserve notice. To these belong, in the first instance, the *Romanovsky* type, originally bred in the province of Yaroslav, but fairly well represented throughout the centre of Russia and highly valued for its excellent, if somewhat heavy, fleece, used for the best sheep-skin coats. Another breed famous in Russia is the *Reshetilovsky*, belonging to the province of Poltava and other parts of the Ukraine, and providing excellent grey skins for hats, collars, etc., whose quality is especially high if they are obtained by crossing the original breed with the *Karakul* (Astrakhan) sheep. The latter, originally bred in Central Asia, became extremely popular in South-Eastern Russia in the course of the last few decades, and was either bred pure or crossed with local varieties of sheep. In the South and South-West, two types of Bessarabian extraction, namely the *Tziganka* (Gipsy) and the *Valashka*, both with much finer fleece than the common sheep, were also bred and crossed with local varieties.

As a general characteristic of the position of Russian sheep-breeding, it may be said that, with the exception of the most noted breeds, the sheep still continued, on the eve of the war, to serve primarily the purposes of immediate consumption in the peasant household, and of all the domestic animals was probably the one in whose breeding the elements of the old régime of self-supplying natural economy had survived to the greatest extent.

The evolution of sheep-breeding in Russia between 1904 and 1914, may be seen from the table on page 315. [1]

Thus, Russian sheep-breeding has been losing heavily all along the line, and the reasons, in almost every locality, were the same. With the increase in population and the extension of arable, facilities for the grazing of sheep were in course of rapid diminution, while the economic evolution of the country generally also tended to have the effect of reducing this branch of stock-farming. Everywhere, indeed, the survivals of the old system of natural economy were

[1] The figures given below include goats, whose total number in Russia, however, did not exceed some 1·5 millions, with the result that their inclusion does not affect the conclusions in any way.

Regions	Number of Sheep 1904	1914	Change per cent.	Per 100 des. of Crops 1904	1914
Northern . . .	711,458	692,052	— 2·8	87	82
North-East . .	2,905,818	2,918,072	0·4	57	53
Petrograd . .	778,509	784,923	0·8	47	49
Moscow . . .	5,256,174	4,305,067	—18·1	69	59
Western . . .	1,568,578	1,560,124	— 0·6	59	58
South-West . .	2,422,724	1,718,141	—29·3	53	35
Ukraine . . .	3,228,107	2,057,775	—36·3	56	32
Central Agricultural .	7,786,603	6,155,231	—21·0	81	62
Middle Volga . .	4,426,640	3,958,502	—10·6	63	54
Eastern . . .	3,629,510	3,409,856	— 6·1	63	50
New Russia . .	3,322,868	1,688,734	—49·2	36	17
South-Eastern . .	7,846,860	5,193,957	—33·9	92	46
Caspian . . .	3,755,408	2,837,441	—24·5	367	177
European Russia .	47,639,257	37,279,875	—21·4	69	49

in full retreat before the advancing tide of commercialization, and the sheep, as, *par excellence*, an animal serving the purpose of immediate consumption in the producer's own household, had to give way either to the cow, or to the extension of arable, according to local conditions.

As the other extensive branches of farming, sheep-breeding, while it diminished in European Russia, tended to increase in Siberia and, especially, in the Asiatic Steppes, which were particularly favourable to its development. In the four provinces of Siberia proper, the number of sheep had increased, between 1904 and 1914, from 3·9 to 4·6 millions, or 15%. In the Asiatic Steppes, namely in the provinces of Akmolinsk and Semipalatinsk, it had increased from 0·6 to 4·1 millions, or 580%. Just before the war attempts were being made by the Colonization Department of the Ministry of Agriculture to encourage the development of sheep-breeding on a large scale in Asiatic Russia. Large tracts of unoccupied land were being offered for lease to sheep-breeders, with a view to helping the revival of extensive merino-breeding, which was so rapidly disappearing in European Russia. This experiment, whose success had always appeared rather doubtful, considering the conditions of the wool market, had been interrupted by the war almost before the first steps had been taken with a view to putting it into practice.

Along with the continual diminution of sheep-breeding, which could not, as yet, find its place in the gradual progress of Russian farming towards more intensive forms, there has been going on a development in another group of live stock, which, being directly dependent on the intensification of farming, was very characteristic of the agricultural evolution of Russia on the eve of the war. The progress of pig-breeding in Russia, indeed, could be looked upon as a sign of the times. Until the close of the last century, this branch of stock-farming in Russia had been in a very backward condition. Pig-breeding had really existed in a few large estates, in which pigs were bred from imported stock, mainly Berkshire and Yorkshire, and in the Western and South-Western provinces, from which live animals and pork were exported to Germany, and where, therefore, there had been an inducement to improve the native stock. Throughout the rest of the country, the number of pigs was relatively very small, and the native animals were small in size and generally of poor quality, though experience had proved that, by crossing with imported breeds, and especially with those of Yorkshire and Berkshire, very fine stock, combining hardiness with excellent flesh and lard and attaining very esteemable size and weight, could be produced. The change in agricultural conditions, which marked the beginning of the twentieth century and made pig-breeding profitable, coupled with the technical progress of farming, and more particularly of root crops, and with the extension of dairy farming, began to attract attention to this, hitherto neglected, branch of the agricultural industry. Pig-breeding, with its characteristic rapidity of development, in many localities expanded in a few years beyond recognition. The most striking example of this expansion was probably that of the province of Tambov and certain districts in its immediate vicinity, in which a complete transformation had been effected, since 1909, by the establishment of the first bacon factory in Russia, engaged in curing bacon for the English market. At first, the new factory found it difficult to obtain suitable animals, but its demand had produced the necessary change in local stock so quickly that by 1912 the breeding of pigs of the required type became one of the principal industries of the local peasantry. In other districts, pig-breeding had been stimulated by the growing demand for pork on the home and foreign markets, and wherever there were distilleries, starch or sugar factories, etc., whose offals could be used for feeding, or where the production of butter left on the hands of the peasants large quantities of butter-milk, this branch of stock-farming was sure, sooner or later, to develop. Accordingly, it increased rapidly in all the more agriculturally advanced districts of European Russia and in those parts of Siberia in which the production of butter was more highly developed.

The following table shows the progress of pig-breeding in various parts of European Russia between 1904 and 1914. It should be borne in mind, however, that since progress in pig-breeding was necessarily expressed rather in the quality of the stock, than in its numbers, the figures not only fail to convey a complete idea of the change in this branch of farming, but, in some cases, as that of the Central Agricultural region, may even create an entirely misleading impression. It will be seen that, in this region, the number of pigs had decreased, though it was there, in the province of Tambov and its neighbourhood, that some of the most rapid progress had been achieved in the development of pig-breeding. This, however, was not necessarily accompanied by an increase in numbers, but rather involved the substitution of a smaller number of better animals for the former multitude of small native pigs. In this case, the need of great circumspection in the use of statistics of live stock, to which I have pointed out, is made very obvious.

Regions	Numbers of Pigs 1904	1914	Change per cent.	Per 100 des. of Crops 1904	1914
Northern	57,706	77,599	34·3	7	9
North-East	392,455	535,019	36·3	8	10
Petrograd	221,038	251,947	13·9	13	16
Moscow	1,077,071	1,069,053	−0·8	14	15
Western	1,289,866	1,313,779	1·8	48	49
South-West	1,440,718	1,573,959	9·2	32	33
Ukraine	1,187,120	1,310,216	10·3	21	20
Central Agricultural	1,168,516	1,110,957	−6·7	12	11
Middle Volga	479,909	508,827	6·0	7	7
Eastern	405,018	407,705	0·6	7	6
New Russia	862,926	943,500	9·3	10	10
South-East	1,169,539	1,211,614	3·5	14	11
Caspian	112,730	138,839	23·1	11	9
European Russia	9,864,612	10,452,944	5·9	14	14

In conclusion, to sum up what has been said here of the development of stock-farming in Russia on the eve of the war, it would appear that, on the whole, it reflected the struggles of the Russian agricultural industry which, at that time, was in the process of transition from the predominance of three-course cropping to improved rotation and the more effective use of the available area. The conditions of all the four branches of stock-farming, dealt with above, bore distinct traces of the influence of a backward system of cultivation, with open fields, compulsory cropping and an increasing scarcity of meadows and pastures. Yet, the observing eye could perceive that, under the stimulus of improved agricultural conditions and of economic progress, involving the growing commercialization of all branches of the agricultural industry, fresh developments constantly gained ground on the old routine and agriculture was being transformed. Year after year, the extensive branches of stock-farming were being shifted to the East beyond the Volga and, farther, beyond the Urals. Intensive branches, on the other hand, were gradually gaining in importance, and though they still lived side-by-side with remnants of a past stage of evolution, and were often hardly perceptible among the vast accumulation of survivals, they were unmistakably forging their way ahead.

CONCLUSION

Looking back at the evolution of Russian farming in the early part of the twentieth century, one cannot fail to perceive the intimate connection which existed between the development of the Russian agricultural industry and the growth in the country of modern industrial capitalism. This connection is clearly exhibited both in space and in time, in the geographical distribution of systems of farming, as well as in the historical evolution of Russian agriculture in all its branches. In the brief sketch of the agricultural map of twentieth-century Russia, given in the opening chapters of the present study, I have endeavoured to show the influence exercised on the organization of farming in various parts of European Russia by its general economic environment, on which depended the purchasing capacity of the local market for agricultural produce. Proceeding to the discussion of Russia's agricultural organization and conditions, in the central part of this work, I had to begin by outlining the principal features of the Russian Agrarian Problem, which had played so outstanding a part in the political and economic history of modern Russia. And, approaching this controversial problem from a purely historical standpoint, as it appears in the light of subsequent evolution, projected against the background of Russia's general economic development, one is enabled to perceive certain features which, not unnaturally, have been liable to escape the eye of the contemporary observer. Considered in the light of later development, from a distance of the last two eventful decades, the Russian agrarian problem, the "land-hunger" among a large section of the peasantry and the generally miserable condition of the Russian countryside would appear to have been due not so much to any specific causes as, in the first instance, to the inherent weakness of the economic system of nineteenth-century Russia. In that system, indeed, capitalistic cells were extremely underdeveloped, not to say practically non-existent. Like a child too big for its age, Russia had vital organs insufficiently large for her huge national body, and suffered accordingly from organic malnutrition. For decades, at the cost of tremendous strain, Russia had been accumulating the means of industrial revolution; and at the close of the last century one could perceive signs of the transformation approaching, if not actually begun. This transformation affected the countryside in two ways: on the one hand, by relieving to some extent the pressure of population on the land in the congested agricultural areas through the rural exodus caused by the demand of the growing industries for additional labour; on the other, by increasing the capacity of the home market for all kinds of agricultural produce and increasingly commercializing agriculture in all its branches and forms. The conditions created by the industrial revolution in Russia, moreover, besides improving the position of the agriculturist, large and small, also provided to the Government the opportunity, which it had hitherto lacked, of active intervention in the countryside, with a view to helping the solution of the agrarian problem and the economic progress of the peasant farmer. Hence, the agrarian policy of the Russian Government in the early part of the twentieth century, which, essentially, could be described as a necessary counterpart of the industrial revolution. The individualist movement in the Russian village, of which enclosures were the manifestation, carried all before it, because it answered the most urgent needs of the peasant farmer. The combination of enclosures with the growth of the co-operative movement pointed to the peasant being conscious of the necessity of adaptation to the requirements of the modern competitive system. By enabling the peasant to enclose his holding, if he so wished; by extending the land-settlement work of the Peasants' Bank and assisting the emigration of peasants to Asiatic Russia on a large scale; finally, by helping the extension of the co-operative movement and the technical progress of peasant husbandry, the Russian Government had taken an active part in the improvement of the condition of the countryside. The State had intervened with wisdom and, on the whole, with outstanding success, in the solution of the agrarian problem; but both the possibility of the Government taking this course in its agrarian policy, and the conspicuous success with which its reforms had met, were entirely due to the change in the general economic situation brought about by the industrial revolution in Russia. Had the agricultural situation remained the same as before; had not the growth of Russian capitalism extended the agricultural market and infused commercialism into the village, and had the latter still been living under conditions of self-sufficient natural economy, all attempts at improving the position of the peasantry by encouraging enclosures, assisting the growth of rural co-operation and helping the technical progress of farming, would have proved futile. It was the commercialization of Russian farming, indeed, that was the keynote and the mainspring of its progress on the eve of the war; and to it have been due all those changes in the Russian agricultural industry, its geographical distribution and its output, with which I have had to deal in the preceding chapters. A vast country of small peasant farmers, with, until the revolution of 1917, a sprinkling of more or less large capitalistically-organized estates, Russia was not in a position to build her economic progress on agricultural exports, in competition with the extensive cereal-growing countries of the New World. Her future lay in the intensification of her farming in all its branches, which could only be based on the extension of her home market for agricultural products; in other words, on the growth within the country of modern industrial capitalism, based on her great, though still undeveloped, natural wealth. On the eve of the war, one was just allowed a glimpse of the beginning of this economic transformation, with its enormous potentialities of progress. Then, as a thunderbolt, came the terrible catastrophe of 1914, and progress changed into destruction.

The influence of the war on the Russian economic system is far too complex a subject to be dealt with in the brief closing chapter of the present study. In a few words, therefore, I shall only try to give here a general outline of its effects on the agricultural situation in Russia. Stated consisely, the principal effect of the war on the Russian agricultural industry was that it had started the process of breaking-up the delicate links which, in the course of the preceding period, had been gradually binding the separate producing units in Russia together into a single economic system: a process of destruction which, later on, since 1917, had been continued and aggravated by the revolution.

Indeed, apart from the withdrawal of millions of men in their prime from the agricultural industry, from the restriction in credit facilities, the depreciation of currency and the difficulty of replacing worn-out implements, the outstanding feature of the agricultural situation during the war was that the facilities for disposing of Russia's agricultural production on the home and foreign markets have been greatly reduced.

The diminution in the volume of Russia's principal agricultural exports during the war is shown on page 322.

Thus, the blow struck by the war to Russia's agricultural exports was simply staggering. The foreign outlets for the products of Russian farming had practically ceased to exist by the close of the second year of the war.

At home, with the progress of the war, the position was being

Changes in the volume of Russia's principal Agricultural Exports during the Great War

	1911–13	1914	1915	1916
	Net Exports in Thousands of Poods			
Grain, various	564,384·4	308,813·8	10,517·9	2,097·1
Other cereals	31,041·2	11,086·7	5,103·5	870·7
Butter	4,551·1	3,212·9	3,232·7	163·4
Eggs	9,370·1	5,621·0	940·3	343·7
Sugar	19,770·6	7,778·4	5,711·2	315·3
Flax	17,952·2	14,067·0	4,845·3	739·4

continually rendered more and more difficult by the growing dislocation of transport and the increasing disorganization of the whole machinery of the internal market. The dislocation of transport may be looked upon as the most fundamental factor of this process of economic disintegration. The claims made by the war on the railways, with their relatively limited rolling-stock which, in years of good crops, had often proved unable quickly to deal with the grain traffic even in peace time, were now taxed beyond their utmost capacity. The trading apparatus, and the grain trade in particular, which had tried to struggle against heavy odds, ultimately broke down under the combined pressure of the transport chaos, on the one hand, and of the Government's policy of food control, on the other. This policy, which originated at the very outbreak of the war, by the close of 1915 had already developed into a far-reaching system of measures, involving local restrictions of export, the control of prices and the extension of the direct intervention of the public authorities in the purchase and distribution of supplies. The credit facilities available to the private trading apparatus have also been in the course of gradual diminution. Co-operative organizations, though their activities had been extending during the war, could not bridge the widening gulf between the agricultural producer and the consuming centres. In the producing localities, enormous stocks of cereals and other agricultural products began to accumulate, without any prospects of ever being moved to market. The peasants, in their bulk, began to relapse into that state of economic isolation, from which the evolution of the last few decades had been helping them gradually to emerge.

At the same time, as the war went on, the costs and difficulties of agricultural production tended to increase. The large farmer's position, in the face of the war-time situation, left him no alternative to the reduction of cultivation. The peasant's position, if less critical, because of his relative independence of hired labour, was also bad enough, mostly owing to the growing dearth of necessaries and to the reduction in marketing facilities. Moreover, in the case of the peasant, the extreme dearth of necessaries, which were either actually unobtainable or unreasonably expensive, combined with the influx of money received by the families of the mobilized, in the way of relief from the State, and with higher wages for any labour the peasant could spare outside his own farm, made the present need for cash much less pressing, thus enabling the peasantry to dispense more easily with those commercial branches of production on which they had hitherto been dependent for ready money. Along with the restriction in marketing facilities, this war-time relative financial independence of the village was also responsible for the tendency towards a return of the peasantry to its pristine state of economic isolation.

The effect of the war on the Russian agricultural industry may be seen from the table below, showing the percentages by which the various crops had been reduced in 1916, as compared with 1913:

Decrease or Increase in the Area under various Crops in 1916, *as compared with* 1913

Regions	Cereals	Potatoes	Flax	Hemp	Sugar-Beet	Total
			Percentages			
Northern	−10	+ 7	+38	—	—	− 8
North-Eastern	− 6	−16	+11	−15	—	− 5
Petrograd	−10	+ 7	+23	+50	—	− 6
Moscow	−17	− 5	+11	+13	−94	−14
Western	+ 3	−29	+32	− 9	—	—
South-Western	− 6	+ 8	−72	−51	−27	− 7
Ukraine	−10	−17	−24	−12	− 6	−11
Central Agricultural	− 6	−25	−42	−12	−24	− 7
Middle Volga	−14	−20	−16	− 6	—	−14
Eastern	− 7	+ 1	−14	− 6	—	− 6
New Russia	− 9	−27	+31	−34	+42	− 9
South-Eastern	−23	−43	+ 1	−53	−17	−23
Caspian	−21	−18	−62	−57	—	−21
European Russia	−11	−16	+10	−11	−21	−11

The above table gives a sufficient idea of the extent to which the agricultural industry of Russia had been reduced by the war. The total area under crops in 1916 was, indeed, some two millions of dessiatins short of that of 1901–5, two years of the war having more than undone all the progress achieved in this respect since the beginning of the twentieth century. Only in flax there has been an increase, as compared with 1913, due to the war-time demand for flax fibre. Sugar-beet, an intensive crop, cultivated mostly in large estates and requiring much labour, had diminished enormously. Next came potatoes, with a 16% decrease in the area, the diminution in this case being probably accounted for, in the

Y*

first instance, by the effect of war-time prohibition, since about 13% of the pre-war output of potatoes in European Russia used to be disposed of to spirit distilleries. The diminution of the total cultivated area was especially heavy in the South-Eastern and Caspian regions (23 and 21% respectively); a phenomenon due partly to these two regions having been largely dependent on the export of grain abroad, partly to the fact that they included three Cossack provinces (the Don, Kuban and Terek), in which nearly all the male population of working age had been called to the Colours by successive mobilizations.

The Agricultural Census of 1916, taken during the summer of that year, had placed on record the agricultural condition of Russia at the close of the second year of the war, and on the eve of the March Revolution of 1917. Beyond 1916, in the revolutionary turmoil into which Russia had been plunged by the strain of the prolonged war, the student can hardly venture as yet. Since 1917, the social, political and economic organization of Russia has been in the melting-pot; and in the cross-currents of various forces striving to build a new system on the ruins of the old, it is, as yet, impossible to discern any clearly-pronounced tendencies of evolution. In the chaos of the last revolutionary twelve years, from the fall of the Imperial Government in March, 1917, to the spring of 1929, one phenomenon would appear, however, to stand out conspicuously, through all the vicissitudes of Russia's recent evolution. This phenomenon, so outstanding in the consistency of its manifestation, is the close dependence of agriculture on the market and on the growth of the other branches of production and trade. Never, indeed, had this intimate connection been exhibited with greater clearness or on more striking examples of almost momentary change, than in the course of the revolutionary period. One of the first acts of the Provisional Government, faced with a food problem growing worse from one day to the next, was the establishment, by the Decree of March 25th, 1917, of the State monopoly of the grain trade. The measure was, as a matter of fact, no more than the carrying to its logical conclusion of the food policy of the Imperial Government, which, to the last, had, however, refrained from so drastic an experiment. What was still left of the private grain trade, had now been done away with, or rather driven underground, thence to fill, by illicit transactions, the gaps left by the State monopoly. By the monopoly and the rapid depreciation of the rouble under the Provisional Government, the market was thrown into a state of utter chaos. Combined with the general condition of unrest throughout the country, the utter disorganization of the agricultural market, aggravated by the collapse of the transport system, brought the destruction of commercial farming in Russia a step farther; and it was in this condition of almost complete relapse into a state of isolated natural economy that the Russian countryside had entered the second stage of the Revolution, inaugurated by the Bolshevik *coup d'état* in November, 1917.

Every student of the Russian Revolution remembers how the peasant farmer, since 1917 the only agricultural producer in Russia, had been affected by the policy of "militant Communism," when the Soviet Government had made an attempt at doing away with all private enterprise and concentrating the production and distribution of goods entirely in the hands of the State. The peasant, though he was left to cultivate his land as before, was not permitted to sell anything on the market, but had to surrender his surplus, over and above a certain minimum allowed for his own use, to the Government, for a fixed price in depreciated money. His reply to this measure was a reduction of the cultivated area to the level of his bare necessities; and the Communist State, after two years of this policy, could get nothing out of the peasant, while the cultivator himself, with no reserve in case of failure of crops, was open to the visitation of famine, should the crops fall below the average yield. This happened in 1920 and, on a larger scale, in 1921, and the terrible famine of the latter year, which had cost Russia millions of lives and untold material losses, will be remembered for generations to come. Seeing the mistake, in the spring of 1921 Lenin inaugurated the New Economic Policy, involving a partial return to competitive conditions in trade and a re-organization of industry on what had since been termed the system of State Capitalism. The peasant farmer was permitted to dispose of his products on the market, and in the course of the next two years, in spite of all obstacles, he succeeded in restoring his cultivation to nearly the pre-revolutionary level. The food situation throughout Russia had improved enormously, and, for two or three years, the Soviet Government had even been able to obtain considerable stocks of grain for export abroad. For some time on the coming in force of the New Economic Policy, it had looked as if, under the régime of State Capitalism, which combined the ownership by the State of the key industries, banks, railways and of large industrial and trading concerns generally, with certain elements of the competitive system in their actual management, a compromise had been achieved which might permit the revival of Russia's economic life. Yet, this illusion was soon dispelled, and by the close of 1923 Russia found herself again faced with a serious economic crisis, due to the failure of the State-owned industries to pay their way and to supply manufactured goods at reasonable prices. The costs of production proved exceedingly high; the industry as a whole was working at a loss, and had to be subsidized either direct from the Budget, or through the banks, whose advances had been swelling by repeated renewals of loans. The situation reacted most unfavourably on the currency and the foreign exchanges, not to speak of the Budget, which had to bear the ultimate cost. At the time, in 1924, the Soviet Government had succeeded in patching-up the somewhat badly shattered system of State Capitalism, mostly by drawing on the peasants, who, in the meantime, had to some extent recovered from the disastrous effects of the undiluted Communist policy of the earlier years. As the only industry in Russia, which, in the Communist State, has been paying its way, peasant farming was called upon, besides paying taxes, to subsidize the State-owned industries by paying monopoly prices for their products, as well as to help in supporting the rate of exchange by supplying grain for export at prices dictated by the purchasing organs of the Commissariat of Trade. For some time, until the peasant farmer realized the position he was in, this device had been effective, and the Soviet Government was enabled to tide over the acute stages of the economic crisis. A policy of this nature, however, was evidently bound to bring its own retribution; and in 1925 the system of State Capitalism was again faced with difficulties. The peasants' demand for manufactured goods began to decline, while it became increasingly difficult for the Government to obtain from the cultivators the quantities of grain scheduled for export abroad. Irrespective of the actual yield of crops, the grain exports, in the course of the last few years, have consistently failed to materialize on the projected scale. The peasant farmer, confronted with a policy of thinly-disguised spoilation, once again sought refuge in retreat from intercourse with the outside world. The bonds connecting the village and the industrial towns have again been loosened greatly, and the expansion and progress of peasant farming, which had been noticeable since 1921, had received a severe check. Indeed, at the time when these pages are being written, Russian towns are again living on rations, while the localities in which the crops proved bad in 1928, are actually suffering from famine. The policy of tampering with the market and the producer is having its disastrous effects.

APPENDIX I

AREA, POPULATION AND DISTRIBUTION OF LAND
according to modes of exploitation

Regions	Area in Sq. Versts	Population[1] per Sq. Verst Total	Rural	Distribution of Land (per cent)[2] Arable	Meadows	Forests	Other Suitable	Waste
Northern	1,207,722·1	1·7	1·6	1·3	2·4	59·2	1·8	35·2
North-East	425,188·4	14·1	13·5	19·5	9·3	54·5	10·8	5·9
Petrograd	181,322·3	25·3	16·7	14·8	14·0	48·8	4·1	18·1
Moscow	374,384·2	37·5	31·4	34·8	17·0	36·9	3·0	7·7
Western	133,927·6	32·4	28·9	26·7	13·0	37·3	3·5	1·8
South-West	109,373·3	72·9	65·5	53·9	9·9	18·8	9·3	8·1
Ukraine	137,771·1	54·9	48·7	66·9	10·0	12·2	4·7	6·2
Central Agr.	232,420·9	47·7	43·8	67·0	9·0	13·6	5·3	4·9
Middle Volga	218,727·3	35·1	31·8	53·1	10·9	25·7	3·1	7·2
Eastern	239,934·2	20·6	19·5	41·4	31·5	23·1	1·8	9·7
New Russia	170,972·6	36·8	28·5	70·3	15·4	3·0	5·0	6·3
South-East	274,983·5	19·5	17·2		No information			
Caspian	270,690·9	6·9	6·2		,,	,,		

[1] Population according to the Census of 1897.
[2] Distribution of the area according to the data of the Survey of 1887.

APPENDIX II

DISTRIBUTION OF THE AREA UNDER CEREALS AND OTHER CROPS

according to the Agricultural Census of 1916

In thousands of Dessiatins and Percentages

Regions	Principal Cereals	Pota toes	Sugar Beet	Flax	Hemp	Grass	Other Crops	Total
Northern	714·2	25·1	—	40·6	1·6	4·4	11·8	797·7
	89·4	3·1	—	5·1	0·2	0·6	1·6	100
North-East	4,801·8	51·3	—	192·5	19·6	97·9	171·8	5,334·9
	90·0	0·9	—	3·6	0·4	1·8	4·3	100
Petrograd	1,170·9	112·2	—	171·0	0·9	158·7	36·5	1,650·2
	71·0	6·7	—	10·4	—	9·6	2·3	100
Moscow	5,057·9	563·1	0·2	475·0	72·0	390·5	274·8	6,833·5
	74·1	8·2	—	6·8	1·0	5·9	4·0	100
Western	2,169·2	248·3	—	115·7	17·8	118·0	208·8	2,877·8
	75·5	8·5	—	4·0	0·7	4·1	7·2	100
South-West	3,278·5	204·2	236·4	2·6	8·2	159·4	716·5	4,605·8
	71·2	4·4	5·1	—	0·2	3·4	15·7	100
Ukraine	4,363·5	214·4	117·6	26·5	75·2	233·6	705·0	5,735·8
	75·7	3·7	2·5	0·4	1·3	4·0	12·4	100
Central Agr.	7,150·6	367·2	86·0	18·2	183·3	175·3	1,479·0	9,459·6
	75·6	3·9	0·9	0·2	1·9	1·9	15·6	100
Middle Volga	5,254·7	135·6	—	50·6	51·9	46·1	836·2	6,375·1
	82·1	2·1	—	0·8	0·8	0·7	13·5	100
Eastern	5,635·5	69·3	—	15·2	22·4	90·0	696·7	6,529·1
	86·3	1·6	—	0·1	0·2	1·4	10·4	100
New Russia	8,286·8	110·8	14·2	61·6	7·1	94·4	611·6	9,186·5
	90·2	1·2	0·2	0·6	0·1	1·0	6·7	100
South-East	8,085·9	71·1	2·5	42·4	6·0	80·8	841·1	9,129·8
	88·6	0·8	—	0·4	—	0·9	9·3	100
Caspian	910·8	21·1	—	8·1	0·8	12·2	354·9	1,307·9
	69·6	1·6	—	0·6	—	0·9	27·3	100
European Russia	56,880·3	2,193·7	456·9	1,220·0	466·8	1,661·3	6,944·7	69,823·7
	81·2	3·1	0·6	1·7	0·7	2·3	10·4	100

APPENDIX III

NUMBERS OF HORNED CATTLE IN VARIOUS REGIONS OF EUROPEAN RUSSIA

according to the Agricultural Census of 1916

Regions	Oxen	Bulls	Cows	Heifers and Bulls over 1½ y.	Young Cattle, 1–1½ y.	Calves	Total	Per 100 dess. Crops
			In thousands of Heads					
Northern	4·4	9·5	700·3	55·5	102·5	355·4	1,227·6	153·7
North-East	21·6	62·3	1,684·4	245·4	402·7	864·6	3,281·0	61·1
Petrograd		14·3	1,042·2	93·0	88·5	478·3	1,716·3	103·5
Moscow	5·3	33·0	2,496·9	195·1	380·0	1,604·2	4,714·4	69·1
Western	47·0	25·0	974·9	160·2	188·1	416·8	1,812·0	65·5
South-West	192·2	28·7	1,164·0	148·8	251·8	817·1	2,602·7	50·6
Ukraine	384·9	37·1	1,144·0	200·1	265·6	833·3	2,865·0	49·7
Central Agr.	216·6	25·4	1,724·4	181·9	378·5	1,212·0	3,738·8	39·5
M. Volga	84·0	21·1	1,269·2	161·9	290·0	282·7	2,648·9	41·4
Eastern	73·5	29·1	1,240·5	210·0	382·1	879·6	2,814·8	43·1
New Russia	176·1	41·4	1,005·2	227·9	363·5	735·1	2,549·2	27·7
South-East	1,211·8	78·2	1,412·5	559·9	666·0	1,189·1	5,117·5	56·0
Caspian	162·1	43·5	839·9	263·3	259·7	647·7	2,216·2	176·9
European Russia	2,579·4	448·8	16,698·3	2,703·3	4,018·9	10,855·7	37·304·3	

[1] The columns and rows in the above table do not total exactly, the discrepancy being due to the figures having been rounded off to thousands.

APPENDIX IV

NUMBERS OF SHEEP AND PIGS IN EUROPEAN RUSSIA

according to the Agricultural Census of 1916

	Sheep		Pigs	
Regions	1,000 heads	Per 100 des. of Crops	1,000 heads	Per 100 des. of Crops
Northern	900·8	11·3	92·6	1·1
North-Eastern	4,897·9	91·8	1,372·7	25·7
Petrograd	1,698·0	102·9	310·1	18·8
Moscow	5,237·6	76·7	1,474·7	21·6
Western	2,230·3	77·5	2,050·5	71·2
South-Western	1,417·9	30·8	1,381·0	29·9
Ukraine	3,143·9	54·6	1,817·7	31·5
Central Agricultural	10,137·8	117·7	1,652·7	17·4
Middle Volga	6,805·8	106·4	801·4	12·5
Eastern	6,398·7	98·0	842·4	12·9
New Russia	2,701·5	29·4	1,556·2	16·9
South-Eastern	7,475·1	81·8	2,173·3	23·7
Caspian	4,439·3	323·7	225·5	17·2
European Russia	57,484·6	—	15,750·8	—

APPENDIX V

PRICES OF LAND IN EUROPEAN RUSSIA, 1906–1910

(based on the records of transactions financed by the State Peasants' Bank)

	Average Prices in Roubles per Dessiatin					
Regions	1906	1907	1908	1909	1910	1906–10
Northern [1]	66	55	32	30	28	42
North-Eastern	27	35	46	45	38	38
Potrograd	74	74	71	67	68	71
Mescow	96	102	104	99	95	99
Western	89	89	99	88	96	93
South-Western	199	202	214	235	232	216
Ukraine	212	207	212	227	227	217
Central Agricultural	162	165	168	179	171	169
Middle Volga	109	98	102	105	112	105
Eastern	77	72	76	79	81	77
New Russia	183	197	210	194	166	190
South-Eastern	133	156	161	179	157	154
Caspian [2]	119	122	127	151	113	126

[1] Exclusive of the province of Archangel.
[2] Province of the Terek only.

APPENDIX VI

DISTRIBUTION OF LANDED PROPERTY IN EUROPEAN RUSSIA IN 1905

Regions	Peasants Nadiel	Peasants Private	Private Proprietors	State	Institutions	Total
	In thousands of dessiatins and percentages					
Northern	7,893·7	880·0	1,182·9	78,382·4	334·6	87,793·6
	8·9	1·0	1·3	88·4	0·4	100
North-East	16,055·5	215·6	9,462·7	16,048·6	345·1	42,127·5
	38·5	0·5	22·1	38·1	1·8	100
Petrograd	5,378·5	3,031·1	5,845·8	2,837·3	363·4	18,456·1
	30·8	17·3	33·5	16·4	2·0	100
Moscow	16,508·7	5,119·0	11,495·0	3,832·6	726·7	37,689·0
	43·9	13·5	30·6	10·1	1·9	100
Western	5,158·4	1,496·6	8,908·1	1,109·0	209·6	16,881·7
	30·6	9·2	53·0	6·8	1·4	100
South-West	6,159·9	824·7	5,713·5	970·5	384·4	14,053·0
	43·8	5·9	40·7	6·9	2·7	100
Ukraine	7,187·8	1,744·3	3,581·4	314·3	251·1	13,078·9
	54·9	13·3	27·3	2·4	2·1	100
Central Agric.	12,873·1	1,785·5	6,195·0	1,295·2	514·5	22,663·3
	56·8	7·8	27·3	5·7	2·4	100
Middle Volga	10,182·9	1,677·4	4,256·4	4,176·4	444·2	20,737·3
	49·0	8·5	20·4	20·0	2·1	100
Eastern	12,964·3	2,408·3	4,300·1	3,725·6	240·3	23,638·6
	54·8	10·1	18·1	15·7	1·3	100
New Russia	6,779·8	3,044·6	6,706·3	739·0	715·6	17,985·3
	37·7	16·8	37·5	4·1	3·9	100
South-East [1]	9,847·4	955·0	1,363·1	1,153·0	1,907·9	15,226·4
	64·6	6·3	8·9	7·6	12·6	100
Caspian [2]	2,354·9	26·2	172·7	836·9	136·8	3,527·5
	66·7	0·7	4·9	23·7	3·9	100

[1] Province of the Don only.
[2] Province of Astrakhan only.

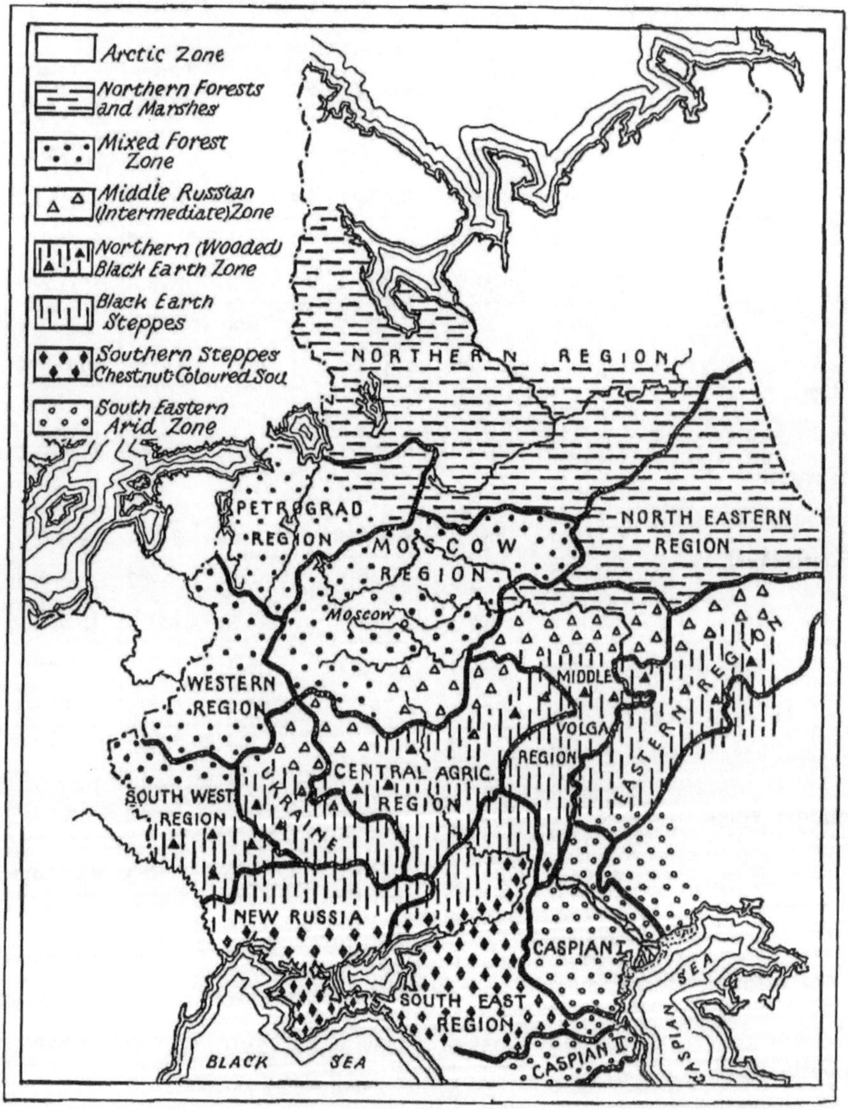

MAP I.—Agricultural Regions and Natural Zones of European Russia

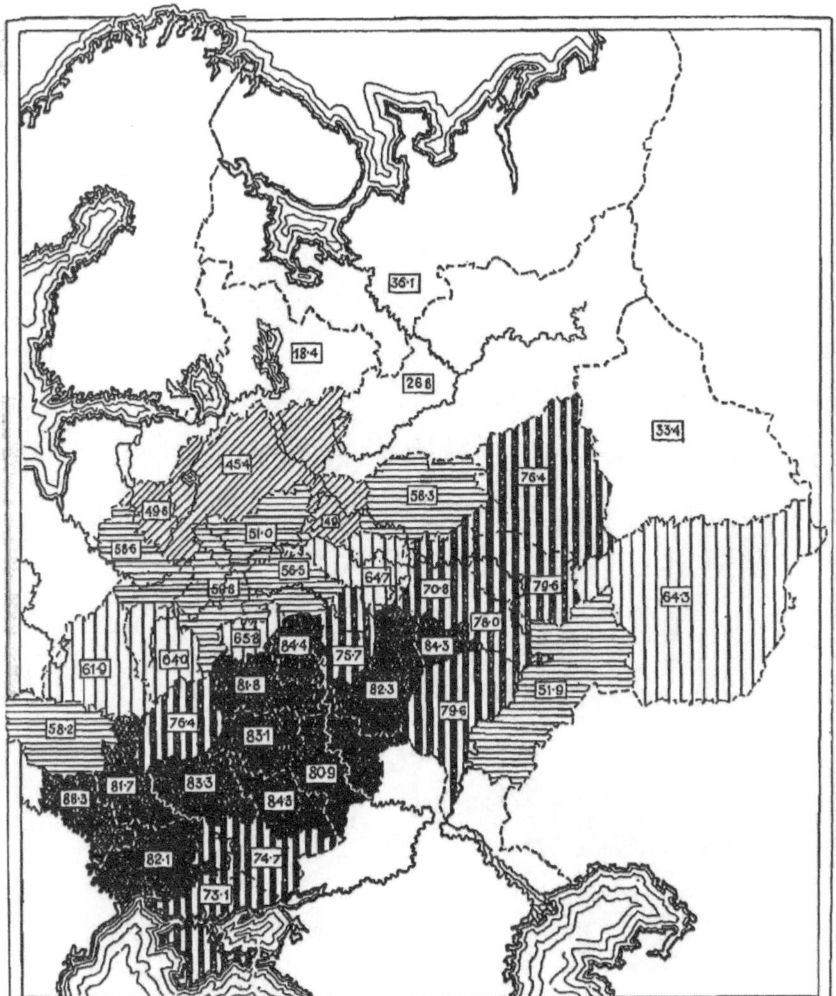

MAP 2.—Percentages of total Agricultural Area actually turned into Arable, according to the Survey of 1887

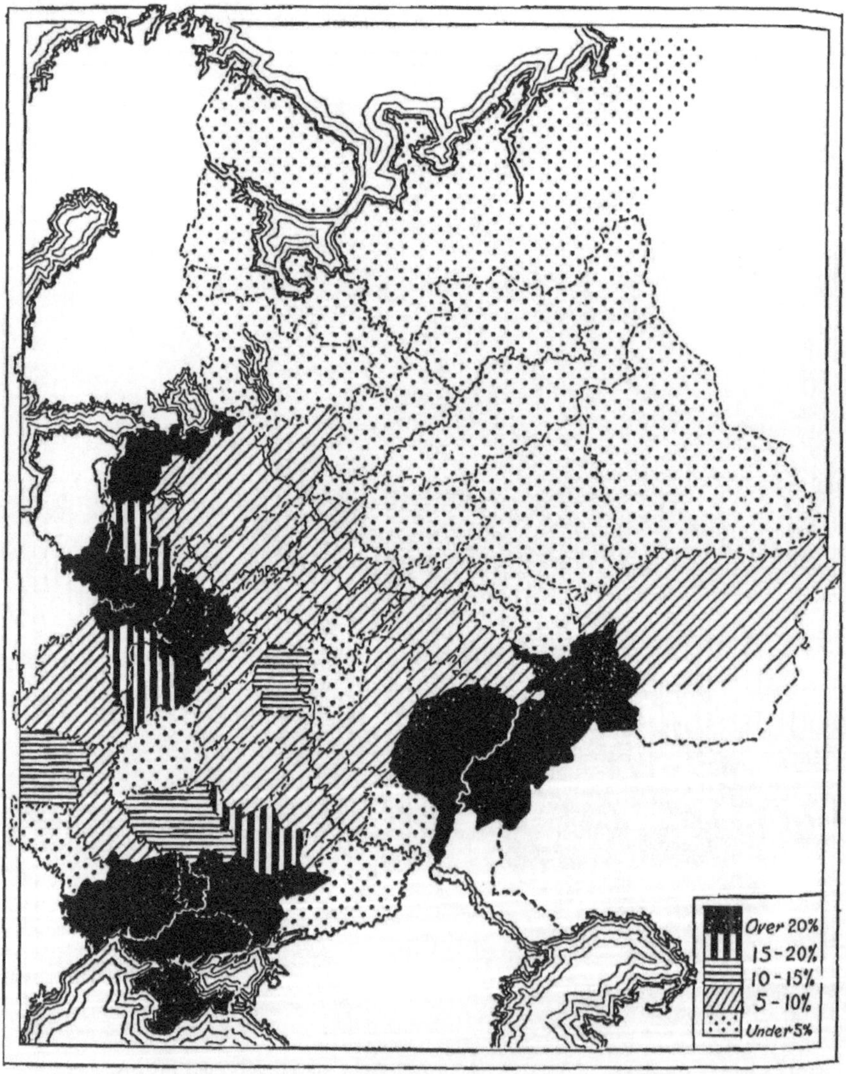

MAP 3.—Peasant Enclosures by the end of 1916

INDEX

Printed in Great Britain by Butler & Tanner Ltd., Frome and London